Polymer Engineering Science and Viscoelasticity
An Introduction

Hal F. Brinson • L. Catherine Brinson

Polymer Engineering Science and Viscoelasticity

An Introduction

 Springer

Hal F. Brinson
University of Houston
Department of Mechanical
  Engineering
4800 Calhoun Road
Houston, TX 77204

L. Catherine Brinson
Northwestern University
Department of Mechanical
  Engineering
Department of Materials
  Science and Engineering
2145 Sheridan Road
Evanston, IL 60208-3111

Library of Congress Control Number: 2007936603

ISBN 978-0-387-73860-4   e-ISBN 978-0-387-73861-1

Printed on acid-free paper.

9 8 7 6 5 4 3 2 1

springer.com

For Clara, Jon, Warren, Max, Casey, Toby and Ellie

Preface

This text is an outgrowth or organized compilation of the notes the authors have used to teach an introductory course on the viscoelasticity of polymers for more than thirty years for the senior author and about fifteen years for the junior author. Originally, the course was taught only to graduate students but in recent years an effort has been made to teach a modification of the course to senior level mechanical engineering students. The authors have long held the view that the lack of knowledge of the fundamental aspects of the time and temperature behavior of polymer materials is a serious shortcoming in undergraduate as well as graduate engineering education. This is especially important in our present society because the use of polymeric materials pervades our experience both in our daily lives and in our engineering profession. Still the basic thrust of undergraduate education and even graduate education to some degree in the areas of mechanical and civil engineering is toward traditional materials of metal, concrete, etc. Until about twenty-five years ago, elementary undergraduate textbooks on materials contained little coverage of polymers. Today many elementary materials texts have several chapters on polymers but, in general, the thrust of such courses is toward metals. Even the polymer coverage that is now included treats stress analysis of polymers using the same procedures as for metals and other materials and therefore often misleads the young engineer on the proper design of engineering plastics. Thus, it is not surprising that some structural products made from polymers are often poorly designed and do not have the durability and reliability of structures designed with metallic materials.

For the above reasons, the view of the authors is that specific courses on polymer materials as well as associated stress analysis and engineering design need to be offered to every engineer. The present text has been developed with this in mind. The intent is to have sufficient coverage for a two semester introductory sequence that would be available to upper class undergraduates and first year graduate students. The level is such that only basic knowledge of solid mechanics and materials science are needed as prerequisites. The book is intended to be self sufficient even for those that have little formal training in solid mechanics and therefore chemical engineers, materials, forestry, chemistry, bio-engineering, etc. students as well as mechanical and civil engineering students can use this text successfully. Similarly, because chemistry background is often weak for non-chemical engineers, introductory material is provided on the chemical basis of polymers, which is essential for proper appreciation of the thermome-chanical response.

Another major objective is for the text to be readable by recent engineering graduates who have not had the advantage of a formal course on polymer science or viscoelasticity. The reason, of course, is that today's engineering curricula, both undergraduate and graduate, have few extra hours such that new courses can be accommodated in degree plans. Therefore, a book such as this one should be of great value to the young engineer who finds him/herself in a position heavily involved with the engineering design and use of polymer based materials. In addition a text such as this should be invaluable to those cross-disciplinary scientists such as biologist, bio-chemist, etc. that need to understand the basic background to rigorous mechanics approaches to the design of structures made with polymer based materials.

The first chapter gives insight to the historical aspects of the subject. A review of basic mechanics of materials (strength of materials) and materials science is given in Chapter 2. Chapter 3 gives an introduction to the mechanical properties of polymers and how they are determined as well as general information on optical, electrical and other properties. Chapter 4 is an introduction to the general character of polymers from a molecular viewpoint and is valuable in assessing the mechanisms associated with viscoelastic deformations. Chapters 5 and beyond speak to the formal mathematics and experimental methods associated with the relationship between stress and strain in viscoelastic solids, both linear and nonlinear, as well as stress analysis and failure.

Acknowledgements

Perhaps the first to receive our grateful acknowledgement are all the authors of books and articles that we have referenced as well as those authors from whom we have gained information that we have internalized so well that the original source of inspiration is forgotten and therefore we cannot properly acknowledge. We know that this text would not have been possible without the guidance from professors and fellow students over our formative years while a student at our respective universities; N.C. State, Northwestern and Stanford (HFB) and Virginia Tech and Caltech (LCB). In addition, we surely have greatly benefited from the council, encouragement and friendship of fellow faculty at the universities where we have had the pleasure to be a faculty member; N.C. State, Virginia Tech, UTSA and University of Houston (HFB) and Northwestern (LCB). Certainly, we have had the privilege to guide and be guided by our graduate students and postdocs over the years who deserve enormous credit for their hard work and the many papers and reports that we have jointly published. As any professor must admit, our success would be small if not for the dedication, energy and resourcefulness of our students – both graduate and under-graduate. Both authors would also like to thank colleagues and staff at University of Poitier in France and the Free University of Brussels in Belgium (HFB) and the Helmut-Schmidt-Universität in Hamburg, Germany (LCB), respectively, where each of us have spent many pleasant hours with colleagues discussing the subject of this book plus many other issues. We especially wish to thank Ms Susanne Pokossi for her skillful, cheerful and tireless assistance in the preparation of the final version of this book.

Foremost of all we gratefully acknowledge those closest to us – our families, our life partners and, of course, our children. We owe them much for their patience, guidance, camaraderie and most of all their love.

Hal F. Brinson   L. Cate Brinson
University of Houston   Northwestern University

Table of Contents

1. Introduction

1.1. Historical Background

Development of synthetic polymers and growth of the polymer industry during the last 70 years has been staggering. The commercial success of polymer-based products has generated a demand such that the total production of plastics (by volume) has exceeded the combined production of all metals for more than 20 years. It has been suggested that Polymer Science evolved from the following five separate technologies: 1. Plastics, 2. Rubbers or Elastomers, 3. Fibers, 4. Surface Finishes, and 5. Protective Coatings, each of which evolved separately to become major industries (Rosen, (1993)). As a result much of the early development of polymers or plastics was focused on these commercial products and other non-structural uses. The need to develop synthetic rubber due to the interruption of trade routes during WW II served as a catalyst to large scale federal funding for polymer research. This increased effort resulted in better understanding of the nature of polymers as well as improved analytical and experimental approaches to their behavior. In more recent years, however, polymers have become an engineering structural material of choice due to low cost, ease of processing, weight savings, corrosion resistance and other major advantages. In fact modern polymeric adhesives and polymer matrix composites (PMC) or fiber-reinforced plastics (FRP) are today being used in many severe structural environments of the aerospace, automotive and other industries.

Not withstanding the recent developments of synthetic structural polymers, naturally occurring polymers have been used for thousands of years and early civilizations understood how to mix fibers (such as wheat flax) with resins to obtain added strength. For example, pottery cemented with natural resins have been found in burial sites that date back to 4000 BC. A cedar chest with extensive glue construction was found in King Tutankhamen's tomb and dates back to 1365 BC. Clegg and Collyer, (1993) report that bitumen, a complex mix of heavier petroleum fractions, is mentioned in the Bible and that amber, a gum-like or brittle fossilized resin from pine trees, was known in ancient Rome. They further note that shellac, a derivative of the lac insect, has been in use since 1,000 BC and was used as late as 1950 for gramophone records.

Rodriguez (1996) notes that the natives of South America made use of the latex derived from trees in the pre-Columbian era. Sometime after the discovery of the Americas, rubber was introduced to Europe and was used as a waterproofing material (McCrum, et al., (1997)). Most authors trace the beginning of polymer science to the development of the vulcanization process by MacIntosh and Hancock in England and Goodyear in the United States in 1839. However, Rodriguez (1996) indicates others had developed applications for rubber as early as the late 18th century. (An excellent time line of polymer science and technology is given by Rodriguez (1996)).

One of the first man-made polymers was Parkesine, so named after its inventor Alexander Parkes. It was introduced in about 1862 but was not a commercial success (Fried, (1995)). However, this early effort led to the development of celluloid (cellulose nitrate) by John Hyatt in 1870 which was a commercial success. The first truly synthetic polymer was a phenol-formaldehyde resin called Bakelite developed in 1907 by Leo Baekeland but it would be two more decades before the nature of the polymerization process would be understood sufficiently to develop polymers based upon a rational process.

While natural polymers had found extensive early use, knowledge of their molecular nature was generally unknown before the middle of the 19th century, when the first speculations about the large molecular weights of polymers were voiced. At that time, the chemical or molecular character of a material's composition was defined in terms of its stoichometric formula and its properties were defined in terms of color, crystal habit, specific gravity, refractive index, melting point, boiling point, solubility, etc. (Tolbolsky and Mark, (1971)). It was only around the turn of the century that concern turned to the chemical structure of materials, which together with advances in measurement techniques led finally to understanding and later acceptance of polymers as consisting of large covalently bonded molecules.

In the late 19th century, materials we now define as molecular high polymers were thought to be composed of large molecules or colloidal aggregates. These colloidal aggregates were said to form from smaller molecules through the action of intermolecular forces of "mysterious origin" which were responsible for the unusual properties such as high viscosity, long-range elasticity, and high strength (Flory, (1953)). Flory attributes the term colloid to Thomas Graham in 1861 and the concept of a colloidal state to Wolfgang Ostwald in about 1907 who suggested that virtually any substance could be in such a state just as in a gas, liquid or solid state.

Until Kekulé in 1877, all geometrical formulas referred to the structure and behavior of small molecules. However, Kekulé in 1877 during a lecture upon becoming the Rector of the U. of Bonn "advanced the hypothesis that the natural organic substances associated with life (proteins, starch, cellulose, etc.) may consist of very long chains and derive their special properties from this peculiar structure" (Tolbolsky and Mark, (1971)).

The fundamental difficulty in evaluating the molecular nature of polymers in the early 20th century was the lack of quantitative characterization methods. Perhaps the greatest limitation resided in the limited means available to accurately measure the high molecular weight of macromolecular materials. The vapor density method which was widely used for low molecular weight materials could not be employed. In 1881 attempts were made to use diffusion rates to distinguish between the molecular weights of starch and the dextrins (Flory, (1953)). Flory further reports that the development of the cryoscopic method for determining molecular weight by Raoult in 1882 and van Hoff's solution laws in 1886 were instrumental in proving the validity of the macromolecular concept. In 1889 Brown and Morris used a freezing point suppression of aqueous solutions method to determine molecular weights as high as 30,000. Rodewald and Kattein in 1900 used osmotic pressure measurements to determine molecular weights as high as 39,700. X-ray procedures were used in 1920 by Polanyi to investigate the nature of cellulose fibers but it was not until Svedberg developed the ultracentrifuge in 1940 that accurate and reproducible measurement of molecular weights from 40,000 to several million was possible. (For an excellent review of ultracentrifugation techniques, see Williams, (1972)).

Dr. Herman Staudinger who was awarded a Nobel Prize in 1953 for his work (see below for other Nobel Prize winners in polymer science) proposed the "macromolecular hypothesis" in the 1920s explaining the common molecular makeup of macromolecular materials. He contradicted the prevalent view of his time that polymeric substances were held together by partial valances and instead proposed the idea of long molecular chains. He accurately gave the proper formulas for polystyrene, polyoxymethylene (paraformaldehyde) and for rubber (Flory, (1953)).

In 1929 W. H. Carothers was the first to clearly define what we know today as the basic parameters of polymers science. Clearly stating his objectives beforehand, he prepared (or synthesized) molecules of definitive structure (mostly condensation polymers such as the polyesters) through established reactions in organic chemistry and proceeded to investigate how the properties of these substances depended on their constitution (Flory, (1953)). Shortly after Carothers definitive work, polymer scientists began to introduce the concepts of statistical thermodynamics to describe the characteristics of long chain polymers of high molecular weight.

More than thirty individuals have been awarded the Nobel prize for Chemistry for their contributions either directly or indirectly to the development of polymer science and associated technology. A few of these are listed below. The interested reader can find more information on all Nobel Laureates at the web address <http://www.nobel.se >.

1939: **Ruzicka Leopold**, Switzerland, Eidgenössische Technische Hochschule, (Federal Institute of Technology), Zurich, (in Vukovar, then Austria-Hungary): "For his work on polymethylenes and higher terpenes"

1953: **Staudinger, Hermann**, Germany, University of Freiburg im Breisgau and Staatliches Institut für makromolekulare Chemie (State Research Institute for Macromolecular Chemistry), Freiburg in Br.: "For his discoveries in the field of macromolecular chemistry."

1963: **Ziegler, Karl**, Germany, Max-Planck-Institut für Kohlenforschung (Max-Planck-Institute for Carbon Research Mülheim/Ruhr; and **Natta, Giulio**, Italy, Institute of Technology, Milan: "For their discoveries in the field of the chemistry and technology of high polymers."

1968: **Onsager, Lars**, U.S.A., Yale University, New Haven, CT: "For the discovery of the reciprocal relations bearing his name, which are fundamental for the thermodynamics of irreversible processes."

1973: **Fischer, Ernst Otto**, Federal Republic of Germany, Technical University of Munich, Munich; and **Wilkinson, Sir Geoffrey**, Great Britain, Imperial College, London: "For their pioneering work, performed independently, on the chemistry of the organometallic, so called sandwich compounds."

1974: **Flory, Paul J.**, U.S.A., Stanford University, Stanford, CA: "For his fundamental achievements, both theoretical and experimental, in the physical chemistry of the macromolecules."

1985: **Hauptman, Herbert A.**, U.S.A., The Medical Foundation of Buffalo, Buffalo, NY,; and **Karle, Jerome**, U.S.A., US Naval Research Laboratory, Washington, DC: "For their outstanding achievements in the development of direct methods for the determination of crystal structures."

1990: **Corey, Elias James**, U.S.A., Harvard University, Cambridge, MA: "For his development of the theory and methodology of organic synthesis"

1991: **Ernst, Richard R.**, Switzerland, Eidgenössische Technische Hochschule Zurich, b. 1933: "For his contributions to the development of the methodology of high resolution nuclear magnetic resonance (NMR) spectroscopy"

2000: **Heeger, Alan J.**, University of California, Alan G. MacDiarmid, University of Pennsylvania, Hideki Shirakawa, University of Tsukuba: For the discovery and development of Conductive Polymers.

The authors realize that many other Nobel Laureates in chemistry have made notable contributions that have impacted the development and understanding of polymer science but those listed here seem to us to be of particular importance. In addition, Nobel Prizes in physics also have encompassed the field of polymer science. Most recently for example, Pierre-Gilles de Gennes received the 1991 Nobel Prize in Physics for his discovery that mathematical methods to describe simple systems can be extended to complex forms of matter including liquid crystals and polymers.

In this brief overview of the historical aspects of polymer science and technology, it would be very inappropriate not to acknowledge the tremendous contributions made by many polymer chemists, polymer physicists, materials scientists and engineers in the last fifty years. The list is so large that it would be impossible to acknowledge everyone. However, all will agree that Paul Flory (Stanford University), Herman Mark (Brooklyn Polytechnic), John Ferry (University of Wisconsin), Turner Alfred (Dow Chemical), Nick Tschoegl (California Institute of Technology), Arthur Tolbolsky (Princeton University), Herbert Leaderman (National Bureau of

Standards, now NIST), Bob Landel (Jet Propulsion Laboratory) and many others have provided the framework and scientific details such that polymers can now be used with confidence as engineering structural materials.

1.1.1. Relation between Polymer Science and Mechanics

As discussed briefly in the next section, polymers have a unique response to mechanical loads and are properly treated as materials which in some instances behave as elastic solids and in some instances as viscous fluids. As such their properties (mechanical, electrical, optical, etc.) are time dependent and cannot be treated mathematically by the laws of either solids or fluids. The study of such materials began long before the macromolecular nature of polymers was understood. Indeed, as will be evident in later chapters on viscoelasticity, James Clerk Maxwell (1831-79), a Scottish physicist and the first professor of experimental physics at Cambridge, developed one of the very first mathematical models to explain such peculiar behavior. Lord Kelvin (Sir William Thomson, (1824-1907)), another Scottish physicist, also developed a similar mathematical model. Undoubtedly, each had observed the creep and/or relaxation behavior of natural materials such as pitch, tar, bread dough, etc. and was intrigued to explain such behavior. Of course, these observations were only a minor portion of their overall contributions to the physics of matter.

Ludwig Boltzmann (1844-1906), an Austrian physicist, correctly conceived the hereditary nature of materials which we now describe as viscoelastic in a series of publications throughout his career. Such ideas were hotly debated at the time by Boltzmann, Ostwald and others but it is now clear that Boltzmann's view was the correct approach. For an excellent discussion of Boltzmann's contributions and their significance, see Markovitz (1975, 1977).

In 1812, even before Maxwell, Kelvin and Boltzman, the Scottish scientist Sir David Brewster (1781-1868) discovered that certain transparent optically isotropic solids (e.g., glass) when loaded developed optical characteristics of natural crystals. That is, he found that such a solid when loaded exhibited birefringence or double refraction and thus behaved as a temporary crystal. His discovery was the beginning of the well-known photoelastic method by which it is possible to experimentally determine the state of stress or strain on the interior of a loaded elastic body using polarized light. Maxwell (as well as F. E. Neumann at an earlier date) also studied the technique and deduced the relationship between stress and the optic effect now known as the Maxwell-Neumann stress-optic law. The importance of these discoveries became apparent during the industrial revolution in the early part of the 20th century when the safe design of precision mechanical parts such as wheels, gears, pushrods, etc. required a stress analysis that was only possible using photoelasticity. As a result, engineers became very interested in finding suitable model materials (polymers) that had desirable characteristics such as good transparency, high stress-optic coefficient, little creep, etc. For this reason, engineers began working with chemists in an effort to create polymers with suitable properties.

Initially, natural crystals (such as mica and quartz) were used to obtain polarized light and model materials were either glass or resins derived from living organisms, e.g., isinglass, a gelatin prepared from the bladder of a sturgeon. The photoelastic procedure was so successful that it led engineers to widely seek more optically sensitive and stable materials. Coker and Filon (1931) in their famous treatise used a number of materials including glass and celluloid. Bakelite, developed at the beginning of the 20th century by L. H. Baekeland, became a favorite photoelastic material for many years. During the 30's a particular form of Bakelite (BT-61-893) was introduced which greatly aided the development of photoelasticity in two and three dimensions. Hetenyi (1938), used this material to develop and explain the so-called "stress-freezing" and slicing method to determine the interior stresses in three-dimensional bodies. CR-39 or Columbia Resin 39 (allyl diglycol carbonate developed by the Columbia Chemical Company in 1945) was also used extensively in the 40's and 50's.

The details of cross-linking were not understood at the time and Bakelite was often termed by engineers as a "heat hardening" resin. Hetenyi (1938) used Houwink's (1937) interpretation of the "micelle" nature of polymers to explain the frozen stresses (photoelastic fringe patterns) inside a body after removable of loading. That is, if a load is applied after the temperature of a birefringent material is raised to a suitable level and then held constant as the temperature is slowly lowered to ambient, a residual fringe (stress) pattern will remain when the load is removed at the lower temperature. The residual pattern was believed to remain due to the network nature of the material and an analogy of a solid network or skeletal phase and a fluid phase in between the network sites was used to explain the frozen stress phenomena. In early photoelastic literature, such polymers were often referred to as di-phase or bi-phase, (i.e., part fluid and part solid), in nature. The specific analogy likened network polymers to a sponge filled with a highly viscous fluid. At low temperatures, the viscous portion would solidify and the network polymer would become a brittle glassy solid with high modulus and high strength. At high temperatures, the viscosity of the fluid phase would decrease sufficiently such that the external load was supported only by the skeletal phase and thus the sample would become a low modulus rubbery material. While this was a useful analogy for the time, it is now more appropriate to explain the phenomena in terms of primary and secondary bonds between molecular chains and, in some sense, one might liken the secondary bonds between network sites as the fluid phase and the primary bonds of the network skeleton as the solid phase.

The engineering use of photoelasticity was greatly aided by the development of polaroid films by E. H. Land (1909 - 1991). This film is a polymer in which the molecular structure has been oriented to cause light to be plane polarized. As plane polarization can also be achieved by reflection (at angles of approximately 57°), the film can be used as a filter to minimize glare as in sunglasses made with polaroid plastics. By careful orientation, the degree of double refraction can be controlled to obtain films with a retardation of a quarter of a wavelength of a particular light. Quarter wave plates used between two oppositely polarized films causes light to be elliptically polarized. The "ellipsometer" often used by polymer chemist is based on such a procedure and, of course, the polarizing microscope uses polaroid films to control the light vector and allows the observation of crystallites in polymers and gives and estimate of their crystalline nature.

In recent years, epoxy resins have become the polymer of choice for three-dimensional photoelastic investigations. Further, the phenomena of birefringence has been used to study plasticity and viscoelasticity effects in materials through the use of extensions to the photoelastic method called photoplasticity and photoviscoelasticity (see Brill (1965) and Brinson (1965, 1968), respectively). Brill used polycarbonate, a thermoplastic polymer, as a model material for his work on photoplasticity and Brinson used an epoxy, a thermosetting polymer, as a model material for his work on photoviscoelasticity. Later, it will become clear why thermoplastic materials are used for photoplasticity while thermosetting materials are used for photoviscoelasticity.

While it is beyond the scope of the discussion here, it can be shown that the stress (strain) tensor, the dielectric tensor and the birefringence tensor are related and, generally, the same types of governing equations apply to each phenomenon. That is, a quadric surface similar to the stress quadric of Cauchy applies to the birefringence tensor and to the dielectric tensor. This knowledge led to the interest of early mechanicians to identify and understand the nature of birefringent materials which, in fact, were natural polymers. As polymer science began to develop, the same group was led to study, understand and use synthetic polymers. For more information on these and other aspects of photoelasticity and experimental mechanics, see Hetenyi (1950) and Kobayashi (1985).

In addition to the use of polymers to study fundamental concepts in mechanics, another driving force for the critical link between polymer science and mechanics has been use of polymers in applications. As the understanding of the physical nature of polymers increased and synthesis techniques matured, many polymers of widespread usage were developed. As these materials were employed in devices and structures, it was essential to analyze and understand from an engineering perspective the response of polymers to load and other environmental variables, such as temperature and moisture. As indicated earlier, today high performance polymer composites are used for critical load bearing applications as diverse as alpine skis and airframe parts, and thus the study of the mechanics of polymers as a structural material is an active and important area of research. Later sections in this text will deal explicitly with the viscoelastic nature of polymeric response and mathematical methods to analyze this behavior.

The fundamental point of the above discussion is that persons interested in theoretical and experimental mechanics of necessity have been aware of and keenly interested in all developments associated with natural and synthetic polymers throughout the history of both natural and synthetic polymers. They have, in some cases, contributed to the general understanding of the properties of polymers and to a high degree have been responsible for their use as engineering materials.

As in the previous section, it would be very inappropriate not to acknowledge the efforts of many who have made outstanding contributions to the development of mathematical and experimental aspects of viscoelasticity which allow the correct interpretation of the mechanical behavior of polymers. The contributions of a few will be discussed in more detail in subsequent chapters but again it should be noted that the number of contributors is so large that it would be impossible to acknowledge everyone However, all will agree that Marcus Reiner, (Technion), R.S. Rivlin (Leheigh University), C. Truesdall (Johns Hopkins University), E.H. Lee (Stanford University), R.H. Schapery (University of Texas), Wolfgang Knauss (California Institute of Technology), M. L. (Max) Williams (University of Pittsburg), Harry Hilton (University of Illinois), R.M. Christensen (Lawrence Livermore Laboratories and Stanford University), J. G. Williams (Imperial College) and many others have contributed to our ability to design safe engineering structures using polymer based materials.

(See Reiner, M., Lectures on Theoretical Rheology, North-Holland, Amsterdam, 1960, for an excellent Bibliography of early contributors).

1.1.2. Perspective and Scope of this Text

Polymers possess many interesting and useful properties that are quite different from those of more traditional engineering materials and these properties cannot be explained or modeled in engineering design situations by traditional approaches. As suggested by Rosen (1993), verification can be observed with three simple experiments.

> **Silly Putty**: This material (polydimethyl siloxane) bounces when dropped but flows when laying stationary and, obviously, has some characteristics of an elastic solid and some characteristics of a viscous fluid.

> **Joule Effect:** A rubber band will contract when heated while a weight is suspended from it. Other materials will undergo the expected thermal expansion.

> **Weissenburg Effect**: When a rod is rotated in a molten polymer or in a concentrated polymer solution, the liquid will rise on rod. For other fluids, the lowest point in fluid will be at rod.

The fundamental difference between polymers and other materials resides in the inherent rheological or viscoelastic properties of polymers. Simply stated, the mechanical (as well as optical, electrical, etc.) properties of polymers such as modulus, strength and Poisson's ratio vary with time. While many materials have properties that vary with time due to creep at high temperature, moisture intrusion, corrosion, and other factors, the time dependent behavior of polymers is due to their unique molecular structure. As will be discussed later, the long chain molecular structure of a polymer gives rise to the phenomena of "fading" memory. It is this fading memory which creates the need to characterize engineering properties in a manner different than those used for traditional structural materials.

One manifestation of the time dependent character of polymers is that they exhibit characteristics of both an elastic solid and that of a viscous fluid as with the example of silly putty above. For this reason, materials such as polymers that exhibit such properties are often said to be viscoelastic. Sometimes the term viscoelastic is used primarily for solid polymers while the term rheologic is for liquid polymers. Fading memory provides the explanation for the three examples mentioned above. To illustrate the point, a demonstration of the Weissenberg effect is given in **Fig. 1.1** which is a schematic of a solid rod being rotated rapidly while being immersed in a viscoelastic liquid. While in a Newtonian fluid, the liquid moves away from the rod due to inertial effects, the liquid polymer will climb the rod due to the combination of elastic and viscous forces in the entangled polymer chains.

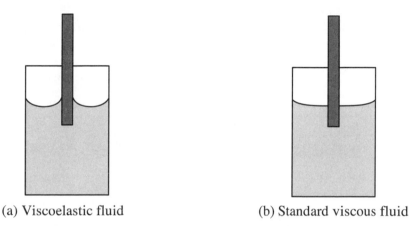

(a) Viscoelastic fluid   (b) Standard viscous fluid

Fig. 1.1 Illustration of the Weissenberg effect due to a rotating rod in a viscoelastic fluid (a) compared to a rod rotating in a Newtonian fluid (b). (See Fredrickson, (1964) for two viscoelastic fluids examples.)

To further illustrate the point of a liquid with both elastic and viscous behavior, the flow of a rheological liquid is shown in **Fig. 1.2**. Here a polymer liquid is in a clear horizontal (to avoid gravity effects) tube and a dark reference mark has been inserted that moves with the fluid. The liquid is unpressurized in frame 1 but a constant pressure has been applied in frames 2 through 5 where motion can be seen to have taken place as time progresses. In frame 6 the pressure has been removed and in frames 7 and 8 the liquid can be seen to partially recover. No recovery would take place if this were an ordinary viscous liquid. This is known as an "elastic after effect" and a similar effect or creep recovery is observed in viscoelastic solids and/or all polymers provided the correct temperature is chosen.

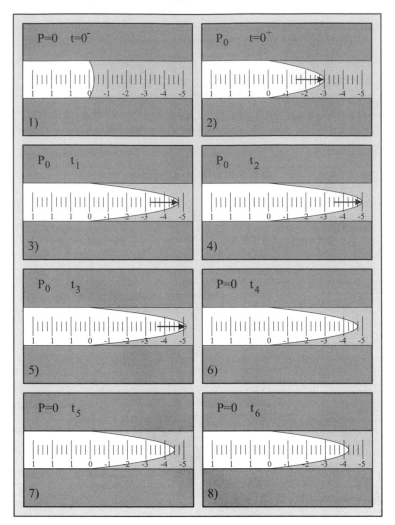

Fig. 1.2 Illustration of recoil in a viscoclastic fluid. (The fluid is on the right in each frame and pressure, indicated by the arrow, is applied from the left.) (Drawn from photograph in Fredrickson, (1964). Original photograph from N.N. Kapoor, MS thesis, U. of Minn.)

The Joule effect arises from thermodynamic (entropy) considerations and will be discussed in sections related to the time and temperature dependent behavior of polymers.

In this text, emphasis will be the on the phenomenological differences between the mechanical behavior of polymers and other materials, rather than their similarities. Emphasis will also be placed on proper procedures to experimentally determine time dependent mechanical properties as well as analytical methods to represent these properties and to use them in the stress analysis of engineering structures.

The classifications of materials used by Fredrickson and given in **Fig. 1.3** is suggestive of the evolution of constitutive (stress-strain) relations and the scientists responsible for their creation. Material rigidity increases from left to right on the top row and the vertical development emphasizes the generality of Boltzman's contributions. The Bingham representation was for a viscous material that displayed a yield point and was originally developed for paint. It is important to understand the connotation of the word "plastic" as used in **Fig. 1.3** and when used to describe a polymer. In fact, the use of the word "plastic" to describe a polymer is unfortunate. This word is best used to denote a type of mechanical behavior associated with unrecoverable deformation or flow, and is misleading when used in a generic way to refer to polymers in general. Certain polymers under favorable conditions will not exhibit any unrecoverable deformation. Hopefully, these distinctions will become clear upon further study of the following chapters.

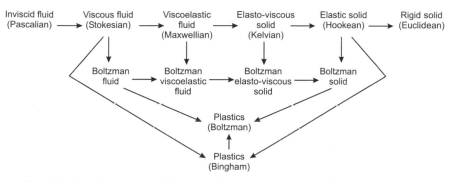

Fig. 1.3 Rheological classification of materials. (After Fredrickson, (1964))

It is recognized that the terms "materials science" and "materials engineering" as applied to the study of materials are generally understood to imply different facets of the same subject. In the same manner, the terms "polymer science" and "polymer engineering" may be interpreted to mean different approaches to the study of polymers. Herein, care has been taken to use the term "polymer engineering science" to imply the study of the nature of polymers which gives rise to their unique engineering properties. Our focus will be on the relation between the molecular structure of polymers, their mechanical properties and the mathematical framework required for the proper stress analysis of polymeric structures. This will, of necessity, entail a general knowledge of polymerization processes, the resulting molecular structure and the relation of each to the final engineering properties. However, no effort will be made to learn how to "engineer" a polymer or, in other words, to learn sufficient chemistry to synthesize a polymer to have specific engineering properties.

Because polymers are, in general, viscoelastic and/or have mechanical properties that are a function of time, the phenomenological and molecular interpretations of viscoelasticity will be covered in detail. While there are approximate methods to design load bearing polymer structures using elementary mechanics of materials principles (some of these will be presented), more precise and more correct procedures will also be discussed at length.

1.2. Review Questions

1.1. Name four naturally occurring polymers.

1.2. Name the earliest polymer mentioned and where it came from.

1.3. Explain the distinction between a mixture, a solution, a suspension and a colloid.

1.4. What was Staudinger's hypothesis?

1.5. Who was the first person to clearly understand the nature of polymers?

1.6. Name the five separate technologies from which Polymer Science is said to have evolved.

1.7. Name four Noble Laureates.

1.8. Name two Scots who did early work on mathematical models for viscoelastic behavior.

1.9. Who correctly conceived the hereditary nature of polymers?

1.10. Who is credited with discovering the phenomena of double refraction?

1.11. Name three materials that have been used in photoelastic analysis.

1.12. Who developed polaroid films?

1.13. Describe the Joule effect.

1.14. Describe the Weissenberg effect.

1.15. Explain differences in the terms "polymer" and "plastic".

1.16. What is a continuum.

1.17. Define viscoelasticity

2. Stress and Strain Analysis and Measurement

The engineering design of structures using polymers requires a thorough knowledge of the basic principles of stress and strain analysis and measurement. Readers of this book should have a fundamental knowledge of stress and strain from a course in elementary solid mechanics and from an introductory course in materials. Therefore, we do not rigorously derive from first principles all the necessary concepts. However, in this chapter we provide a review of the fundamentals and lay out consistent notation used in the remainder of the text. It should be emphasized that the interpretations of stress and strain distributions in polymers and the properties derived from the standpoint of the traditional analysis given in this chapter are approximate and not applicable to viscoelastic polymers under all circumstances. By comparing the procedures discussed in later chapters with those of this chapter, it is therefore possible to contrast and evaluate the differences.

2.1. Some Important and Useful Definitions

In elementary mechanics of materials (Strength of Materials or the first undergraduate course in solid mechanics) as well as in an introductory graduate elasticity course five fundamental assumptions are normally made about the characteristics of the materials for which the analysis is valid. These assumptions require the material to be,

- **Linear**
- **Homogeneous**
- **Isotropic**
- **Elastic**
- **Continuum**

Provided that a material has these characteristics, be it a metal or polymer, the elementary stress analysis of bars, beams, frames, pressure vessels, columns, etc. using these assumptions is quite accurate and useful. However, when these assumptions are violated serious errors can occur if the same analysis approaches are used. It is therefore incumbent upon engineers to thoroughly understand these fundamental definitions as well as how to determine if they are appropriate for a given situation. As a result, the reader is encouraged to gain a thorough understanding of the following terms:

Linearity: Two types of linearity are normally assumed: Material linearity (Hookean stress-strain behavior) or linear relation between stress and strain; Geometric linearity or small strains and deformation.

Elastic: Deformations due to external loads are completely and instantaneously reversible upon load removal.

Continuum: Matter is continuously distributed for all size scales, i.e. there are no holes or voids.

Homogeneous: Material properties are the same at every point or material properties are invariant upon translation.

Inhomogeneous or Heterogeneous: Material properties are not the same at every point or material properties vary upon translation.

Amorphous: Chaotic or having structure without order. An example would be glass or most metals on a macroscopic scale.

Crystalline: Having order or a regular structural arrangement. An example would be naturally occurring crystals such as salt or many metals on the microscopic scale within grain boundaries.

Isotropic: Materials which have the same mechanical properties in all directions at an arbitrary point or materials whose properties are invariant upon rotation of axes at a point. Amorphous materials are isotropic.

Anisotropic: Materials which have mechanical properties which are not the same in different directions at a point or materials whose properties vary with rotation at a point. Crystalline materials are anisotropic.

Plastic: The word comes from the Latin word plasticus, and from the Greek words plastikos which in turn is derived from plastos (meaning molded) and from plassein (meaning to mold). Unfortunately, this term is often used as a generic name for a polymer (see definition below) probably because many of the early polymers (cellulose, polyesters, etc.) appear to yield and/or flow in a similar manner to metals and could be easily molded. However, not all polymers are moldable, exhibit plastic flow or a definitive yield point.

Viscoelasticity or Rheology: The study of materials whose mechanical properties have characteristics of both solid and fluid materials. Viscoelasticity is a term often used by those whose primary interest is solid mechanics while rheology is a term often used by those whose primary interest is fluid mechanics. The term also implies that mechanical properties are a function of time due to the intrinsic nature of a material and that the material possesses a memory (fading) of past events. The latter separates such materials from those with time dependent properties due primarily to changing environments or corrosion. All polymers (fluid or solid) have time or temperature domains in which they are viscoelastic.

Polymer: The word Polymer originates from the Greek word "polymeros" which means many-membered, (Clegg and Collyer 1993). Often the word polymer is thought of as being composed of the two words; "poly" meaning many and "mer" meaning unit. Thus, the word polymer means many units and is very descriptive of a polymer molecule.

Several of these terms will be reexamined in this chapter but the intent of the remainder of this book is to principally consider aspects of the last three.

2.2. Elementary Definitions of Stress, Strain and Material Properties

This section will describe the most elementary definitions of stress and strain typically found in undergraduate strength of materials texts. These definitions will serve to describe some basic test methods used to determine elastic material properties. A later section will revisit stress and strain, defining them in a more rigorous manner.

Often, stress and strain are defined on the basis of a simple uniaxial tension test. Typically, a "dogbone" specimen such as that shown in **Fig. 2.1(a)** is used and material properties such as Young's modulus, Poisson's ratio, failure (yield) stress and strain are found therefrom. The specimen may be cut from a thin flat plate of constant thickness or may be machined from a cylindrical bar. The "dogbone" shape is to avoid stress concentrations from loading machine connections and to insure a homogeneous state of stress and strain within the measurement region. The term homogeneous here indicates a uniform state of stress or strain over the measurement region, i.e. the throat or reduced central portion of the specimen. **Fig. 2.1(b)** shows the uniform or constant stress that is present and that is calculated as given below.

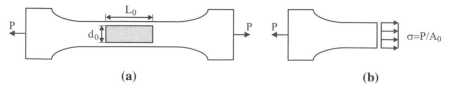

Fig. 2.1 "Dogbone" tensile specimen.

The engineering (average) stress can be calculated by dividing the applied tensile force, **P**, (normal to the cross section) by the area of the original cross sectional area A_0 as follows,

$$\sigma_{av} = \frac{P}{A_0} \qquad (2.1)$$

The engineering (average) strain in the direction of the tensile load can be found by dividing the change in length, ΔL, of the inscribed rectangle by the original length L_0,

$$\varepsilon_{av} = \int_{L_0}^{L} \frac{dL}{L_0} = \frac{\Delta L}{L_0} = \frac{L - L_0}{L_0} \qquad (2.2)$$

or

$$\varepsilon_{av} = \frac{L}{L_0} - 1 = \lambda - 1 \qquad (2.3)$$

The term λ in the above equation is called the extension ratio and is sometimes used for large deformations such as those which may occur with low modulus rubbery polymers.

True stress and strain are calculated using the instantaneous (deformed at a particular load) values of the cross-sectional area, **A**, and the length of the rectangle, **L**,

$$\sigma_t = \frac{F}{A} \qquad (2.4)$$

and

$$\varepsilon_t = \int_{L_0}^{L} \frac{dL}{L} = \ln \frac{L}{L_0} = \ln(1 + \varepsilon) \qquad (2.5)$$

Hooke's law is valid provided the stress varies linearly with strain and Young's modulus, **E**, may be determined from the slope of the stress-strain curve or by dividing stress by strain,

$$E = \frac{\sigma_{av}}{\varepsilon_{av}} \qquad (2.6)$$

or

$$E = \frac{P/A_0}{\Delta L/L_0} \qquad (2.7)$$

and the axial deformation over length **L$_0$** is,

$$\delta = \Delta L = \frac{PL_0}{A_0 E} \qquad (2.8)$$

Poisson's ratio, **ν**, is defined as the absolute value of the ratio of strain transverse, ε_y, to the load direction to the strain in the load direction, ε_x,

$$\nu = \frac{\varepsilon_y}{\varepsilon_x} \qquad (2.9)$$

The transverse strain ε_y, of course can be found from,

$$\varepsilon_y = \frac{d - d_0}{d_0} \qquad (2.10)$$

and is negative for an applied tensile load.

Shear properties can be found from a right circular cylinder loaded in torsion as shown in **Fig. 2.2**, where the shear stress, **τ**, angle of twist, **θ**, and shear strain, **γ**, are given by,

$$\tau = \frac{Tr}{J} \ , \ \theta = \frac{TL}{JG} \ , \ \gamma = \frac{\delta}{L} = \frac{r\theta}{L} \qquad (2.11)$$

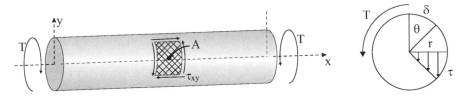

Fig. 2.2 Typical torsion test specimen to obtain shear properties.

Herein, **L** is the length of the cylinder, **T** is the applied torque, **r** is the radial distance, **J** is the polar second moment of area and **G** is the shear modulus. These equations are developed assuming a linear relation between shear stress and strain as well as homogeneity and isotropy. With these assumptions, the shear stress and strain vary linearly with the radius and a pure shear stress state exists on any circumferential plane as shown on the surface at point **A** in **Fig. 2.2**. The shear modulus, **G**, is the slope of the shear stress-strain curve and may be found from,

$$G = \frac{\tau}{\gamma} \tag{2.12}$$

where the shear strain is easily found by measuring only the angular rotation, **θ**, in a given length, **L**. The shear modulus is related to Young's modulus and can also be calculated from,

$$G = \frac{E}{2(1+\nu)} \tag{2.13}$$

As Poisson's ratio, **ν**, varies between **0.3** and **0.5** for most materials, the shear modulus is often approximated by, **G ~ E/3**.

While tensile and torsion bars are the usual methods to determine engineering properties, other methods can be used to determine material properties such as prismatic beams under bending or flexure loads similar to those shown in **Fig. 2.3**.

The elementary strength of materials equations for bending (flexural) stress, σ_x, shear stress, τ_{xy}, due to bending and vertical deflection, **v**, for a beam loaded in bending are,

$$\sigma_x = \frac{M_{zz}y}{I_{zz}} , \tau_{xy} = \frac{VQ}{I_{zz}b} , \frac{d^2v}{dx^2} = \frac{M_{zz}}{EI_{zz}} \tag{2.14}$$

where **y** is the distance from the neutral plane to the point at which stress is calculated, M_{zz} is the applied moment, I_{zz} is the second moment of the cross-sectional area about the neutral plane, **b** is the width of the beam at the point of calculation of the shear stress, **Q** is the first moment of the area about the neutral plane (see a strength of materials text for a more explicit definition of each of these terms), and other terms are as defined previously.

For a beam with a rectangular cross-section, the bending stress, σ_x, varies linearly and shear stress, τ_{xy}, varies parabolically over the cross-section as shown in **Fig. 2.4**.

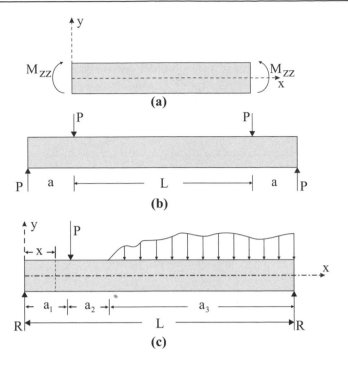

Fig. 2.3 Beams in bending

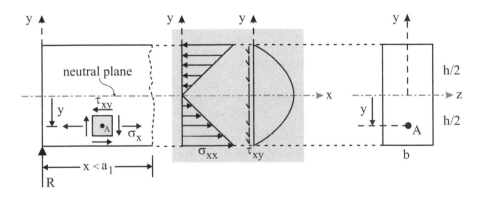

Fig. 2.4 Normal and shear stress variation in a rectangular beam in flexure.

Using **Eq. 2.14**, given the applied moment, **M**, geometry of the beam, and deflection at a point, it is possible to calculate the modulus, **E**. Strictly speaking, the equations for bending stress and beam deflections are only valid for pure bending as depicted in **Fig. 2.3(a-b)** but give good approximations for other types of loading such as that shown in **Fig. 2.3(c)** as long as the beam is not very short. Very short beams require a shear correction factor for beam deflection.

As an example, a beam in three-point bending as shown in **Fig. 2.5** is often used to determine a "flex (or flexural) modulus" which is reported in industry specification sheets describing a particular polymer.

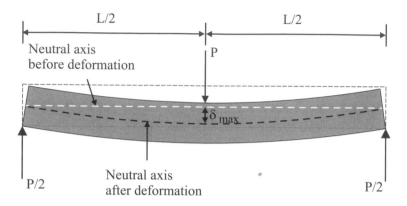

Fig. 2.5 Three-point bend specimen.

The maximum deflection can be shown to be,

$$\delta_{max} = \frac{PL^3}{48EI} \tag{2.15}$$

from which the flexural (flex) modulus is found to be,

$$E = \frac{PL^3}{48I} \frac{1}{\delta_{max}} \tag{2.16}$$

Fundamentally, any structure under load can be used to determine properties provided the stress can be calculated and the strain can be measured at the same location. However, it is important to note that no method is available to measure stress directly. Stresses can only be calculated through the determination of forces using Newton's laws. On the other hand, strain can be determined directly from measured deformations. That is, displacement or motion is the physically measured quantity and force (and hence stress) is a defined, derived or calculated quantity. Some might argue that photoelastic techniques may qualify for the direct measurement of stress but it can also be argued that this effect is due to interaction of light on changes in the atomic and molecular structure associated with a birefringent material, usually a polymer, caused by load induced displacements or strain.

It is clear that all the specimens used to determine properties such as the tensile bar, torsion bar and a beam in pure bending are special solid mechanics boundary value problems (BVP) for which it is possible to determine a "closed form" solution of the stress distribution using only the loading, the geometry, equilibrium equations and an assumption of a linear relation between stress and strain. It is to be noted that the same solutions of these BVP's from a first course in solid mechanics can be obtained using a more rigorous approach based on the Theory of Elasticity.

While the basic definitions of stress and strain are unchanged regardless of material, it should be noted that the elementary relations used above are not applicable to polymers in the region of viscoelastic behavior. For example, the rate of loading in a simple tension test will change the value measured for E in a viscoelastic material since modulus is inherently a function of time.

2.3. Typical Stress-Strain Properties

Properties of materials can be determined using the above elementary approaches. Often, for example, static tensile or compression tests are performed with a modern computer driven servo-hydraulic testing system such as the one shown in **Fig. 2.6**. The applied load is measured by a load cell (shown in (a) just above the grips) and deformation is found by either an extensometer (shown in (b) attached to the specimen) or an electrical resistance strain gage shown in (c). The latter is glued to the specimen and the change in resistance is measured as the specimen and the gage elongate. (Many additional methods are available to measure strain, including laser extensometers, moiré techniques, etc.) The cross-sectional area of the specimen and the gage length are input into the computer and the stress strain diagram is printed as the test is being run or can be stored for later use. The reason for a homogeneous state of stress and strain is now obvious. If a homogeneous state of stress and strain do not exist, it is only possible to determine the average strain value over the gage length region with this procedure and not the true properties of the material at a point.

Typical stress-strain diagrams for brittle and ductile materials are shown in **Fig. 2.7**. For brittle materials such as cast iron, glass, some epoxy resins, etc., the stress strain diagram is linear from initial loading (point 0) nearly to rupture (point B) when average strains are measured. As will be discussed subsequently, stress and strain are "point" quantities if the correct mathematical definition of each is used. As a result, if the strain were actually measured at a single point, i.e., the point of final failure, the stress and strain at failure even for a brittle material might be slightly higher than the average values shown in **Fig. 2.7**.

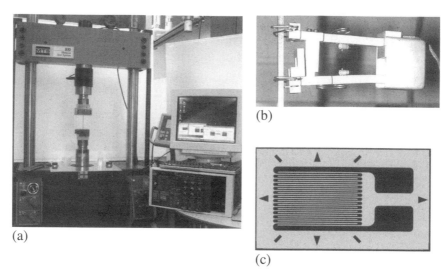

Fig. 2.6 (a) Servo-hydraulic testing system:
(b) extensometer
(c) electrical resistance strain gage.

For ductile materials such as many aluminum alloys, copper, etc., the stress-strain diagram may be nonlinear from initial loading until final rupture. However, for small stresses and strains, a portion may be well approximated by a straight line and an approximate proportional limit (point A) can be determined. For many metals and other materials, if the stress exceeds the proportional limit a residual or permanent deformation may remain when the specimen is unloaded and the material is said to have "yielded". The exact yield point may not be the same as the proportional limit and if this is the case the location is difficult to determine. As a result, an arbitrary "0.2% offset" procedure is often used to determine the yield point in metals. That is, a line parallel to the initial tangent to the stress-strain diagram is drawn to pass through a strain of 0.002 in./in. The yield point is then defined as the point of intersection of this line and the stress-strain diagram (point C in **Fig. 2.7**). This procedure can be used for polymers but the offset must be much larger than 0.2% definition used for metals. Procedures to find the yield point in polymers will be discussed in Chapter 3 and 11.

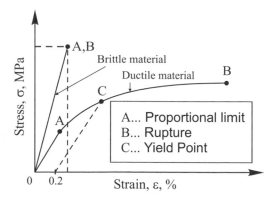

Fig. 2.7 Stress-strain diagrams for brittle and ductile materials

An approximate sketch of the stress-strain diagram for mild steel is shown in **Fig. 2.8(a)**. The numbers given for proportional limit, upper and lower yield points and maximum stress are taken from the literature, but are only approximations. Notice that the stress is nearly linear with strain until it reaches the upper yield point stress which is also known as the elastic-plastic tensile instability point. At this point the load (or stress) decreases as the deformation continues to increase. That is, less load is necessary to sustain continued deformation. The region between the lower yield point and the maximum stress is a region of strain hardening, a concept that is discussed in the next section. Note that if true stress and strain are used, the maximum or ultimate stress is at the rupture point.

The elastic-plastic tensile instability point in mild steel has received much attention and many explanations. Some polymers, such as polycarbonate, exhibit a similar phenomenon. Both steel and polycarbonate not only show an upper and lower yield point but visible striations of yielding, plastic flow or slip lines (Luder's bands) at an approximate angle of 54.7° to the load axis also occur in each for stresses equivalent to the upper yield point stress. (For a description and an example of Luder's band formation in polycarbonate, see **Fig. 3.7(c)**). It has been argued that this instability point (and the appearance of an upper and lower yield point) in metals is a result of the testing procedure and is related to the evolution of internal damage. That this is the case for polycarbonate will be shown in Chapter 3. For a discussion of these factors for metals, see Drucker (1962) and Kachanov (1986).

If the strain scale of **Fig. 2.8(a)** is expanded as illustrated in **Fig. 2.8(b)**, the stress-strain diagram of mild steel is approximated by two straight lines; one for the linear elastic portion and one which is horizontal at a stress level of the lower yield point. This characteristic of mild steel to "flow", "neck" or "draw" without rupture when the yield point has been exceeded has led to the concepts of plastic, limit or ultimate design. That is, just because the yield point has been exceeded does not mean that the material cannot support load. In fact, it can be shown that economy of design and weight savings can be obtained using limit design concepts. Concepts of plasticity and yielding date back to St. Venant in about 1870 but the concepts of plastic or limit design have evolved primarily in the last 50 years or so (see Westergaard (1964) for a discussion of the history of solid mechanics including comments on the evolution of plasticity). Computational plasticity has its origins associated with calculations of deformations beyond the yield point for stress-strain diagrams similar to that of mild steel and will be briefly discussed in Chapter 11 in the context of polymers.

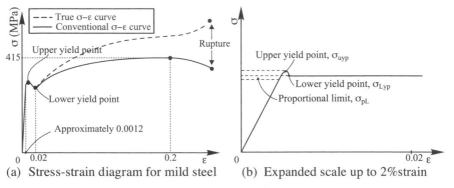

(a) Stress-strain diagram for mild steel   (b) Expanded scale up to 2%strain

Fig. 2.8 Typical tensile stress-strain diagrams (not to scale).

As will be discussed in Chapter 3, the same procedures discussed in the present chapter are used to determine the stress-strain characteristics of polymers. If only a single rate of loading is used, similar results will be obtained. On the other hand, if polymers are loaded at various strain rates, the behavior varies significantly from that of metals. Generally, metals do not show rate effects at ambient temperatures. They do, however, show considerable rate effects at elevated temperatures but the molecular mechanisms responsible for such effects are very different in polymers and metals.

It is appropriate to note that industry specification sheets often give the elastic modulus, yield strength, strain to yield, ultimate stress and strain to failure as determined by these elementary techniques. One objective of this text is to emphasize the need for approaches to obtain more appropriate specifications for the engineering design of polymers.

2.4. Idealized Stress-Strain Diagrams

The stress-strain diagrams discussed in the last section are often approximated by idealized diagrams. For example, a linear elastic perfectly brittle material is assumed to have a stress-strain diagram similar to that given in **Fig. 2.9(a).** On the other hand, the stress-strain curve for mild steel can be approximated as a perfectly elastic-plastic material with the stress-strain diagram given in **Fig. 2.9(b).** Metals (and polymers) often have nonlinear stress-strain behavior as shown in **Fig. 2.10(a).** These are sometimes modeled with a bilinear diagram as shown in **Fig. 2.10(b)** and are referred to as a perfectly linear elastic strain hardening material.

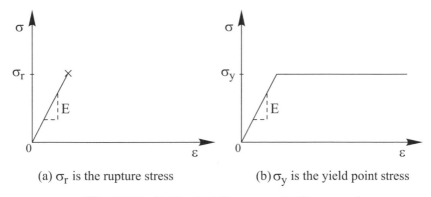

(a) σ_r is the rupture stress   (b) σ_y is the yield point stress

Fig. 2.9 Idealized uniaxial stress-strain diagrams:
(a) Linear elastic perfectly brittle.
(b) Linear elastic perfectly plastic.

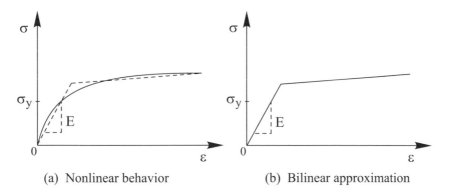

(a) Nonlinear behavior   (b) Bilinear approximation

Fig. 2.10   Nonlinear stress-strain diagram with linear elastic strain hardening approximation (σ_y is the yield point stress).

2.5. Mathematical Definitions of Stress, Strain and Material Characteristics

The previous sections give a brief review of some elementary concepts of solid mechanics which are often used to determine basic properties of most engineering materials. However, these approaches are sometimes not adequate and more advanced concepts from the theory of elasticity or the theory of plasticity are needed. Herein, a brief discussion is given of some of the more exact modeling approaches for linear elastic materials. Even these methods need to be modified for viscoelastic materials but this section will only give some of the basic elasticity concepts.

Definition of a Continuum: A basic assumption of elementary solid mechanics is that a material can be approximated as a continuum. That is, the material (of mass ΔM) is continuously distributed over an arbitrarily small volume, ΔV, such that,

$$\lim_{\Delta V \to 0} \frac{\Delta M}{\Delta V} = \frac{dM}{dV} = \text{const.} = \rho = (\text{density at a point}) \qquad (2.17)$$

Quite obviously such an assumption is at odds with our knowledge of the atomic and molecular nature of materials but is an acceptable approximation for most engineering applications. The principles of linear elasticity, though based upon the premise of a continuum, have been shown to be useful in estimating the stress and strain fields associated with dislocations and other non-continuum microstructural details.

Physical and Mathematical Definition of Normal Stress and Shear Stress: Consider a body in equilibrium under the action of external forces F_i as shown in **Fig. 2.11(a).** If a cutting plane is passed through the body as shown in **Fig. 2.11(b)**, equilibrium is maintained on the remaining portion by internal forces distributed over the surface **S**.

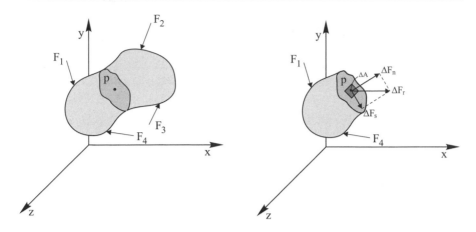

Fig. 2.11 Physical definition of normal force and shear force.

At any arbitrary point **p**, the incremental resultant force, $\Delta \mathbf{F_r}$, on the cut surface can be broken up into a normal force in the direction of the normal, **n**, to surface **S** and a tangential force parallel to surface **S**. The normal stress and the shear stress at point **p** is mathematically defined as,

$$\sigma_n = \lim_{\Delta A \to 0} \frac{\Delta F_n}{\Delta A} \quad \tau_s = \lim_{\Delta A \to 0} \frac{\Delta F_s}{\Delta A} \tag{2.18}$$

where $\Delta \mathbf{F_n}$ and $\Delta \mathbf{F_s}$ are the normal and shearing forces on the area $\Delta \mathbf{A}$ surrounding point **p**.

Alternatively, the resultant force, $\Delta \mathbf{F_r}$, at point **p** can be divided by the area, $\Delta \mathbf{A}$, and the limit taken to obtain the stress resultant σ_r as shown in **Fig. 2.12**. Normal and tangential components of this stress resultant will then be the normal stress σ_n and shear stress τ_s at point **p** on the infinitesimal area $\Delta \mathbf{A}$.

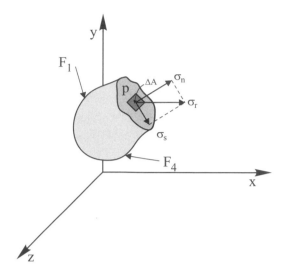

Fig. 2.12 Stress resultant definition.

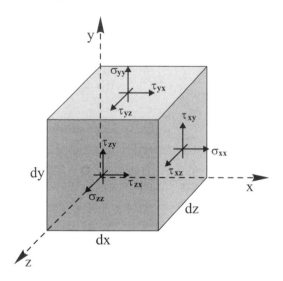

Fig. 2.13 Cartesian components of internal stresses.

If a pair of cutting planes a differential distance apart are passed through the body parallel to each of the three coordinate planes, a cube will be identified. Each plane will have normal and tangential components of the stress resultants. The tangential or shear stress resultant on each plane can further be represented by two components in the coordinate directions. The internal stress state is then represented by three stress components on each coordinate plane as shown in **Fig. 2.13**. (Note that equal and opposite components will exist on the unexposed faces.) Therefore at any point in a body there will be nine stress components. These are often identified in matrix form such that,

$$\sigma_{ij} = \begin{pmatrix} \sigma_{xx} & \tau_{xy} & \tau_{xz} \\ \tau_{yx} & \sigma_{yy} & \tau_{yz} \\ \tau_{zx} & \tau_{zy} & \sigma_{zz} \end{pmatrix} \tag{2.19}$$

Using equilibrium, it is easy to show that the stress matrix is symmetric or,

$$\tau_{xy} = \tau_{yx}, \qquad \tau_{xz} = \tau_{zx}, \qquad \tau_{yz} = \tau_{zy} \tag{2.20}$$

leaving only six independent stresses existing at a material point.

Physical and Mathematical Definition of Normal Strain and Shear Strain: If a differential element is acted upon by stresses as shown in **Fig. 2.14(a)** both normal and shearing deformations will result. The resulting deformation in the x-y plane is shown in **Fig. 2.14(b)**, where **u** is the displacement component in the x direction and **v** is the displacement component in the y direction.

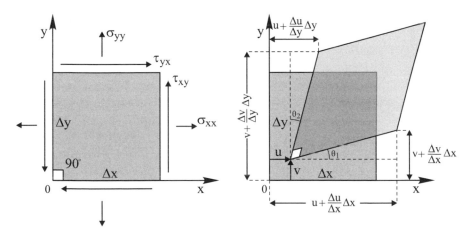

Fig. 2.14 Definitions of displacements **u** and **v** and corresponding shear and normal strains.

The unit change in the $\mathbf{\Delta x}$ dimension will be the strain ε_{xx} and is given by,

$$\varepsilon_{xx} = \lim_{\Delta x \to 0} \left\{ \frac{\left(u + \dfrac{\Delta u}{\Delta x} \Delta x \right) - u}{\Delta x} \right\} \tag{2.21}$$

with similar definitions for the unit change in the y and z directions. (The assumption of small strain and linear behavior is implicit here with the assumption that θ is small and thus its impact on $\mathbf{\Delta u}$ is ignored.) Therefore the normal strains in the three coordinate directions are defined as,

$$\varepsilon_{xx} = \lim_{\Delta x \to 0} \frac{\Delta u}{\Delta x} = \frac{\partial u}{\partial x} \;, \quad \varepsilon_{yy} = \lim_{\Delta y \to 0} \frac{\Delta v}{\Delta y} = \frac{\partial v}{\partial y} \;,$$

$$\varepsilon_{zz} = \lim_{\Delta z \to 0} \frac{\Delta w}{\Delta z} = \frac{\partial w}{\partial z} \tag{2.22}$$

where $\mathbf{u}$, $\mathbf{v}$ and $\mathbf{w}$ are the displacement-components in the three coordinate directions at a point. Shear strains are defined as the distortion of the original 90° angle at the origin or the sum of the angles $\theta_1 + \theta_2$. That is, again using the small deformation assumption,

$$\tan\left(\theta_1 + \theta_2\right) \approx \left(\theta_1 + \theta_2\right) = \lim_{\Delta x, \Delta y \to 0} \left[\frac{\left(v + \dfrac{\Delta v}{\Delta x} \Delta x \right) - v}{\Delta x} + \frac{\left(u + \dfrac{\Delta u}{\Delta y} \right) - u}{\Delta y} \right] \tag{2.23}$$

which leads to the three shear strains,

$$\gamma_{xy} = \left(\frac{\partial v}{\partial x} + \frac{\partial u}{\partial y} \right) \;, \quad \gamma_{xz} = \left(\frac{\partial w}{\partial x} + \frac{\partial u}{\partial z} \right) \;, \quad \gamma_{yz} = \left(\frac{\partial w}{\partial y} + \frac{\partial v}{\partial z} \right) \tag{2.24}$$

Stresses and strains are often described using tensorial mathematics but in order for strains to transform as tensors, the definition of shear strain must be modified to include a factor of one half as follows,

$$\varepsilon_{xy} = \frac{1}{2}\left(\frac{\partial v}{\partial x} + \frac{\partial u}{\partial y} \right) \;, \quad \varepsilon_{xz} = \frac{1}{2}\left(\frac{\partial w}{\partial x} + \frac{\partial u}{\partial z} \right) \;, \quad \varepsilon_{yz} = \frac{1}{2}\left(\frac{\partial w}{\partial y} + \frac{\partial v}{\partial z} \right) \tag{2.25}$$

The difference between the latter two sets of equations can lead to very erroneous values of stress when attempting to use an electrical strain gage rosette to determine the state of stress experimentally. In **Eqs. 2.25** the traditional symbol ε with mixed indices has been used to identity tensorial shear strain. The symbol γ with mixed indices will be used to describe non-tensorial shear strain, also called engineering strain.

In general, as with stresses, nine components of strain exist at a point and these can be represented in matrix form as,

$$\epsilon_{ij} = \begin{pmatrix} \epsilon_{xx} & \epsilon_{xy} & \epsilon_{xz} \\ \epsilon_{yx} & \epsilon_{yy} & \epsilon_{yz} \\ \epsilon_{zx} & \epsilon_{zy} & \epsilon_{zz} \end{pmatrix} \qquad (2.26)$$

Again, it is possible to show that the strain matrix is symmetric or that,

$$\epsilon_{xy} = \epsilon_{yx} , \qquad \epsilon_{xz} = \epsilon_{zx} , \qquad \epsilon_{yz} = \epsilon_{zy} \qquad (2.27)$$

Hence there are only six independent strains.

Generalized Hooke's Law: As noted previously, Hooke's law for one dimension or for the condition of uniaxial stress and strain for elastic materials is given by $\sigma = E\,\epsilon$. Using the principle of superposition, the generalized Hooke's law for a three dimensional state of stress and strain in a homogeneous and isotropic material can be shown to be,

$$\epsilon_{xx} = \frac{1}{E}\left[\sigma_{xx} - \nu\left(\sigma_{yy} + \sigma_{zz}\right)\right], \quad \gamma_{xy} = \frac{\tau_{xy}}{G}$$

$$\epsilon_{yy} = \frac{1}{E}\left[\sigma_{yy} - \nu\left(\sigma_{xx} + \sigma_{zz}\right)\right], \quad \gamma_{yz} = \frac{\tau_{yz}}{G} \qquad (2.28)$$

$$\epsilon_{zz} = \frac{1}{E}\left[\sigma_{zz} - \nu\left(\sigma_{xx} + \sigma_{yy}\right)\right], \quad \gamma_{xz} = \frac{\tau_{xz}}{G}$$

where **E, G** and ν are Young's modulus, the shear modulus and Poisson's ratio respectively. Only two are independent and as indicated earlier,

$$G = \frac{E}{2(1 + \nu)} \qquad (2.29)$$

The proof for **Eq. 2.29** may be found in many elementary books on solid mechanics.

Other forms of the generalized Hooke's law can be found in many texts. The relation between various material constants for linear elastic materials are shown below in **Table 2.1** where **E, G** and ν are previously defined and where **K** is the bulk modulus and λ is known as Lame's constant.

Table 2.1 Relation between various elastic constants. λ and G are often termed Lame' constants and K is the bulk modulus.

$$\dagger \ A \equiv \sqrt{(E+\lambda)^2 + 8\lambda^2}$$

	Lamé's Modulus, λ	Shear Modulus, G	Young's Modulus, E	Poisson's Ratio, ν	Bulk Modulus, K
λ,G			$\dfrac{G(3\lambda+2G)}{\lambda+G}$	$\dfrac{\lambda}{2(\lambda+G)}$	$\dfrac{3\lambda+2G}{3}$
λ,E		$\dfrac{A^\dagger+(E-3\lambda)}{4}$		$\dfrac{A^\dagger-(E+\lambda)}{4\lambda}$	$\dfrac{A^\dagger+(3\lambda+E)}{6}$
λ,ν		$\dfrac{\lambda(1-2\nu)}{2\nu}$	$\dfrac{\lambda(1+\nu)(1-2\nu)}{\nu}$		$\dfrac{\lambda(1+\nu)}{3\nu}$
λ,K		$\dfrac{3(K-\lambda)}{2}$	$\dfrac{9K(K-\lambda)}{3K-\lambda}$	$\dfrac{\lambda}{3K-\lambda}$	
G,E	$\dfrac{(2G-E)G}{E-3G}$			$\dfrac{E-2G}{2G}$	$\dfrac{GE}{3(3G-E)}$
G,ν	$\dfrac{2G\nu}{1-2\nu}$		$2G(1+\nu)$		$\dfrac{2G(1+\nu)}{3(1-2\nu)}$
G,K	$\dfrac{3K-2G}{3}$		$\dfrac{9KG}{3K+G}$	$\dfrac{3K-2G}{2(3K+G)}$	
E,ν	$\dfrac{\nu E}{(1+\nu)(1-2\nu)}$	$\dfrac{E}{2(1+\nu)}$			$\dfrac{E}{3(1-2\nu)}$
E,K	$\dfrac{3K(3K-E)}{(9K-E)}$	$\dfrac{3EK}{9K-E}$		$\dfrac{3K-E}{6K}$	
ν,K	$\dfrac{3K\nu}{1+\nu}$	$\dfrac{3K(1-2\nu)}{2(1+\nu)}$	$3K(1-2\nu)$		

Hooke's law is a mathematical statement of the linear relation between stress and strain and usually implies both small strains ($\varepsilon^2 \ll \varepsilon$) and small deformations. It is also to be noted that in general elasticity solutions in two and three dimensions, the displacement, stress and strain variables are functions of spatial position, x_i. This will be handled more explicitly in Chapter 9.

Again, it is important to note that stress and strain are point quantities, yet methods for strain measurement are not capable of measuring strain at an infinitesimal point. Thus, average values are measured and moduli are obtained using stresses calculated at a point. For this reason, strains are best measured where no gradients exist or are so small that an average is a good approximation. One approach when large gradients exist is to try to measure the gradient and extrapolate to a point. The development of methods to measure strains within very small regions has become a topic of great importance due to the development of micro-devices and machines. Further, such concerns as interface or interphase properties in multi-phase materials also creates the need for new micro strain measurement techniques.

Indicial notation and compact form of generalized Hooke's Law: Because of the cumbersome form of the generalized Hooke's Law for material constitutive response in three dimensions (**Eq. 2.28**), a shorthand notation referred to as indicial or index notation is extensively used. Here we provide a brief summary of indicial notation and further details may be found in many books on continuum mechanics (e.g., Flügge, 1972). In indicial notation, the subscripts on tensors are used with very precise rules and conventions and provide a compact way to relate and manipulate tensorial expressions.

The conventions are as follows:
- Subscripts indicating coordinate direction (**x, y, z**) can be generally represented by a roman letter variable that is understood to take on the values of **1, 2**, or **3**. For example, the stress tensor can be written as σ_{ij} which then gives reference to the entire 3x3 matrix. That is the stress and strain matrices given by **Eqs. 2.19** and **2.26** become,

$$\sigma_{ij} = \begin{pmatrix} \sigma_{11} & \sigma_{12} & \sigma_{13} \\ \sigma_{21} & \sigma_{22} & \sigma_{23} \\ \sigma_{31} & \sigma_{32} & \sigma_{33} \end{pmatrix} \quad \varepsilon_{ij} = \begin{pmatrix} \varepsilon_{11} & \varepsilon_{12} & \varepsilon_{13} \\ \varepsilon_{21} & \varepsilon_{22} & \varepsilon_{23} \\ \varepsilon_{31} & \varepsilon_{32} & \varepsilon_{33} \end{pmatrix} \qquad (2.30)$$

- Summation convention: if the same index appears twice in any term, summation is implied over that index (unless suspended by the phrase "no sum"). For example,

$$\sigma_{ii} = \sigma_{11} + \sigma_{22} + \sigma_{33} \qquad (2.31)$$

- Free index: non-repeated subscripts are called free subscripts since they are free to take on any value in 3D space. The count of the free indices on a variable indicates the order of the tensor. e.g. F_i is a vector (first order tensor), σ_{ij} is a second order tensor.
- Dummy index: repeated subscripts are called dummy subscripts, since they can be changed freely to another letter with no effect on the equation.
- Rule 1: The same subscript cannot appear more than twice in any term.

- Rule 2: Free indices in each term (both sides of the equation) must agree (all terms in an equation must be of the same order).
  Example of valid expression: $v_i = a_{ij}u_j - \lambda e_{kl}d_{ikl}$
- Rule 3: Both free and repeated indices may be replaced with others subject to the rules.
  Example of valid expression: $a_{ij}u_j + d_i = a_{ik}u_k + d_i$
- Unlike in vector algebra, the order of the variables in a term is unimportant, as the bookkeeping is done by the subscripts. For example consider the inner product of a second order tensor and a vector:

$$A_{ij}u_j = u_j A_{ij} \qquad (2.32)$$

- Differentiation with respect to spatial coordinates is represented by a comma, for example

$$\frac{dv_i}{dx_j} = v_{i,j} \qquad (2.33)$$

- The identity matrix is also referred to as the Kronecker Delta function and is represented by

$$\delta_{ij} = \begin{cases} 1, \text{if } i = j \\ 0, \text{if } i \neq j \end{cases} \qquad (2.34)$$

The properties of δ_{ij} are thus

$$\begin{aligned} \delta_{ii} &= 3 \\ \delta_{ij}v_j &= v_i \\ \delta_{ij}\delta_{jk} &= \delta_{ik} \\ \delta_{ij}\sigma_{jk} &= \sigma_{ik} \end{aligned} \qquad (2.35)$$

Although the conventions listed above may seem tedious at first, with a little practice index notation provides many advantages including easier manipulations of matrix expressions. Additionally, it is a very compact notation and the rules listed above can often be used during manipulation to reduce errors in derivations.

The generalized Hooke's Law from **Eq. 2.28** may be rewritten to relate tensorial stress and strain in index notation as follows:

$$\varepsilon_{ij} = \frac{1+\nu}{E}\sigma_{ij} - \frac{\nu}{E}\sigma_{kk}\delta_{ij} \qquad (2.36)$$

or

$$\sigma_{ij} = 2G\varepsilon_{ij} + \lambda\varepsilon_{kk}\delta_{ij} \tag{2.37}$$

Additionally, the strain-displacement relations, **Eqs. 2.22** and **2.25**, can be written as

$$\varepsilon_{ij} = \frac{1}{2}\left(u_{i,j} + u_{j,i}\right) \tag{2.38}$$

where u_i are the three displacement components, represented as **u**, **v**, and **w** earlier (e.g., $u_2=v$).

These expressions will be used extensively later in Chapter 9 when dealing with viscoelasticity problems in two and three dimensions.

Consequences of Homogeneity and Isotropy Assumptions: It is interesting to examine the consequences if a material is linearly elastic but not homogeneous or isotropic. For such a material, the generalized Hooke's law is often expressed using index notation as,

$$\sigma_{ij} = E_{ijkq}\varepsilon_{kq} \tag{2.39}$$

For a material that is nonhomogeneous, the material properties are a function of spatial position and E_{ijkq} becomes $E_{ijkq}(x,y,z)$. The nonhomogencity for a particular material determines exactly how the moduli vary across the material. The geometry of the material on an atomic or even microscale determines symmetry relationships that govern the degree of anisotropy of the material. Without regard to symmetry constraints, **Eq. 2.39** could have 81 independent proportionality properties relating stress components to strain components.

The complete set of nine equations (one for each stress) each with nine coefficients (one for each strain term) can be found from **Eq. 2.39**. This is accomplished using the summation convention over repeated indices. That is, **Eq. 2.39** is understood to be a double summation as follows,

$$\sigma_{ij} = \sum_{k=1}^{3}\sum_{q=1}^{3} E_{ijkq}\varepsilon_{kq} \tag{2.40}$$

(The expansion is left as an exercise for the reader. See problem 2.4.)

If a material is nonlinear elastic as well as heterogeneous and anisotropic, **Eq. 2.39** becomes,

$$\sigma_{ij} = E_{ijkl}(x,y,z)\varepsilon_{kl} + E'_{ijkl}(x,y,z)\varepsilon_{kl}^{2} + \cdots \tag{2.41}$$

Again each term on the right hand side of **Eq. 2.40** represents a double summation and each coefficient of strain is an independent set of material parameters. Thus, many more than 81 parameters may be required to represent a nonlinear heterogeneous and anisotropic material. Further, for viscoelastic materials, these material parameters are time dependent. The introduction of the assumption of linearity reduces the number of parameters to 81 while homogeneity removes their spatial variation (i.e., the E_{ijkq} parameters are now constants). Symmetry of the stress and strain tensors (matrices) reduces the number of constants to 36. The existence of a strain energy potential reduces the number of constants to 21. Material symmetry reduces the number of constants further. For example, an orthotropic material, one with three planes of material symmetry, has only 9 constants and an isotropic material, one with a center of symmetry, has only two independent constants (and **Eq. 2.39** reduces to **Eq. 2.28**). Now it is easy to see why the assumptions of linearity, homogeneity and isotropy are used for most engineering analyses.

A plane of material symmetry exists within a material when the material properties (elastic moduli) at mirror imaged points across the plane are identical. For example, in the sketch given in **Fig. 2.15**, the yz plane is a plane of symmetry and the elastic moduli would be the same at the material points A and B.

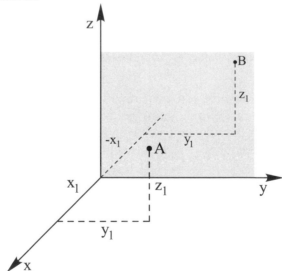

Fig. 2.15 Definition of a plane of material symmetry.

Experimentation is needed to determine if a material is homogeneous or isotropic. One approach is to cut small tensile coupons from a three-dimensional body and perform a uniaxial tensile or compressive test as well as a torsion test for shear. Obviously, to obtain a statistical sample of specimens at a single point would require exact replicas of the same material or a large number of near replicas. Assuming that such could be accomplished for a body with points A and B as in **Fig. 2.15**, the following relationships would hold for homogeneity,

$$E_{xxxx}\big|_A = E_{xxxx}\big|_B \qquad E_{yyyy}\big|_A = E_{yyyy}\big|_B \qquad E_{zzzz}\big|_A = E_{zzzz}\big|_B \qquad (2.42)$$

That is, the modulus components are invariant (constant) for all directions at a point. (See Problem 2.5.)

The above measurement approach illustrates the influence of heterogeneity and anisotropy on moduli but is not very practical. A sonic method of measuring properties, though not as precise as tensile or torsion tests, is often used and is based upon the fact that the speed of sound, v_s, in a medium is related to its modulus of elasticity, E, and density, ρ, such that (Kolsky, 1963),

$$v_s = \sqrt{\frac{E}{\rho}} \qquad (2.43)$$

The above is adequate for a thin long bar of material but for three-dimensional bodies the velocity is related to both dilatational (volume change - see subsequent section for definition) and shear effects as well as geometry effects, etc.

It is to be noted that the condition of heterogeneity and anisotropy are confronted when considering many materials used in engineering design. For example, while many metals are isotropic on a macroscopic scale, they are crystalline on a microscopic scale. Crystalline materials are at least anisotropic and may be heterogencous as well. Wood is both heterogeneous and anisotropic as are many ceramic materials. Modern polymer, ceramic or metal matrix composites such as fiberglass, etc. are both heterogeneous and anisotropic. The mathematical analysis of such materials often neglects the effect of heterogeneity but does include anisotropic effects. (See Lekhnitskii, (1963), Daniel, (1994)).

2.6. Principal Stresses

In the study of viscoelasticity as in the study of elasticity, it is mandatory to have a thorough understanding of methods to determine principal stresses and strains. Principal stresses are defined as the normal stresses on the planes oriented such that the shear stresses are zero - the maximum and minimum normal stresses at a point are principal stresses. The determination of stresses and strains in two dimensions is well covered in elementary solid mechanics both analytically and semi-graphically using Mohr's circle. However, practical stress analysis problems frequently involve three dimensions. The basic equations for transformation of stresses in three-dimensions, including the determination of principal stresses, will be given and the interested reader can find the complete development in many solid mechanics texts.

Often in stress analysis it is necessary to determine the stresses (strains) in a new coordinate system after calculating or measuring the stresses (strains) in another coordinate system. In this connection, the use of index notation is very helpful as it can be shown that the stress σ'_{ij} in a new co-ordinate system, x'_i, can be easily obtained from the σ_{ij} in the old coordinate system, x_i, by the equation,

$$\sigma'_{ij} = a_{ik} a_{jq} \sigma_{kq} \tag{2.44}$$

where the quantities a_{ij} are the direction cosines between the axes x'_i and x_i and may be given in matrix form as,

$$a_{ij} = \begin{pmatrix} a_{11} & a_{12} & a_{13} \\ a_{21} & a_{22} & a_{23} \\ a_{31} & a_{32} & a_{33} \end{pmatrix} \tag{2.45}$$

In **Eq. 2.44**, the repeated indices on the right again indicate summation over the three coordinates, **x,y,z** or the indices **1,2,3**. It is left as an exercise for the reader to show that this process leads to the familiar two-dimensional expressions found in the first course in solid mechanics (see Problem 2.6.),

$$\sigma'_x = \sigma_x \cos^2 \vartheta + \sigma_y \sin^2 \vartheta + 2\tau_{xy} \sin \vartheta \cos \vartheta \tag{2.46a}$$

or

$$\sigma'_x = \frac{\sigma_x + \sigma_y}{2} + \frac{\sigma_x - \sigma_y}{2} \cos 2\vartheta + \tau_{xy} \sin 2\vartheta \tag{2.46b}$$

$$\tau'_{xy} = -\left(\sigma_x - \sigma_y\right)\sin\vartheta\cos\vartheta + \tau_{xy}\left(\cos^2\vartheta - \sin^2\vartheta\right) \qquad (2.47a)$$

or

$$\tau'_{xy} = -\frac{\sigma_x - \sigma_y}{2}\sin 2\vartheta + \tau_{xy}\cos 2\vartheta \qquad (2.47b)$$

Using **Eq. 2.44** it is possible to show that the three principal stresses (strains) can be calculated from the following cubic equation,

$$\sigma_i^3 - I_1\sigma_i^2 + I_2\sigma_i - I_3 = 0 \qquad (2.48)$$

where the principal stresses, σ_i, are given by one of the three roots σ_1, σ_2 or σ_3 and,

$$I_1 = \sigma_{xx} + \sigma_{yy} + \sigma_{zz} = \sigma_1 + \sigma_2 + \sigma_3$$

$$I_2 = \sigma_{xx}\sigma_{yy} + \sigma_{yy}\sigma_{zz} + \sigma_{xx}\sigma_{zz} - \sigma_{xy}^2 - \sigma_{yz}^2 - \sigma_{xz}^2 = \sigma_1\sigma_2 + \sigma_2\sigma_3 + \sigma_3\sigma_1 \qquad (2.49)$$

$$I_3 = \sigma_{xx}\sigma_{yy}\sigma_{zz} - \sigma_{xx}\sigma_{yz}^2 - \sigma_{yy}\sigma_{xz}^2 - \sigma_{zz}\sigma_{xy}^2 + 2\sigma_{xy}\sigma_{yz}\sigma_{zx} = \sigma_1\sigma_2\sigma_3$$

The quantities I_1, I_2, and I_3 are the same for any arbitrary coordinate system located at the same point and are therefore called invariants.

In two-dimensions when $\sigma_{zz} = 0$ and a state of plane stress exists, **Eq. 2.48** reduces to the familiar form,

$$\sigma_{1,2} = \frac{\sigma_{xx} + \sigma_{yy}}{2} \pm \sqrt{\left(\frac{\sigma_{xx} - \sigma_{yy}}{2}\right)^2 + \left(\tau_{xy}\right)^2} \qquad (2.50)$$

where the comma does not indicate differentiation in this case, but is here used to emphasize the similarity in form of the two principle stresses by writing them in one equation. The proof of **Eq. 2.50** is left as an exercise for the reader (see Problem 2.7).

The directions of principal stresses (strains) are also very important. However, the development of the necessary equations will not be presented here but it might be noted that the procedure is an eigenvalue problem associated with the diagonalization of the stress (strain) matrix.

2.7. Deviatoric and Dilatational Components of Stress and Strain

A general state of stress at a point or the stress tensor at a point can be separated into two components, one of which results in a change of shape (deviatoric) and one which results in a change of volume (dilatational). Shape changes due to a pure shear stress such as that of a bar in torsion given in **Fig. 2.2** are easy to visualize and are shown by the dashed lines in **Fig. 2.16(a)** (assuming only a horizontal motion takes place).

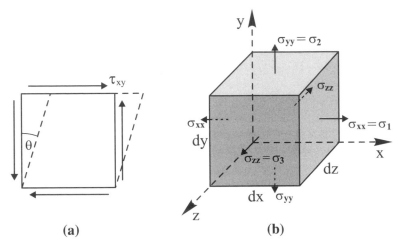

(a)   **(b)**

Fig. 2.16   (a) Shape changes due to pure shear.
(b) Normal stresses leading to a pure volume change.

Shear Modulus: Because only shear stresses and strains exist for the case of pure shear, the shear modulus can easily be determined from a torsion test by measuring the angle of twist over a prescribed length under a known torque, i.e.,

$$T = \theta \frac{JG}{L} \tag{2.51}$$

where all terms are as previously defined in **Eq. 2.11**.

Bulk Modulus: Volume changes are produced only by normal stresses. For example, consider an element loaded with only normal stresses (principal stresses) as shown in **Fig. 2.16(b)**. The change in volume can be shown to be (for small strains),

$$\frac{\Delta V}{V} = \varepsilon_{xx} + \varepsilon_{yy} + \varepsilon_{zz} \tag{2.52}$$

Substituting the values of strains from the generalized Hooke's law, **Eq. 2.28**, gives,

$$\frac{\Delta V}{V} = \frac{1-2v}{E}\left(\sigma_{xx} + \sigma_{yy} + \sigma_{zz}\right) \tag{2.53}$$

If Poisson's ratio is $v = 0.5$, the change in volume is zero or the material is incompressible. Here it is important to note that Poisson's ratio for metals and many other materials in the linear elastic range is approximately 0.33 (i.e., $v \sim 1/3$). However, near and beyond the yield point, Poisson's ratio is approximately 0.5 (i.e., $v \sim 1/2$). That is, when materials yield, neck or flow, they do so at constant volume.

In the case when all the stresses on the element in **Fig. 2.16(b)** are equal ($\sigma_{xx} = \sigma_{yy} = \sigma_{zz} = \sigma$), a spherical state of stress (hydrostatic stress) is said to exist and,

$$\frac{\Delta V}{V} = \frac{1-2v}{E}\left(3\sigma\right) \tag{2.54}$$

By equating **Eqs. 2.52** and **2.54** the Bulk Modulus can be defined as the ratio of the hydrostatic stress, σ, to volumetric strain or unit change in volume ($\Delta V/V$),

$$K = \frac{E}{3\left(1-2v\right)} \tag{2.55}$$

Notice that the bulk modulus becomes infinite, $K \sim \infty$, if the material is incompressible and Poisson's ratio is, $v \sim 1/2$.

Obviously, one method for obtaining the bulk modulus of a material would be to create a hydrostatic compression (or tension) state of stress and measure the resulting volume change.

Dilatational and Deviatoric Stresses for a General State of Stress: For a general stress state, the dilatational or volumetric component is defined by the mean stress or the average of the three normal stress components shown in **Fig. 2.13**,

$$\sigma = \sigma_m = \frac{\sigma_{xx} + \sigma_{yy} + \sigma_{zz}}{3} = \frac{1}{3}\sigma_{kk} \tag{2.56}$$

In **Eq. 2.56** care has been taken to provide three different symbolic ways of indicating the volumetric stress, σ, σ_m, or $\sigma_{kk}/3$ to emphasize the many notations found in the literature. Since the sum of the normal stresses is the first Invariant, I_1, the mean stress, σ_m, will be the same for any axis orientation at a point including the principal axes as shown in **Eq. 2.56**. Thus, independent of axis orientation the general stress state can be separated into a volumetric component plus a shear component as shown in **Fig. 2.17**. That is, if the stresses responsible for volumetric changes are subtracted from a general stress state, only stresses responsible for shape changes remain. This statement can be expressed in matrix form as,

$$
\begin{pmatrix} \sigma_{xx} & \tau_{xy} & \tau_{xz} \\ \tau_{yx} & \sigma_{yy} & \tau_{yz} \\ \tau_{zx} & \tau_{zy} & \sigma_{zz} \end{pmatrix} = \begin{pmatrix} \sigma_m & 0 & 0 \\ 0 & \sigma_m & 0 \\ 0 & 0 & \sigma_m \end{pmatrix} + \begin{pmatrix} s_{xx} & s_{xy} & s_{xz} \\ s_{yx} & s_{yy} & s_{yz} \\ s_{zx} & s_{zy} & s_{zz} \end{pmatrix} \quad (2.57)
$$

or in index notation as

$$
\sigma_{ij} = \frac{1}{3}\sigma_{kk}\delta_{ij} + s_{ij} \quad (2.58)
$$

where s_{ij} are the deviator (shape change) components of stress and δ_{ij} is the Kronecker Delta function as defined earlier (**Eq. 2.34**).

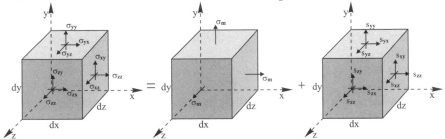

Fig. 2.17 Separation of a general stress state into dilatational and deviator stresses.

Since the trace of the first two matrices in **Eq. 2.52** are the same, i.e.,

$$
\sigma_{kk} = \sigma_{xx} + \sigma_{yy} + \sigma_{zz} = 3\sigma_m \quad (2.59)
$$

the trace of the third matrix is zero, i.e.,

$$
s_{kk} = s_{xx} + s_{yy} + s_{zz} = 0 \quad (2.60)
$$

Using **Eq. 2.60**, the deviator matrix can be separated into five simple shear stress systems,

$$
\begin{pmatrix} s_{xx} & s_{xy} & s_{xz} \\ s_{yx} & s_{yy} & s_{yz} \\ s_{zx} & s_{zy} & s_{zz} \end{pmatrix} = \begin{pmatrix} 0 & s_{xy} & 0 \\ s_{yx} & 0 & 0 \\ 0 & 0 & 0 \end{pmatrix} + \begin{pmatrix} 0 & 0 & 0 \\ 0 & 0 & s_{yz} \\ 0 & s_{zy} & 0 \end{pmatrix}
$$

$$+\begin{pmatrix} 0 & 0 & s_{xz} \\ 0 & 0 & 0 \\ s_{zx} & 0 & 0 \end{pmatrix}+\begin{pmatrix} s_{xx} & 0 & 0 \\ 0 & -s_{xx} & 0 \\ 0 & 0 & 0 \end{pmatrix}+\begin{pmatrix} 0 & 0 & 0 \\ 0 & -s_{zz} & 0 \\ 0 & 0 & s_{zz} \end{pmatrix} \qquad (2.61)$$

That the stress states given by the first three matrices on the right side of **Eq. 2.61** are pure shear states is obvious. The last two are also pure shear states but at 45° to the indicated axis as shown in **Fig. 2.18**.

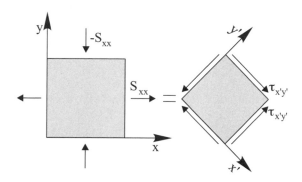

Fig. 2.18 Pure shear state.

Therefore each term in **Eq. 2.61** represents a pure shear state and results in only shape changes with no volume change.

Strains can also be separated into dilatational and deviatoric components and the equation for strain analogous to **Eq. 2.58** is,

$$\varepsilon_{ij} = e_{ij} + \varepsilon_m \delta_{ij} \quad \text{or} \quad \varepsilon_{ij} = e_{ij} + \frac{1}{3}\varepsilon_{kk}\delta_{ij} \qquad (2.62)$$

where e_{ij} are the deviatoric strains and $e_m = \frac{1}{3}\varepsilon_{kk}$ is the dilatational component. The trace of the strain tensor analogous to **Eq. 2.59** can also clearly be written.

The generalized Hooke's law given by **Eq. 2.28** or **Eq. 2.36** can now be written in terms of deviatoric and dilatational stresses and strains using the equations above as well as **Eqs. 2.52-2.55**

$$s_{ij} = 2Ge_{ij}$$
$$\sigma_{kk} = 3K\varepsilon_{kk} \qquad (2.63)$$

The importance of the concept of a separating the stress (and strain) tensors into dilatational and deviatoric components is due to the observation that viscoelastic and/or plastic (meaning yielding, not polymers) deformations in materials are predominately due to changes in shape. For this reason, volumetric effects can often be neglected and, in fact, the assumption of incompressibility is often invoked. If this assumption is used, the solution of complex boundary value problems (BVP) are often greatly simplified. Such an assumption is often made in analyses using the theory of plasticity and theory of viscoelasticity and each will be discussed in later chapters.

Further, the observation that deformations in viscoelastic materials such as polymers is more related to changes of shape than changes of volume suggest that shear and volumetric tests may be more valuable than the traditional uniaxial test.

It can be shown that additional invariants exist for both dilatational and deviatoric stresses. For a derivation and description of these see Fung (1965) and Shames, et al. (1992). The invariants for the deviator state will be used briefly in Chapter 11 and are therefore given below.

$$\mathbf{J}_1 = \sigma_1 + \sigma_2 + \sigma_3 = 0$$
$$\mathbf{J}_2 = 3\sigma_m^2 - \mathbf{I}_2 \qquad (2.64)$$
$$\mathbf{J}_3 = \mathbf{I}_3 - \mathbf{J}_2\sigma_m - \sigma_m^3$$

All invariants have many different forms other than those given herein.

2.8. Failure (Rupture or Yield) Theories

Simply stated, failure theories are attempts to have a method by which the failure of a material can be predicted and thereby prevented. Most often the physical property to be limited is determined by experimental observations and then a mathematical theory is developed to accommodate observations. To date, no universal failure criteria have been determined which is suitable for all materials. Because of the large interest in light weight but strong materials such as polymer, metal and ceramic matrix composites (PMC, MMC and CMC respectively) that will operate at high temperatures or under other adverse conditions there has been much activity in developing special failure criteria appropriate for individual materials. As a result, the number of failure theories now is in the hundreds. Here we will only give the essential features of the classical theories, which were primarily developed for metals. For this reason, it is suggested that the reader keep an open mind and be extremely careful when investigating the behavior of polymers using these traditional methods. It is virtually certain that actual behavior will not always be well represented using any of the following theories due to the time dependent nature of polymer based materials. The same statement is likely true for most of the current popular theories used for composites.

Ductile materials often have a stress-strain diagram similar to that of mild steel shown in **Fig. 2.8** and can be approximated by a linear elastic-perfectly plastic material with a stress-strain diagram such as that given in **Fig. 2.9(b)**. Failure for ductile materials is assumed to occur when stresses or strains exceed those at the yield point. Materials such as cast iron, glass, concrete and epoxy are very brittle and can often be approximated as perfectly linear elastic-perfectly brittle materials similar to that given in **Fig. 2.9(a)**. Failure for brittle materials is assumed to occur when stresses or strains reach a value for which rupture (separation) will occur.

The following are the simple statements and expressions for three well known and often used failure theories. They are described in terms of principal stresses, where $\sigma_1 > \sigma_2 > \sigma_3$, and a failure stress in a uniaxial tensile test, $\sigma_f|_{tensile}$, which is either the rupture stress or the yield stress as appropriate for the material. Typically, tensile and compression properties as found in a uniaxial test are assumed to be the same.

Maximum normal stress theory (Lame-Navier): Failure occurs when the largest principal stress (either tension or compression) is equal to the maximum tensile stress at failure (rupture or yield) in a uniaxial tensile test.

$$\sigma_1 = \sigma_f|_{tensile} \qquad (2.65)$$

Maximum shear stress theory (Tresca): Failure occurs when the maximum shear stress at–an arbitrary point in a stressed body is equal to the maximum shear stress at failure (rupture or yield) in a uniaxial tensile test.

$$\tau_{max} = \frac{\sigma_1 - \sigma_3}{2} = \tau_{max}|_{tensile} = \frac{\sigma_f|_{tensile}}{2} \qquad (2.66)$$

$$\sigma_1 - \sigma_3 = \sigma_f|_{tensile}$$

Maximum distortion energy (or maximum octahedral shear stress) theory (von Mises): Failure occurs when the maximum distortion energy (or maximum octahedral shear stress) at an arbitrary point in a stressed medium reaches the value equivalent to the maximum distortion energy (or maximum octahedral shear stress) at failure (yield) in simple tension

$$\sigma_1^2 + \sigma_2^2 + \sigma_3^2 - \left(\sigma_1\sigma_2 + \sigma_2\sigma_3 + \sigma_3\sigma_1\right) = 2\sigma_f^2\Big|_{\text{tensile}} \tag{2.67}$$

Development of the octahedral shear stress can be found in many texts and will not be given here. However, it is appropriate to note the geometry of the octahedral plane. That is, if a diagonal plane is identified for stressed element as shown in **Fig. 2.19(a)** such that the normal to the diagonal plane makes an angle of 54.7°, the stress state will be as shown in **Fig. 2.19(b)**. The resultant shear stress on this octahedral plane, so named because there are eight such planes at a point, is the octahedral shear stress.

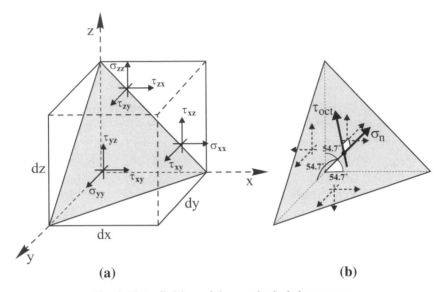

(a)   **(b)**

Fig. 2.19 Definition of the octahedral shear stress.

Comparison Between Theory and Experiment: Comparisons between theory and experiment have been made for many materials. Shown in **Fig. 2.20** are the graphs in stress space for the equations for the three theories given above. Also shown is experimental data on five different metals as well as four different polymers. It will be noted that cast iron, a very brittle material agrees well with the maximum normal stress theory while the ductile materials of steel and aluminum tend to agree best with the von Mises criteria. Polymers tend to be better represented by von Mises than the other theories.

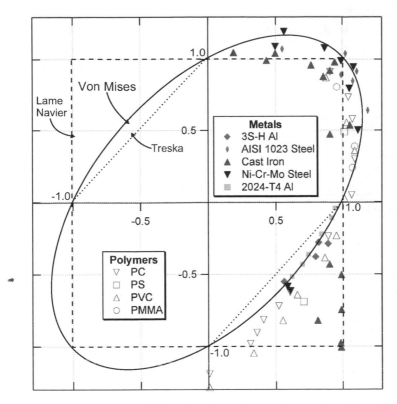

Fig. 2.20 Comparison between failure theories and experiment. (Data from Dowling, (1993): metal p. 252, polymer p. 254)

2.9. Atomic Bonding Model for Theoretical Mechanical Properties

Materials scientists and engineers have long sought methods to determine the mechanical properties of materials from knowledge of the bonding properties of individual atoms, which, of course, hold materials together. Observation of elastic behavior suggests the existence of both attractive and repulsive forces between individual atoms. Stretching an elastic bar in tension, stretches the atomic bonds and release of the load allows the bonds to return to their original equilibrium positions. Likewise, compression causes atoms to move closer together and release of the load allows the atoms to return to their equilibrium position. A hypothetical tensile (or compressive) bar composed of perfectly packed atoms is shown in **Fig.**

2.21. The distances between the centers of four neighboring atoms, mnpq, form a rhombus. When stretched, the strains in the vertical and horizontal directions, ε_x and ε_y, can be calculated from geometrical changes in the position of the spheres and the ratio can be shown to give a Poisson's ratio of $\nu = 1/3$, which is close to the measured value for metals and many materials. The proof is left as an exercise for the reader (see Problem 2.9). This simple calculation tends to give some confidence in the use of an atomic model to represent mechanical behavior.

Now consider just two atoms in equilibrium with each other as shown in **Fig. 2.22**. Application of a tensile force, F_T, will induce an attractive force, F_A, between the two atoms in order to maintain equilibrium. Application of a compressive force will induce a repulsive force, F_R, between the two atoms to maintain equilibrium. These attractive and repulsive forces will vary depending upon the separation distance. It is to be noted that the attractive forces in interatomic bonds are largely electrostatic in nature. For example, Coulomb's law for electrostatic charges indicates that the force is inversely proportional to the square of the spacing. The repulsive forces are caused by the interactions of the electron shells of the atoms and is somewhat difficult to estimate directly.

The variation of attractive and repulsive forces and energies with separation distance are given in **Figs. 2.22(d-e)**, where r_0 is the equilibrium spacing. The forms of the equations agree with physical observations but the values of the constants α, β, **m** and **n** vary for different materials. Obviously, the effect of dislocations, vacancies, grain boundaries, etc. complicates the picture in metals and the long molecular chains, entanglements and other defects complicate the picture in polymers. The energy equations and diagrams given in **Fig. 2.22** can be simply calculated from the force diagram using the basic definitions of work an energy given in elementary mechanics. This proof is left as an exercise for the reader.

Obviously, if the tensile forces are large enough, the distance between atoms can become so great that the attractive force will tend to zero and no force would be required for the atom to be in equilibrium. On the other hand, the application of a compressive forces can not force the two atoms to merge and the repulsive force will increase without bound. For this reason, it should be possible to calculate the theoretical strength of a material if sufficient information is known about the bonding forces in atoms of a particular material. This interpretation has been used by many (see for example, (Courtney, (1990), McClintock and Argon, (1966), Richards, (1961), Shames and Cozzarelli, (1992)) to formulate nonlinear stress-strain relations, laws for creep, plasticity effects, etc. However, as far as is known by the authors, no direct experimental verification has ever been made and, at best, such deduction must be termed empirical.

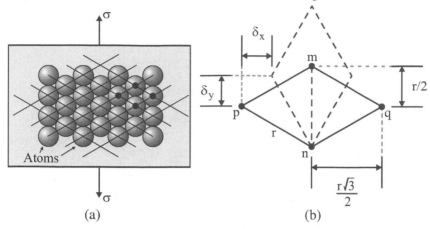

(a)

Close packed crystal structure in a material subject to tensile stress.

(b)

Elongation and contraction of centers due to tensile loading.

Fig. 2.21 Atomic deformations in a material composed of perfectly packed atoms.

Not withstanding the empirical nature of the force and energy variations in **Fig. 2.22**, this approach does give insight to the strength limitations of materials. For example, by examination of **Fig. 2.22(d)** it can be shown that for a perfect crystalline arrangement of atoms as in **Fig. 2.21** that the strength of a material should be the same order of magnitude as its elastic modulus (see (Richards, 1961)). The fact that no material has such high strength properties is an indication of weaknesses caused by imperfections in their molecular structure (e.g. imperfections such as dislocations, vacancies, etc.). Even near perfect crystalline materials do not have such high strength properties. On the other hand, it has been recognized that it is possible to increase strength properties drastically by developing processing approaches to create more nearly perfect crystalline structure and to minimize imperfections in molecular structure. Most of these processing improvements (directional solidification, powder metallurgy, etc.) are used for metals and ceramic type materials. Indeed, it is recognized that the large number of secondary bonds as opposed to primary bonds in polymers gives rise to their relatively modest properties when compared with most metals. Never-the-less, as will be noted in the following chapters, the properties of polymers can also be improved greatly by increasing crystallinity, using additives and developing improved processing techniques.

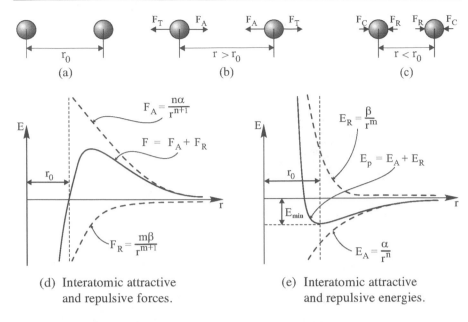

(d) Interatomic attractive
    and repulsive forces.

(e) Interatomic attractive
    and repulsive energies.

Fig. 2.22   Attractive and repulsive forces and energies between atoms.

2.10. Review Questions

2.1. Name five assumption that are normally made to solve problems in elementary solid mechanics.

2.2. Name two types of nonlinearities encountered in solid mechanics.

2.3. Describe a heterogeneous or an inhomogeneous material. Name several materials that are inhomogeneous

2.4. Describe an anisotropic material. Name several materials that are anisotropic.

2.5. Give a mathematical definition for a continuum.

2.6. Define crystallinity, amorphous, anisotropic and material symmetry.

2.7. Define true stress and true strain and write an appropriate equation for each.

2.8. Discuss the characteristics one would seek in developing a test specimen to determine material properties.

2.9.   What is a Luder's band? At what angle do they occur? Name two materials in which they are known to occur.

2.10. Explain the difference between engineering shear strain and the tensorial alternative.

2.11. How many material constants are needed to characterize a linear elastic homogeneous isotropic material? How many material constants are needed to characterize a linear elastic homogeneous anisotropic material?

2.12. Describe a plane of material symmetry. What type of symmetry does an isotropic material possess?

2.13. Define a stress invariant and give the proper expression for the first invariant of stress.

2.14. Define deviatoric and dilatational stresses.

2.15. Give a definition for the classical failure theories of Tresca and von Mises.

2.16. A brittle material is likely to follow which failure theory? On what plane would a brittle material tested in uniaxial tension fail?

2.17. A ductile material is likely to follow which failure theory?

2.18. What is the octahedral shear stress.

2.19. At what angle does a slip band form for a Tresca material tested in uniaxial tension.

2.20. At what angle does a slip band form for a von Mises material tested in uniaxial tension.

2.21. The strength of a material for a perfect arrangement of atoms might be expected to be on the order of what other material parameter?

2.22 Poisson's ratio can be shown to be equal to what value for a perfect arrangement of atoms?

2.11. Problems

2.1. If the engineering strain in a tensile bar is 0.0025 and Poisson's ratio is 0.33, find the original length and the original diameter if the length and diameter under load are 2.333 ft. and 1.005 in. respectively.

2.2. Find the true strain for the circumstances described in problem 2.1.

2.3. A circular tensile bar a ductile material with an original cross-sectional area of 0.5 in.^{2} is stressed beyond the yield point until a neck is formed. The area of the neck is 0.25 in.^{2} Find the average engineering strain in the necked region. The true strain. (Hint: Assume yielding occurs with no volume change.)

2.4. The generalized Hooke's law in tensor (matrix) notation is given as $\sigma_{ij} = E_{ijkq}\,\varepsilon_{kq}$. Expand and find the algebraic expansion for σ_{12}.

2.5. From a thin plate of material small tensile coupons are cut at points A, B and C as shown and the following moduli properties are determined

$$E_x\big|_A, E_x\big|_B, E_x\big|_C, E_y\big|_A, E_y\big|_B, E_y\big|_C, E_\theta\big|_A, E_\theta\big|_B, E_\theta\big|_C$$

Give a correct relationship among the moduli for a homogeneous material. Give a correct relationship among the moduli for n anisotropic material.

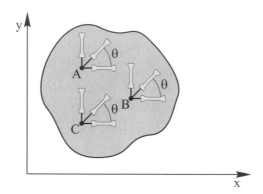

2.6. Show that the tensorial transformation relation given by $\sigma'_{ij} = a_{ik} a_{jq} \sigma_{kq}$ reduces to the form

$$\sigma'_x = \sigma_x\,\cos^2\theta + \sigma_y\,\sin^2\theta + 2\tau_{xy}\,\sin\theta\cos\theta$$

2.7. Expand **Eq. 2.58** and show that the matrix given below is recovered.

$$s_{ij} = \begin{pmatrix} \sigma_{xx} - \sigma_m & \sigma_{xy} & \sigma_{xz} \\ \sigma_{yx} & \sigma_{yy} - \sigma_m & \sigma_{yz} \\ \sigma_{zx} & \sigma_{zy} & \sigma_{zz} - \sigma_m \end{pmatrix}$$

2.8. Using the geometry given in **Fig. 2.21** show that the ratio of lateral to longitudinal strain is 1/3. (Hint: spheres at **m** and **n** that are initially in contact stretch vertically when a stress is applied resulting in a separation of the spheres at **m** and **n**. Also, spheres at **p** and **q** will move inward to maintain contact with spheres at **m** and **n**.)

3. Characteristics, Applications and Properties of Polymers

Many materials found in nature are polymers. In fact, the basic molecular structure of all plant and animal life is similar to that of a synthetic polymer. Natural polymers include such materials as silk, shellac, bitumen, rubber, and cellulose. However, the majority of polymers or plastics used for engineering design are synthetic and often they are specifically formulated or "designed" by chemists or chemical engineers to serve a specific purpose. Other engineers (mechanical, civil, electrical, etc.) typically design engineering components from the available materials or, sometimes, work directly with chemists or chemical engineers to synthesize a polymer with particular characteristics. Some of the useful properties of various engineering polymers are high strength or modulus to weight ratios (light weight but comparatively stiff and strong), toughness, resilience, resistance to corrosion, lack of conductivity (heat and electrical), color, transparency, processing, and low cost. Many of the useful properties of polymers are in fact unique to polymers and are due to their long chain molecular structure. These issues will be discussed at length in the next chapter. In this chapter, focus will be on general characteristics, applications and an introduction to the mechanical behavior including elementary concepts of their inherent time dependent or viscoelastic nature.

3.1. General Classification and Types of Polymers

There are a variety of ways to classify polymers according to their molecular structure and these will be covered in more detail later in Chapter 4. However, there are two general types that should be mentioned here. Most polymers can be broadly classified as either thermoplastics or thermosets. The fundamental physical difference between the two has to do with the bonding between molecular chains - thermoplastics have only secondary bonds between chains, while thermosets also have primary bonds between chains. The names are not only associated with the chemical structure of each but their general thermal and processing characteristics as well since this basic structural difference greatly impacts material properties. Thermoplastic polymers can be melted or molded while thermosetting polymers cannot be melted or molded in the general sense of the term. Thermoplastic or thermosetting polymers are sometimes identified by other names such as "linear" and "cross-linked" respectively. It should be noted that the term linear here applies to molecular structure and not to mechanical (stress-strain) characteristics.

As will be discussed later, a polymer can be a hard and stiff glass-like solid, a soft and flexible elastomeric rubber, or a viscous liquid depending only on the use temperature as compared to two reference temperatures identified as the glass-transition temperature, T_g, and the melt temperature, T_m. All thermoplastic materials may exist in one of these three phases upon changes in the use temperature, while thermosetting polymers generally exist only in the first two phases. The glass-transition and melt temperatures for different polymers range from well below to well above ambient and therefore a particular polymer may be either glassy, elastomeric or liquid at room temperature depending only on its chemical composition. These reference or transition temperatures as well as thermal effects will be thoroughly discussed in later chapters.

Thermoplastic Polymers: Thermoplastic polymers may be either amorphous or crystalline. Crystallinity (or morphology) will be discussed in more detail in the next chapter but it is important to point out here that the degree of crystallinity is low by standards for crystalline metals, ceramics and other materials. That is, polymers are rarely over 50 % crystalline. Crystalline polymers are often more dense than amorphous polymers due to closer packing of their long chain molecules and, in general, the following properties are enhanced.

> **Hardness**
> **Friction and wear**
> **Less creep or time dependent behavior**
> **Corrosion resistance and/or resistance to environmental stress cracking**

An example of a much-used crystalline thermoplastic polymer is polyethylene. LDPE (low density polyethylene) is considered to be semicrystalline while HDPE (high density polyethylene) or UHDPE (ultra high density polyethylene) are considered to be highly crystalline. LDPE is one of the most widely used plastics accounting for more than 20% of the total polymer market and is used extensively for milk containers and other packaging operations. HDPE and UHDPE are used extensively in water and gas (natural) pipelines. Other typical crystalline thermoplastics used in engineering design include LLDPE (linear low density polyethylene) and the following;

    Polypropylene   **Polyamides (nylon)**
    Acetals   **Polytetrafluoroethylene (PTFE)**
    Polyesters   **Polyetheretherketone (PEEK)**

Amorphous thermoplastics (those with no regular molecular structure) are;

    Polyvinyl Chloride (PVC)   **Polymethyl Methacrylate (PMMA)**
    Polystyrene (PS)   **Acrylonitrile-butadiene-styrene (ABS)**
    Polycarbonate   **Polyethersulphone**

In general thermoplastic polymers are easier to produce and cost less than thermosets. Information on the volume of sales in the US and basic costs of a few thermoplastics is given in **Table 3.1**.

Thermosetting Polymers: In general thermosetting polymers are used where high thermal and dimensional stability are required. Applications include use as electrical and thermal insulation materials, adhesives, high performance composites and especially where high strength and modulus are required. Some examples of thermosetting polymers are.

    Aminos   **Phenolics (Bakelite)**
    Polyurethanes   **Polyesters**
    Epoxides

Information on the volume of sales in the US and basic costs of the major thermosetting and thermoplastic polymers is given in **Table 3.1** and the volume distribution by products in **Fig. 3.1.**

Table 3.1 US polymer production.

Production volume data from *American Plastics Council* for 2001: Canadian and Mexican production data included in some categories; dry-weight basis except phenolic resins. Pricing data from *Plastics News* for Feb. 2003. See current data on respective websites.

Resin	US Production $(x10^6 \text{ lbs})$	US Production $(x10^6 \text{ kg})$	% Total production	Price (US$/lb)
Epoxy	601	273	0.59%	$1.12
Urea and Melamine	3,040	1382	3.01%	$0.82
Phenolic	4,362	1983	4.31%	$0.80
Total Thermosets	**8003**	**3638**	**7.92%**	
LDPE	7,697	3499	7.61%	$0.59
LLDPE	10,272	4669	10.16%	$0.53
HDPE	15,284	6947	15.12%	$0.54
PP	15,934	7243	15.76%	$0.47
ABS	1,217	553	1.20%	$0.77
SAN	127	58	0.13%	$1.04
Other Styrenics	1,517	690	1.50%	
PS	6,114	2779	6.05%	$0.73
Nylon	1,139	518	1.13%	$1.14
PVC	14,257	6480	14.10%	$0.45
Thermoplastic Polyester	6,898	3135	6.82%	$1.04
Total Thermoplastics	**80,456**	**36571**	**79.57%**	
Engineering Resins*	2,542	1155	2.51%	
All Other#	10,108	4595	10.00%	
Total Other	12,650	5750	12.51%	
Grand Total	**101,109**	**45959**	**100%**	

*includes acetal, granular fluoropolymers, polyamide-imide, polycarbonate, thermoplastic polyester, polyimide, modified polyphenylene oxide, polyphenylene sulfide, polysulfone, polyetherimide and liquid crystal polymers; #includes polyurethanes (TDI, MDI and polyols), unsaturated (thermoset) polyester, and other resins.

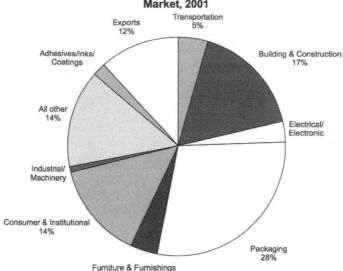

Total Sales & Captive Use of Selected Thermoplastic Resins by Major Market, 2001

Fig. 3.1 Major markets for thermoplastic resins. Data compiled by VERIS Consulting, LLC and reported on the web by APC's Plastics Industry Producers' Statistics Group.

Additives: A "pure" synthetic polymer may not have the desirable characteristics for a particular application. However, through the use of additives (or fillers) various properties can often be modified to fill a particular need. For example, many "structural plastics" often contain additives to enhance their properties for a special application. As a result, commercial plastics may be very different from those of the base polymer even though they may have the same basic chemistry. Some typical additives are (See Crawford (1992) for a discussion of each):

Antistatic Agents	**Coupling Agents**
Lubricants	**Flame Retardants**
Plasticizers	**Pigments**
Stabilizers	**Reinforcements (alumina, fibers, etc.)**

A good example for additives is the inclusion of rubber particles to increase the toughness of otherwise brittle polymers. In the case of epoxy adhesives, the fracture toughness can be significantly improved by the addition of microscopic rubber particles. These particles form a second phase and normally are not covalently bonded to the matrix phase. A photomicrograph of a rubber-toughened polymer, high impact polystyrene (HIPS), is given **Fig. 3.2**. Under sufficient external loading the rubber inclusions become highly stressed and cavitation of these particles occur (rupture) absorbing energy and enhancing toughness.

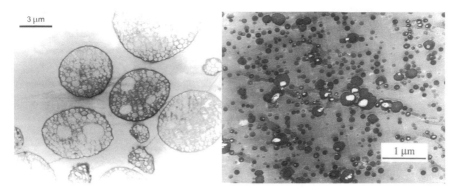

Fig. 3.2 Example of a rubber toughened polymers: HIPS (left, Serpooshan et al. (2007)) and ABS (right, Bucknall et al. (2000)). Reprinted with permission of John Wiley and Sons, Inc and Elsevier.

Blends or Alloys: Sometimes two or more plastics are mixed or "alloyed" to achieve special properties and are known as polyblends. ABS (acrylonitrile, butadiene and styrene) and PBT (polybutylene terephthalate) are often used in engineering applications with polycarbonate, polysulphone, etc. Several combinations and their improved features are given below (see Crawford (1992) for a more complete discussion of alloys).

Alloy	Characteristics
Polycarbonate/ABS	Good heat and impact resistance.
Polycarbonate/ PBT	High toughness.
PVC/acrylic	Good chemical and flame resistance.
PVC/ABS	Good processability, flame resistance and impact strength.

The fundamental point is that many structural plastics are, in fact, composites composed of combinations of several materials. As a result, mechanical and other properties are influenced by each component and it is most appropriate for design engineers to have a familiarity with the effect of various additives. Often, manufacturers change additives or blend ratios from time to time to enhance certain properties for a large volume customer or for enhanced and more economical processing, etc. Changing additives or the introduction of new additives may change one or more engineering properties and creates the need for continual testing to evaluate commercial polymers.

While additives, fillers and blends do alter a polymer and, in effect, may cause a polymer to be both heterogeneous and anisotropic most testing programs to determine mechanical properties are performed under the assumption of homogeneity and isotropy. As a result, industrial test programs to measure stress, strain, modulus and strength and other properties given in "specification" sheets are very similar to those described in Chapter 2 for metals or other time independent materials. Such information may not be adequate to evaluate the long-term structural performance of a polymer used in engineering design.

3.2. Typical Applications

Polymers are widely used in the automotive industry, aerospace industry, computer industry, building trades and many other applications. For example, automobile bumpers are now made with a polymer blend that has sufficient toughness to meet state and federal standards. This has resulted in a significant weight saving and the conversion from metal has also been cost effective due to decreased energy costs and the ability to easily recycle the polymer blend from older cars to manufacture bumpers for new vehicles.

The above illustrates, interestingly, that the cost to produce polymers is sometimes less than the cost to produce certain metals. Crawford, 1992, gives data on the relative energy required to manufacture thin sheets of various polymers and metals including the proportion of energy related to the feedstock, fuel and processing. Since this data is not for the a uniform sheet thickness the data has been divided by the sheet thickness and normalized with respect to the energy required to produce mild steel. The result is given in **Table 3.2** and indicates that aluminum requires approximately 11% more total energy to manufacture than steel while the polymers cited require at least 50% less energy to manufacture than similar thin sheets of metal. This gives a good indication why polymer products are replacing such items as aluminum foil food wraps, soft drink containers, computer housings, etc. Of course this substitution of polymer for metal occurs mostly for non-structural products. Because the modulus and strength of structural metals such as aluminum and steel are much greater than the modulus and strength of polymers, the latter cannot perform as well in structural circumstances. The exception is for fiber reinforced polymers but then the production cost is often much higher.

Table 3.2 Relative energy required to manufacture various sheet materials normalized relative to steel.

Aluminum	1.11
Steel	1.00
PC	0.49
Acrylic	0.47
Nylon	0.52
LDPE	0.26
HDPE	0.29
Polystyrene	0.34
Polypropylene	0.24
PVC	0.26

Fiber Reinforced Plastics: Fiber reinforced plastics (FRP) or polymer matrix composites (PMC) are now frequently used in automotive, aerospace, boating, sporting goods, construction and other applications. Unfortunately, these products go by many different names. For example the FRP materials made with glass fibers often are call glass reinforced polymers (GRP) or simply fiberglass.

FRP or PMC materials are made by a number of processes. For example, the materials used in many applications (bathtubs, boats, auto hoods, etc.) are formed by compression molding a polymer containing chopped glass fibers (usually about 1 in. long) in a polyester matrix to form what is known as a sheet-molding compound (e.g. SMC-25, sheet molding compound with 25% fiber). FRP or PMC composites used for water, oil or gas pipelines are formed by a filament winding process using continuous glass fibers which are first passed through a polymer (e.g., polyester, epoxy, etc.) bath to coat the fiber prior to winding.

Advanced composites (so called due to the extremely high mechanical properties of the fibers) used in the aerospace industry and for certain sports equipment (e.g. skis, tennis rackets, golf clubs, etc.) are made with continuous carbon fibers in a polymer matrix (e.g., epoxy, PEEK, etc.) and are most often laminated, vacuum bagged and cured under high heat and pressure.

All composites are in general inhomogeneous, anisotropic and cannot be considered a continuum at a local or microscopic level. Therefore special testing programs are normally required to determine mechanical properties. The assumptions of a continuum, homogeneity and isotropy are often made and may give estimates of behavior that can be used in engineering design though this should only be done with extreme care.

Adhesives: Nearly all adhesives are polymers and are used extensively to connect structural components made of wood, composites, metals, polymers, and other materials. Though the amount of adhesive needed for a particular application is small, the cost of a polymer adhesive is high compared to other applications. For example, it is not unusual for an adhesive to cost on the order of $1.00 or more per ounce while general use polymers of the same type might cost less than $1.00 per pound (see **Table 3.1**). For this and other reasons, the world market for adhesives is in excess of five billion dollars per year. As mentioned earlier, adhesives often contain elastomeric particles to enhance their fracture toughness. In addition, many adhesives contain alumina or other metallic particles for increased tensile and shear strength and in such cases are in reality particulate composites.

Insulation Applications: One of the earliest uses for polymeric materials was for the insulation of electrical cables for power lines, etc. due to their low conductivity. In addition, polymers are now being used as thermal insulation in buildings, automobiles, etc. A few polymers (e.g., polybenzimidazole) have such high thermal resistance that they are used as fabrics for clothing of firefighters who must deal with very intense heat such as that in fires in buildings and oil wells.

The relative insulation characteristics of polyurethane foam and polystyrene foam as compared to brick and wood is given in **Fig. 3.3**. Thermal conductivity coefficients, thermal expansion coefficients and dielectric constants for various polymers and other materials are given in **Table 3.3**.

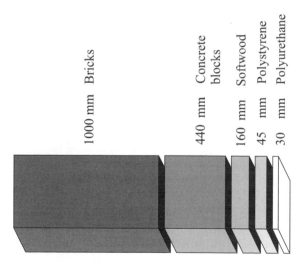

Fig. 3.3 Comparsion of relative thermal insulation characteristics of various materials.

Optical Applications: Typically, amorphous polymers are transparent unless fillers or other additives are used that cause them to be opaque, while crystalline polymers are translucent or opaque. For this reason, amorphous thermoplastic polymers are often used in optical applications, the most prominent of which is lenses to enhance vision. Polymer lenses are lighter, tougher and have better fracture resistance than regular glass or silicon oxide based lens. Further they have a higher refractive index and hence transmit more light than ordinary glass. The refractive index, light transmission and dispersive properties of several polymers are given in **Table 3.4**.

Amorphous polymers, notably, polycarbonate, are often used as windows where enhanced fracture resistance is needed. (Parenthetically, those familiar with baseball in the 60's may recall a TV commercial of Sandy Kofax attempting unsuccessfully to break a substitute school window made of polycarbonate with a baseball.) Polycarbonate and PET are often used as a glazing material for high performance windows in automobiles, airplanes and elsewhere.

Table 3.3 Thermal and electrical properties of polymers

Material	Conductivity (W/m-K)	Coefficient of Thermal Expansion (10^{-6}/K av. @RT)	Dielectric Constant (Average)
ABS	0.33	-	2.8
Epoxy	0.20	-	3.5
Phenolic	0.20	-	-
Polypropylene	0.22	160	2.3
Polyethylene	0.33	150	2.3
Polycarbonate	0.20	-	3.0
Polystyrene	0.19	130	
Teflon	0.20	170	2.1
Polyurethane foam	0.020		-
Polystyrene foam	0.037	-	-
Aluminium	216	24	-
Copper	394	-	-
Steel	67	12	-
Brick	0.70	-	-
Concrete	1.1	10	-
Oak	0.19	-	-
Pine	0.16	-	-
Glass	0.80	8	9.0
Air	0.03	-	1.0

Table 3.4 Typical optical properties

Material	Refractive Index	Light Transmission
Acrylic	1.49	92
Polycarbonate	1.59	89
Polystyrene	1.57	88
PMMA	1.49	-
Glass	1.5	-

Fibers

One of the major applications of polymers is for use as fibers in clothing, ropes, rugs or tapestries and many other household or commercial purposes. Natural fibers such as flax for garments date back to prehistoric times. Plant derived natural fibers such as cotton, flax, and rayon are based on the polymer cellulose, while animal based fibers such as wool and silk are polyamides. Synthetic fibers such as polyesters and polyamides (including nylon and Kevlar) account for the majority of the fiber market today and are used in textiles and in high performance applications (e.g., space suits and reinforcements in polymer matrix composites). Both natural and synthetic polymeric fibers are semi-crystalline, with significant molecular orientation in both the crystalline and noncrystalline domains. In typical manufacturing processes for fibers, this molecular orientation is achieved through spinning and drawing steps. Increasing the degree of molecular orientation in polymeric fibers leads to superior strength and stiffness characteristics.

A very important fiber used in high performance polymer matrix composites is the carbon or graphite fiber. Carbon fibers were first made by a complicated heat (or pyrolysis) treatment of rayon fibers but are now primarily made by pyrolyzing either a PAN (polyacrlonitrile) or pitch based fiber. The resulting fiber consists of layers of graphene sheets oriented predominantly along the fiber axis and provides extremely high strength to weight ratios. Polymer composites incorporating carbon fibers have excellent mechanical properties and are used in aircraft/spacecraft structural components, sporting equipment and now even as handles in builder's tools. See Hyer (1998) for an excellent description of various fiber types used in composites as well as their microstructure.

The tensile modulus of a few fibers is given in **Table 3.5**.

Table 3.5 Tensile modulus of select fibers (Warner (1995))

Material	Tensile Modulus
Cotton	8.1 GPa
Rayon	8.2 GPa
Nylon66	2.3 GPa
Kevlar49	125 GPa
Carbon (IM)	250 GPa

3.3. Mechanical Properties of Polymers

The mechanical properties of polymers are most often obtained using a uniaxial tensile test at a constant rate of strain or head motion similar to those used for metals and other materials. Schematic stress-strain diagrams characteristic of those found for the indicated types of solid polymers is shown in **Fig. 3.4**. Curve 1 represents a linear elastic and brittle material like an epoxy, polystyrene, etc. Curve 2 is similar to that of a semi-ductile material like PMMA. Curve 3 is similar to that of a ductile material like PET or polycarbonate. Curve 4 is similar to that of a typical elastomer such as a flexible urethane. Elastic modulus, Poisson's ratio, failure stress and strain are defined as given in Chapter 2 but the 0.2% offset method to determine yield stress cannot be used as strains in polymers are quite large compared to structural metals such as steel and aluminum. The yield stress of a ductile material is often assumed to be equal to the proportional limit stress or the first peak in the stress strain diagram (termed the intrinsic yield point) as indicated in **Fig. 3.5**. It is to be noted that many approaches to determining the yield point are used, although the intrinsic yield point is the most common. One method due to Considere is shown in **Fig. 3.5** (see Ward and Hadley, (1993) for reference). With this method, the extrinsic yield point is the point of tangency of a line drawn from a point on the strain axis of -1.0 to the stress-strain diagram. Both true stress and true strain are normally used but in **Fig. 3.5** true stress and nominal or average strain is used. A comparison of tensile modulus, strength and strain at break (yield), and impact strength of a number of polymers developed using elementary test procedures is given in **Table 3.6.**

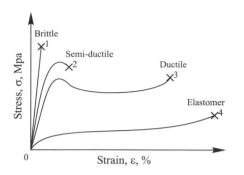

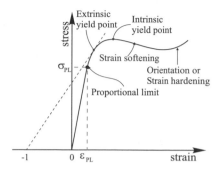

Fig. 3.4 Typical stress-strain (load-elongation) diagrams of various polymer types

Fig. 3.5 Considere's definition of yielding for polymers. (After Kinloch and Young (1983), p. 108)

If tests are performed at different constant strain rates or temperatures, stress-strain response similar to that shown in **Fig. 3.6** is obtained for many polymers. Notice that modulus and intrinsic yield point vary with both rate and temperature. Also, the stress-strain response appears to be nonlinear even at low stress levels. However, caution on the interpretation of the information obtained from such elementary tests is suggested, as it will be shown in a later section that linearity as well as other essential mechanical properties should be deduced from isochronous stress-strain diagrams.

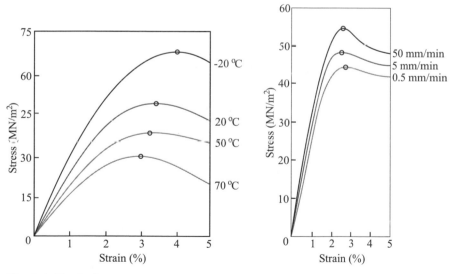

Fig. 3.6 Typical temperature and rate dependent stress-strain response. Instrinsic yield points indicated by circles.

Table 3.6 Comparsion of mechanical properties of selected polymers
(Except as noted, data are average values taken form Billmeyer (1984), p. 470-480)

Polymer*	Tensile Modulus GPa (ksi)	Tensile Strength MPa (ksi)	Elongation %	Impact Strength J/m (ft-lb/in) (notched)
Cellulose Acetate	1.59 (230)	37.6 (5.45)	38	150 (2.8)
Nylon 66	2.07 (300)	72.4 (10.5)	180	80 (1.5)
Polycarbonate	2.41 (350)	60.7 (8.8)	115	790 (14.8)
Polyethylene (LD)	0.39 (57)	20.1 (3.0)	570	260 (5.0)
Poly(ethylene terephthalate)	3.55 (500)	65.5 (9.5)	175	24 (0.45)
Poly(methylmethacrylate)	3.10 (450)	62.1 (9.0)	6.0	320 (0.4)
Polypropylene	1.38 (200)	33.8 (4.9)	450	70 (1.3)
Polysulfone	2.48 (360)	70.3 (10.2)	75	64 (1.2[a])
Polyimide	3.10 (450)	72.4 (10.5)	6.0	59 (1.1)
Poly(vinylchloride) (rigid)	3.31 (480)	48.2 (7.0)	21	545 (10.2)
Polyurethane (rigid)	3.55 (500[b])	72.4 (10.5)	4.5	320 (0.4)
Epoxy (cast)	2.41 (350)	58.6 (8.5)	4.5	32 (0.6)

[a]From Rodriguez (1996), p. 701   [b]from plasticssusa.com

*Note: Property values such as those listed in this table vary widely and should not be used for design purposes without validating by testing the exact polymer to be used. ASTM Standard testing procedures offer reliable experimental protocols for such experiments. Mechanical properties of polymers can also be found in reference handbooks such as The Polymer Handbook (2006) and other textbooks such as Rodriguez, 1996 (p. 696-710) as well as various online databases such as plasticssusa.com. Variability of polymer properties can be seen for example in Fig. 3.7, where the true stress and strain at rupture for polycarbonate differ from the values tabulated here.

3.3.1. Examples of Stress-Strain Behavior of Various Polymers

From an engineering design standpoint, a fundamental question to ask about the stress-strain diagrams found in the literature and from industry specification sheets is: how was the strain measured? Was it measured by,

- Machine head motion divided by the length of the specimen
- An extensometer
- An electrical strain gage

or some other method? The reason for asking such a question is to know whether the strain truly represents the behavior occurring at a material point and, therefore, if stress-strain equations developed therefrom are accurate and justified. If not, the stress analysis used for design may only be approximate and not a good predictor of actual service performance. In the following discussion, details of strain measurement for polycarbonate will serve as an example of possible differences using different strain measurement methods. In addition, experimental data for stress-strain response of polycarbonate, polypropylene, and epoxy will serve to illustrate the differences in ductile and brittle polymers, as well as to point out important factors affecting stress-strain results for polymers (e.g., strain measure, strain rate, loading mode, temperature.)

Stress-Strain Behavior of Polycarbonate: Specimens of polycarbonate are shown in **Fig. 3.7** together with the stress-strain properties obtained by three different methods: 1) electrical strain gages, 2) use of an effective gage length, and 3) thickness changes measured in the necked region. In the first method, strain was measured with electrical resistance strain gages attached to the specimen with an adhesive and the change in resistance monitored with deformation under load. The axial and transverse strains are directly related to the changes in resistance resulting from deformations in the respective directions. The second approach determined an average strain by dividing the machine head motion by the total length of specimen between the grips. Due to the dogbone shape of the specimen and because of stress concentration factors at the grips, the average strain determined in this manner is not accurate. However, the average strain obtained using the total length between grips can be corrected through the use of a proportionality factor found by comparing the electrical resistance strain gage measurements for very small strain levels to that obtained by using machine head motion. In this manner, reasonably accurate strains can be determined from the machine head motion prior to neck formation. The third technique was used to determine strains after formation of the neck and involved micrometer measurements of the thickness in the neck area as the neck propagated and conversion to axial strain via assuming a Poisson's ratio of 0.4.

It is appropriate to note that electrical resistance gages must be used with care as polymers, in general, are poor conductors of heat. As a result, the electrical current in the strain gage can cause local heating of the material under the gage and thereby appreciably soften the material giving rise to erroneous measures of strain. Further, as the strain gage (a metal) is much stiffer than the polymer, appreciable reinforcement can occur for thin polymers. These effects can be minimized using sufficiently thick specimens and by pulsing the current to minimize local heating. Errors due to these sources were negligible for the data shown.

As the electrical strain gage is located outside of the neck, the strain measurement given by curve 1 in **Fig. 3.7(d)** does not provide a useful measure of strain beyond the point of Luder's band formation. The average strain (effective gage length) technique, curve 2, does account for the neck in the material, but is inaccurate after neck formation since the proportionality factor used is strictly applicable only in the small/linear deformation region. Only the third method, curve 3, of directly measuring thickness changes in the specimen in the neck area properly represents the local strain in the material after yielding. Note that the first two methods should provide essentially equivalent strain measures prior to yielding, and that either of them can be combined with the results of the third method to provide an accurate picture of material response up to failure.

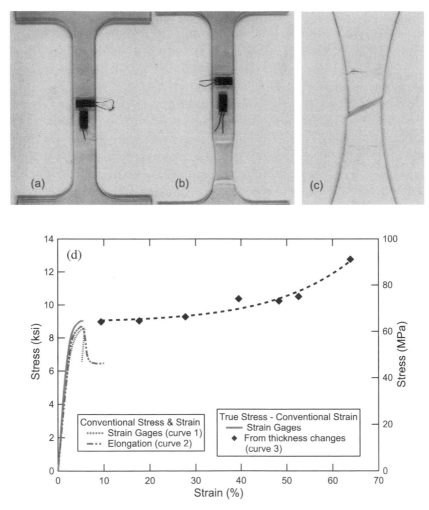

Fig. 3.7 Example of stress-strain response of polycarbonate.

Both engineering and true stresses are given in **Fig. 3.7(d)** with engineering (conventional) strain measurements. Curve 1 shows the results of the electrical strain gage measurement from zero load to Luder's band formation or the initiation of yielding. The strain reaches only a maximum of about 5% and both strain and stress appear to decrease after this point. Curve 2 represents the strain determined by the effective gage length method as described above and the maximum strain reaches about 10%. The strain does not reach a larger value because the length of the neck is only a small portion of the total specimen length between grips. The engineering stress decreases after the neck forms in both 1 and 2, and this decrease is not seen when true stress is used. Curve 3 is based on true stress (using the cross sectional area of the necked region) and using electrical resistance strain gage measurements up to ~5% strain with thickness measurements above this level. Note again that curves 1 and 2 provide misleading information on the stress-strain response of the material after yield and only curve 3 is representative of local material response through the necking stage.

The left photograph, **Fig. 3.7(a),** is prior to neck formation, the center photograph **Fig. 3.7(b),** is after the neck has formed and drawing begins and the right photograph **Fig. 3.7(c)** is a close-up of a fully formed Luder's band from a different specimen. The Luder's band begins to form near the point of maximum stress and shortly thereafter a prominent slip band or neck appears similar to that shown in **Fig. 3.7(c)** which is from a separate test of a thinner specimen. Note the initial Luder's band angle for a thin specimen is about 57.3^0 with the vertical while the angle in **Fig. 3.7(c)** is more than 60°. The reason, of course, is that the slip band angle gradually increases to 90° when the full neck is formed. A more descriptive discussion of these results may be found in Brinson (1972) and Brinson and Das Gupta (1974). It is also to be noted that the stress-strain results agree with the very careful optical strain measurements for polycarbonate by Brill (1965) using a very fine inscribed grid.

Notice the similarity between the stress-strain behavior of polycarbonate given by curve 2 and that for mild steel in **Fig. 2.7**. Both show an elastic-plastic tensile instability point or a decrease of stress with increasing strain after the first peak in the stress-strain diagram is reached. An upper and lower yield point can be defined for polycarbonate as for mild steel providing the intrinsic yield point, engineering stress and the approximate measure of strain after neck formation are used. Although the similarity to mild steel is perhaps useful, such a stress-strain diagram as given by curve 2 for polycarbonate is not truly indicative of the local stress-strain response of the material through necking and to failure.

The stress-strain response of polycarbonate is a function of test rate as is shown in **Fig. 3.8**. Little rate effect is observed for low stress levels but a very significant effect is observed for higher levels. The intrinsic yield stress is clearly rate dependent and should the tests have been carried to rupture a drawing behavior similar to that shown in **Fig. 3.7** would have occurred for each rate. These results suggest the need to include rate and/or time in developing yield criteria for polymers. This will be discussed more fully in Chapter 10.

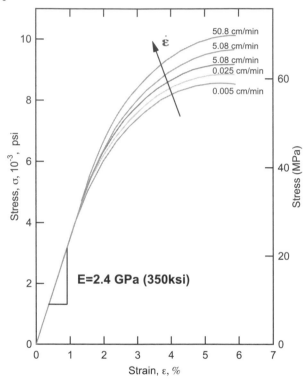

Fig. 3.8 Constant strain-rate behavior of a thermoplastic polymer (polycarbonate). (Data from Brinson (1973))

Stress-Strain Behavior of Polypropylene: Both tensile and compressive stress-strain response of polypropylene is shown in **Fig. 3.9**. Quite obviously, the behavior in tension and compression are quite different for stresses above about 2,000 psi. This indicates that care must be used in analysis where the behavior in tension and compression are assumed to be the same. (See Rybicky and Kanninen (1973) for an example of the difference on the analysis of a beam in 3-point bending.)

Stress-Strain Response of Epoxy: The constant strain-rate stress-strain response as a function of temperature of an unmodified epoxy is given in **Fig. 3.10**. The initial portion of each curve is linear and for room temperature (not shown) the material was linear up to the fracture or rupture point. The data presented suggests a brittle to ductile transition might be defined but the transition is merely the transition from glassy to rubbery behavior previously discussed.

The behavior of a modified or rubber toughened epoxy is shown in **Fig. 3.11** as a function of strain rate at room temperature. Comparison with **Fig. 3.10** indicates the significance of adding rubber tougheners to dramatically alter the ductility of the material.

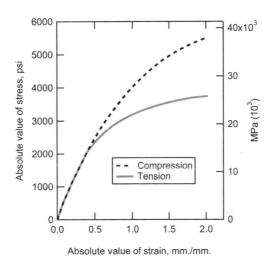

Fig. 3.9 Stress-strain behavior of polypropylene. (Data from E.F. Rybicky and M.F. Kanninen, (1973).)

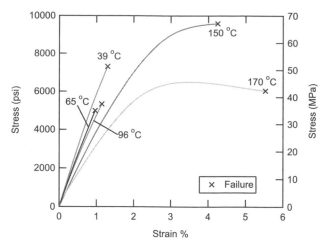

Fig. 3.10 Temperature dependent stress-strain response of a typical brittle epoxy. (Data from Hiel, et al. (1983))

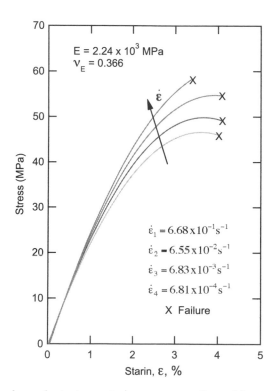

Fig. 3.11 Rate dependent stress-strain response of a rubber toughened epoxy, room temperature. (Data from Brinson, et al. (1975))

3.4. An Introduction to Polymer Viscoelastic Properties and Characterization

The fact that the response of polymer based materials is time dependent and/or viscoelastic has been mentioned in previous sections. Further, it has been indicated that this time dependence is inherent to polymeric materials due to their unique molecular structure and is quite different from time dependence induced in other materials such as metals by fatigue, moisture, corrosion or other environmental factors. In fact, these same environmental factors also affect polymers but manifest themselves differently than in other materials due to the intrinsic viscoelastic nature of the molecular structure.

For the above reason, unique tests and analysis approaches must be adopted for polymer-based materials to determine the manner in which properties vary with time. The following sections introduce the necessary terms, definitions and general behavior which will be useful in the more advanced approaches in later chapters.

3.4.1. Relaxation and Creep Tests

One of the fundamental methods used to characterize the viscoelastic time-dependent behavior of a polymer is the relaxation test. In a relaxation test, a constant strain is applied quasi-statically to a uniaxial tensile (or compression or torsion) bar at zero time. That is, the bar is suddenly stretched to a new position and rigidly fixed such that the strain remains constant for the duration of the test. The sudden strain must not induce any dynamic or inertia effects (which explains the term quasi-static, i.e., the loading motion is sufficiently slow that inertia effects can be ignored).

In a relaxation test, it is also normal to assume that the material has no previous stress or strain history or if one did exist, the effect has been nullified in some way. One method to accomplish this for polymers is to anneal the sample at a suitable temperature sufficient to remove any previous history and then to cool very slowly. The nature of such a process will become obvious in later sections.

If a polymer is loaded in the described manner, the stress needed to maintain the constant strain will decrease with time. Eventually, the stress will go to zero for an ideal thermoplastic polymer but will decrease to a constant value for a crosslinked polymer. The strain input and the stress output for typical thermoset and thermoplastic materials in a relaxation test is shown in **Fig. 3.12**.

Obviously, if the stress is a function of time and the strain is constant, the modulus will also vary with time. The modulus so obtained is defined as the relaxation modulus of the polymer and is given by,

$$E(t) = \frac{\sigma(t)}{\varepsilon_0} = \text{Relaxation Modulus} \qquad (3.1)$$

or

$$\sigma(t) = \varepsilon_0 E(t) \qquad (3.2)$$

The latter equation is the uniaxial stress-strain relation for a polymer analogous to Hooke's law for a material that is time independent but is valid only for the case of a constant input of strain. The relaxation test provides the defining equation for the material property identified as the relaxation modulus. More general differential and integral stress-strain relations for an arbitrary loading will be developed in later Chapters.

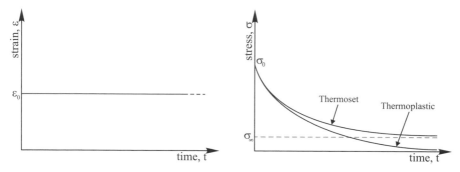

Fig. 3.12 Relaxation test: strain input (left) and qualitative stress output (right).

The limiting moduli at **t = 0** and at **t = ∞** for a crosslinked material are defined as,

$$E(t = 0) = \frac{\sigma(t = 0)}{\varepsilon_0} = E_0 = \text{Initial Modulus} \qquad (3.3)$$

$$E(t = \infty) = \frac{\sigma(t = \infty)}{\varepsilon_0} = E_\infty = \text{Equilibrium Modulus} \qquad (3.4)$$

In addition to the relaxation test, another fundamental characterization test for viscoelastic materials is the creep test in which a uniaxial tensile (or compression or torsion) bar is loaded with a constant stress at zero time as shown in **Fig. 3.13**. Again, the load is applied quasi-statically or in such a manner as to avoid inertia effects and the material is assumed to have no prior history. In this case, the strain under the constant load increases with time and the test defines a new quantity called the creep compliance,

$$D(t) = \frac{\varepsilon(t)}{\sigma_0} = \text{Creep Compliance} \tag{3.5}$$

In this case,

$$\varepsilon(t) = \sigma_0 D(t) \tag{3.6}$$

In a creep test, the strain will tend to a constant value after a long time for a thermoset while the strain will increase without bound for a thermoplastic. Initial and equilibrium compliances similar to initial and equilibrium modulus can also be defined for thermosetting materials.

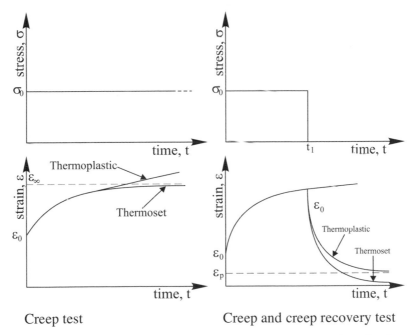

Creep test   Creep and creep recovery test

Fig. 3.13 Creep and creep recovery tests: stress input (above) and qualitative material strain response (below).

An equally important facet of a constant stress test is to understand the resulting strain variation if the stress is removed. This is referred to as a creep-recovery test and is also shown in **Fig. 3.13**. For an ideal thermoset material, the strain will decay to zero after a sufficient time interval which may be quite long compared to the time of loading. For an ideal thermoplastic material, a residual deformation or permanent strain will remain even after a very long (or infinite) time.

The deformation mechanisms associated with relaxation and creep are related to the long chain molecular structure of the polymer. Continuous loading gradually induces strain accumulation in creep as the polymer molecules rotate and unwind to accommodate the load. Similarly, in relaxation at a constant strain, the initial sudden strain occurs more rapidly than can be accommodated by the molecular structure. However, with time the molecules will again rotate and unwind so that less stress is needed to maintain the same strain level. It is also clear from these tests that polymers have some characteristics of a solid and some characteristics of a fluid. In a relaxation test, the ratio of the initial stress and strain is,

$$E(t = 0) = \frac{\sigma_0}{\varepsilon_0} \tag{3.7}$$

and in a creep test,

$$D(t = 0) = \frac{\varepsilon_0}{\sigma_0} \tag{3.8}$$

which is analogous to the behavior of an elastic solid. On the other hand in a creep test the rate of change of strain (or slope) for a thermoplastic material is,

$$\frac{d\varepsilon(t = \infty)}{dt} = \text{constant} \tag{3.9}$$

after a sufficiently long period of time which is characteristic of a fluid. The flow characteristics of a thermoplastic are due to the lack of primary bonds between molecular chains and the solid characteristics of a thermoset are due to entanglements and the primary bonds between individual chains. In both thermosets and thermoplastics, creep (which is also viscous like), is related to the motion of molecules between entanglements, while the mechanisms for creep are further limited to motion between crosslinking sites for thermosets. The initial and equilibrium moduli of a thermoset are solid like with the former being due to both entanglements and crosslinks and the latter being principally due to crosslinks.

3.4.2. Isochronous Modulus vs. Temperature Behavior

The variation of modulus with temperature can be determined from relaxation tests conducted at different temperatures. In a relaxation test (**Fig. 3.14**) conducted at a constant temperature, the ratio of stress to strain at a given instant in time of 10 seconds, one minute, or another suitable time, is identified as the ten second modulus, E(10), or one minute modulus, E(1), etc.

$$E_{10} = E(t = 10 \text{ sec.}) = \frac{\sigma(t = 10 \text{ sec.})}{\varepsilon_0} = 10 \text{ sec. Relaxation Modulus} \qquad (3.10)$$

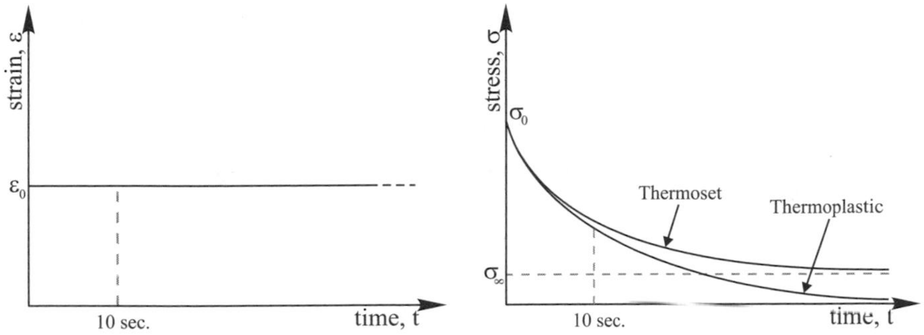

Fig. 3.14 Definition of the 10 second relaxation modulus for an isothermal test.

The variation of the 10 second relaxation modulus with temperature for amorphous, crystalline and crosslinked polystyrene is shown in **Fig. 3.15** (after Tobolsky (1962)). Similar curves are shown for polyblends in **Fig. 3.16**. As may be observed, there are five regions of viscoelastic behavior. These are the glassy, transition ("leathery"), rubbery plateau, rubbery flow and liquid flow regions. In some texts, only four regions are identified with the rubbery flow region not being identified separately from the liquid flow region. Thermoset materials do not show a liquid flow region though if the temperature is very high for a prolonged period, degradation can take place and give the appearance of a flow region. Also, the color of the polymer will darken and degradation will be obvious. An example will be given later. The transition region is suppressed in crystalline materials as shown in **Fig. 3.15**.

Two very important temperatures are indicated in **Fig. 3.15** and are the melt temperature (or first order transition temperature), T_m, and the glass transition (or second order transition temperature) T_g. The T_m and T_g can only be determined approximately from isochronous modulus-temperature data similar to that given in **Fig. 3.15**. Often, manufacturers specification sheets will define a softening temperature which is not clearly defined as either the T_m or the T_g but is somewhere in between the two. The T_g is also frequently determined approximately from DMA (dynamic mechanical analysis – see Chapter 5) but the most accurate procedure to determine both T_m and T_g is through specific or relative volume measurements as obtained from a dilatometer. Typically the relative or specific volume of amorphous or crystalline polymers varies with temperature as shown in **Fig. 3.17**. The T_m is identified as the temperature at which a discontinuous change in relative volume takes place while the T_g is the temperature at which a discontinuous change in the slope of the relative volume takes place. These concepts are discussed in more detail in Chapter 4 for crystalline polymers and in Chapter 7 for concepts of polymer aging.

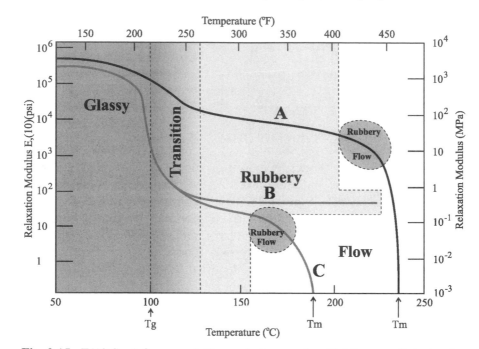

Fig. 3.15 E(10 Sec.) for a crystalline polystyrene, A, a lightly cross-linked polystyrene, B, and amorphous polystyrene, C.

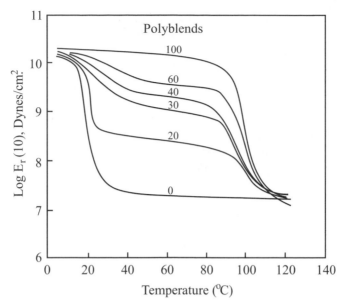

Fig. 3.16 E(10 Sec.) for polyblends as the phase fraction of the two polymers varies from 0 to 100%. (Data from Tobolsky (1962)).

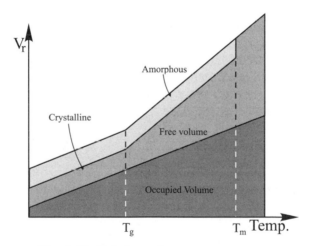

Fig. 3.17 Relative volume vs. temperature.

In **Fig. 3.17**, both occupied and free volume regions are indicated. The occupied volume is the portion of polymer containing molecular mass and the free volume represents the region within the polymer that is not occupied by molecular mass. As a rule of thumb, the free volume at the T_g is approximately 2.5% of the total volume. The variation of free volume gives an interpretation of the molecular mechanisms associated with the five regions of viscoelastic behavior shown in **Fig. 3.15**. Below the T_g, the amount of free volume is small and there is little room for molecular motion. The vibratory motion of individual atoms is suppressed, bond angles are frozen, and the rotation of the backbone bonds and relative motion between chains is inhibited when the polymer is stressed. At the T_g, however, the amount of free volume begins to increase dramatically as temperature is increased affording extra room for change in mobility and hence viscoelastic or time dependent behavior. At the T_m secondary bonds become ineffective and chains are able to move relative to each other freely.

3.4.3. Isochronous Stress-Strain Behavior – Linearity

For many applications and analysis methods, it is very important to determine if the polymer mechanical response under specific conditions is linear or nonlinear. This can only be accomplished by rigorously determining if the creep compliance (or relaxation modulus) is independent of stress (or strain). One method to determine linearity is by conducting creep (or relaxation) tests at different stress levels (at least three levels as shown in **Fig. 3.18**) and obtaining the creep compliance (or relaxation modulus) at constant times as well as the "isochronous" stress-strain diagram. If this isochronous variation of stress vs. strain at any given time is linear as shown in the lower diagram **Fig. 3.18**, the material is linear. If the variation is nonlinear, the material is nonlinear. Linearity of the isochronous stress-strain plot derives from the fact that the ratio of the strain to stress at a given time, t_i, from each stress level must be identical if the material is to be linear. That is for time t_1 we have,

$$D(t=t_1)=\frac{\varepsilon_a(t=t_1)}{\sigma_0\big|_a}=\frac{\varepsilon_b(t=t_1)}{\sigma_0\big|_b}=\frac{\varepsilon_c(t=t_1)}{\sigma_0\big|_c} \qquad (3.11)$$

which means that the compliance $D(t=t_1)$ is independent of stress level.

Similarly,

$$D(t=t_2)=\frac{\varepsilon_a(t=t_2)}{\sigma_0\big|_a}=\frac{\varepsilon_b(t=t_2)}{\sigma_0\big|_b}=\frac{\varepsilon_c(t=t_2)}{\sigma_0\big|_c} \qquad (3.12)$$

and

$$D(t=t_3)=\frac{\varepsilon_a(t=t_3)}{\sigma_0\big|_a}=\frac{\varepsilon_b(t=t_3)}{\sigma_0\big|_b}=\frac{\varepsilon_c(t=t_3)}{\sigma_0\big|_c} \qquad (3.13)$$

Note that the conditions above can be deduced from the requirement that the creep compliance is only a function of time (**D(t)**), and not a function of stress level (**D(t,σ)**), for a linear material:

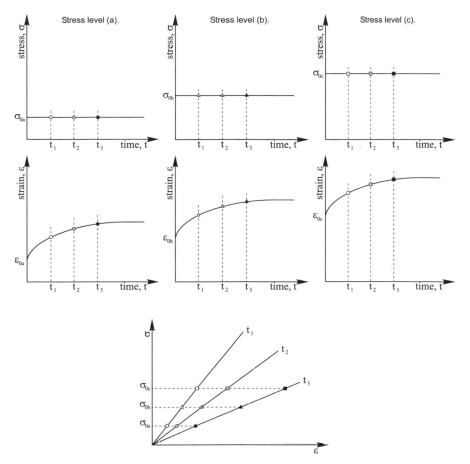

Fig. 3.18 Linearity as indicated by isochronous stress-strain data at constant times from independent creep tests.

$$D(t) = \frac{\varepsilon_a(t)}{\sigma_0\big|_a} = \frac{\varepsilon_b(t)}{\sigma_0\big|_b} = \frac{\varepsilon_c(t)}{\sigma_0\big|_c} \tag{3.14}$$

Relaxation tests may be used in the same manner to determine linearity. The above discussion focuses on stress linearity. For viscoelastic materials there is another important linearity which will be discussed at length in Chapter 6, that of translational linearity with time.

Before discussing mechanical models or other mathematical representation of viscoelastic behavior, it is very important to note that the preceding section deals only with observed behavior or the experimental response of polymers under laboratory conditions. That is, the viscoelastic properties are defined from observations of real behavior and need not be defined by a particular mathematical model. Mathematical models are developed for the simple purpose of understanding and describing observed behavior. Also, as will be evident later, other loading modes such as constant strain rate and steady state oscillation, etc. can be used to determine viscoelastic properties.

3.5. Phenomenological Mechanical Models

In this section, elementary mechanical models that can describe some aspects of viscoelastic polymeric behavior are presented. Although these simple models cannot represent the behavior of real polymers over their complete history of use, they are very helpful to gain physical understanding of the phenomena of creep, relaxation and other test procedures and to better understand the relationship between stress and strain for a viscoelastic material. Undoubtedly, the first models were developed on the basis of observations and not just as a mathematical exercise. Generalized mechanical models are presented later in Chapter 5.

The simplest mechanical models for viscoelastic behavior consist of two elements: a spring for elastic behavior and a damper for viscous behavior. First it is convenient to introduce the model of a linear spring to represent a Hookean bar under uniaxial tension where the spring constant is the modulus of elasticity. As indicated in **Fig. 3.19** the spring constant can be replaced by Young's modulus if the stress replaces **P/A** and strain replaces **δ/L**.

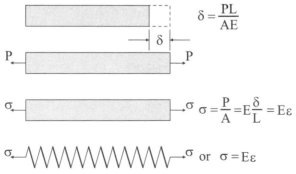

$$\delta = \frac{PL}{AE}$$

$$\sigma = \frac{P}{A} = E\frac{\delta}{L} = E\varepsilon$$

$$\text{or} \quad \sigma = E\varepsilon$$

Fig. 3.19 Linear elastic spring analog for a Hookean elastic tensile bar.

Consider a semi-infinite fluid as shown in **Fig. 3.20**. If a flat plate at the top of the fluid is moved with a velocity, **V = du/dt**, and if the fluid is Newtonian the shear deformation varies linearly from top to bottom assuming a no slip boundary condition between the fluid and the plate as well as between the fluid and the container.

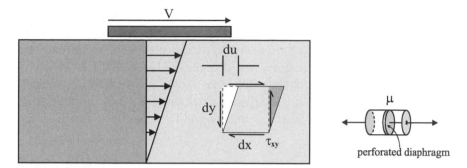

Fig. 3.20 Linear viscous damper analog for a Newtonian viscous fluid.

The strain on a differential element of the fluid is given by du/dy. The Newtonian law of viscosity for the shear process shown in **Fig. 3.20** may thus be expressed as,

$$\tau = \mu \frac{d}{dt}\left(\frac{du}{dy}\right) = \mu \frac{d\gamma}{dt} = \mu\dot{\gamma} \tag{3.15}$$

where μ is viscosity. A linear viscous damper (or dashpot) also shown in **Fig. 3.20** will be used to model a Newtonian fluid such that it can form a uniaxial fluid analogue to a tensile bar. The housing of the damper contains a fluid with a viscosity μ. The diaphragm is perforated and when it is pulled through the fluid by an applied force, motion occurs according to the Newtonian law of viscosity given above.

Spring and damper elements can be combined in a variety of arrangements to produce a simulated viscoelastic response. Early models due to Maxwell and Kelvin combine a linear spring in series or in parallel with a Newtonian damper as shown in **Fig. 3.21**. Other basic arrangements include the three-parameter solid and the four-parameter fluid as shown in **Fig. 3.22**.

These models are very useful in understanding the physical relation between stress and strain that occurs in polymers and other viscoelastic materials. For example, if suddenly a constant stress is applied as in a creep test, each model with a free spring will have a sudden increase in strain.

The Kelvin will not have a sudden increase in strain as the damper will not allow a sudden jump in strain. Under the condition of constant stress, each model with a free damper (Maxwell and four parameter fluid) will have an ever-increasing creep strain and will be similar to the response for a thermoplastic polymer described in **Fig. 3.13**. Those with a free spring (Kelvin and three parameter solid) will creep to a limiting constant strain and will be similar to the response of thermoset polymers described in **Fig. 3.13**. In relaxation, the stress will decay to zero for those models with a free damper (Maxwell and four parameter fluid) and the stress will decay to a limiting value for those without a free damper (Kelvin and three parameter solid) as shown in **Fig. 3.12** for thermoplastic and thermosetting materials respectively. Note that a simple stress relaxation test is not possible for a Kelvin model as the damper will prohibit a sudden increase in strain.

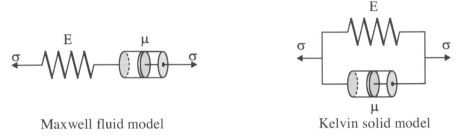

Maxwell fluid model   Kelvin solid model

Fig. 3.21 Spring and damper arrangements for Maxwell and Kelvin models.

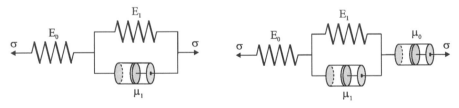

Three parameter solid model   Four parameter fluid model

Fig. 3.22 Spring and damper arrangements for three and four element models.

3.5.1. Differential Stress-Strain Relations and Solutions for a Maxwell Fluid

The models described in the preceding section are useful in developing mathematical relations between stress and strain in viscoelastic polymers and in giving insight to their response to creep, relaxation and other types of loading. Consider again the Maxwell fluid from **Fig. 3.21**,

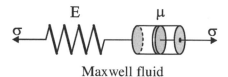

Maxwell fluid

An equation between stress and strain can be obtained for any mechanical model by using equilibrium and kinematic equations for the system and constitutive equations for the elements. For a Maxwell fluid, equilibrium gives,

$$\sigma = \sigma_s = \sigma_d \qquad (3.16)$$

where σ is the applied stress, σ_s is the stress in the spring and σ_d is the stress in the damper. The kinematic condition is,

$$\varepsilon = \varepsilon_s + \varepsilon_d \qquad (3.17)$$

where ε is the total strain in the Maxwell element, ε_s is the strain in the spring and ε_d is the strain in the damper. The constitutive equations are,

$$\sigma_s = E\,\varepsilon_s = \sigma \qquad (3.18)$$

and

$$\sigma_d = \mu \frac{d\varepsilon_d}{dt} = \mu \dot{\varepsilon}_d = \sigma \qquad (3.19)$$

Differentiating **Eq. 3.17** and replacing the strain rates of the spring and damper using **Eqs. 3.18** and **3.19** gives after rearrangement,

$$\dot{\sigma} + \frac{E}{\mu}\sigma = E\dot{\varepsilon} \qquad (3.20)$$

The result indicates that the relation between stress and strain for a material that is Maxwellian in behavior is a differential equation which must be solved for particular cases of applied stresses or strains. In viscoelastic literature, it is usual to write the differential equation in a standard form with ascending derivatives from right to left on both sides of the equation. Hence,

$$\sigma + \frac{\mu}{E}\dot{\sigma} = \mu\dot{\varepsilon} \qquad (3.21)$$

or

$$\sigma + p_1\dot{\sigma} = q_1\dot{\varepsilon} \qquad (3.22)$$

Differential equations for all mechanical models can be found using the same procedure. In this form the inverse of the coefficient of the stress rate is defined as the relaxation time, i.e. $\tau = \mu/E$.

To obtain the solution of **Eq. 3.20** for the case of creep note the applied stress is constant and can be written as,

$$\sigma(t) = \sigma_0 H(t) \tag{3.23}$$

where **H(t)** is the Heavyside function and is defined as,

$$H(t) = 1 \ \ \text{for} \ \ t > 0$$
$$H(t) = 0 \ \ \text{for} \ \ t < 0 \tag{3.24}$$

In other words the stress is constant for time greater than zero. With this input the solution of **Eq. 3.20** is,

$$\varepsilon(t) = \sigma_0 \left(\frac{1}{E} + \frac{t}{\mu} \right) \tag{3.25}$$

or

$$\varepsilon(t) = \sigma_0 D(t) \tag{3.26}$$

where

$$D(t) = \left(\frac{1}{E} + \frac{t}{\mu} \right) \tag{3.27}$$

is the creep compliance.

The creep and creep recovery behavior for a Maxwell fluid is shown in **Fig. 3.23(a)** and agrees with the description of a thermoplastic materials given in **Fig. 3.13**.

The solution of **Eq. 3.20** for relaxation is obtained using a step input in strain,

$$\varepsilon(t) = \varepsilon_0 H(t) \tag{3.28}$$

with the resulting stress output of,

$$\sigma(t) = \varepsilon_0 E e^{-t/\tau} \tag{3.29}$$

where

$$E(t) = E e^{-t/\tau} \tag{3.30}$$

is the relaxation modulus. The relaxation behavior for a Maxwell fluid is shown in **Fig. 3.23(b)** and agrees with the description of thermoplastic materials given in **Fig. 3.12.**

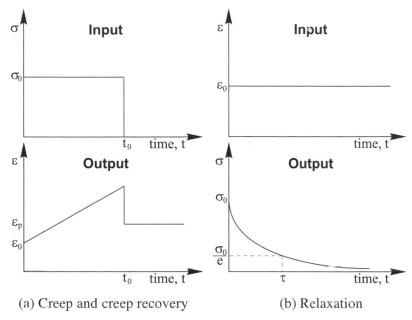

(a) Creep and creep recovery (b) Relaxation

Fig. 3.23 Creep, creep recovery and relaxation response of a Maxwell fluid.

From **Eq. 3.29**, the stress at a time equal to the relaxation time is,

$$\sigma(t = \tau) = \sigma_0 / e \tag{3.31}$$

This result provides a general definition of the relaxation time of a polymer and allows the relaxation time to be found easily from experimental data without recourse to a mechanical model. It can be used as a material property to give an indication of the time scale associated with viscoelastic response in a polymer and is indicative of the intrinsic viscosity of the polymer. It should again be noted that the relaxation time for a Maxwell model is related to the viscosity through the equation, $\tau = \mu/E$. In a sense, the Maxwell model provides a defining relationship for the viscosity of a material. It will be shown later that a polymer possesses a distribution of relaxation times and that an individual chain can be thought of as having various relaxation times.

It is instructive to consider the response to a Maxwell fluid under a constant strain rate loading as shown in **Fig. 3.23a**. For a constant strain-rate,

$$\varepsilon = Rt \quad \text{and} \quad d\varepsilon/dt = R = \text{constant} \tag{3.32}$$

The differential equation then becomes

$$\sigma + \frac{\mu}{E}\dot{\sigma} = \mu R \tag{3.33}$$

and the solution can be shown to be,

$$\sigma(t) = \tau ER\left(1 - e^{-t/\tau}\right) \tag{3.34}$$

or, since **t= ε / R**

$$\sigma(\varepsilon) = \tau ER\left(1 - e^{-\varepsilon/\tau R}\right) \tag{3.35}$$

For various constant strain rates, several results are plotted in **Figs. 3.23(b)** and **3.23(c)**. Note that the time scale and the strain scale in these two figures are related by the constant rate of each test and obviously the abscissa can be interpreted as only strain. While the stress versus time curves would be linear for a single spring (a pure elastic material), the result for the Maxwell element appears nonlinear since the damper continuously relaxes some of the stress as time increases. The apparent stress-strain behavior (or plot of **Eq. 3.35**) is therefore as shown in **Figure 3.24(c)**. That is, the stress-strain response might be mistakenly interpreted as nonlinear, even though the Maxwell model is composed only of linear elements. The reason, of course, is due to the simple relationship between strain and time. If isochronous stress-strain curves were constructed for a Maxwell model using creep or relaxation data or the constant strain rate data of **Fig. 3.24**, a linear stress-strain response would be obtained. Also the construction of isochronous stress-strain curves from constant strain rate tests as given in **Fig. 3.23** would be linear (see problem 3.5).

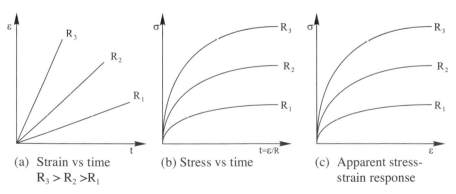

(a) Strain vs time
     $R_3 > R_2 > R_1$

(b) Stress vs time

(c) Apparent stress-
     strain response

Fig. 3.24 Stress response of a Maxwell model in a constant strain-rate test.

The results shown in **Fig. 3.23(c)** are very similar to those for the polymers illustrated in **Fig. 3.6**. From this example, it is now clear that the apparent nonlinear stress-strain response displayed in **Fig. 3.6** may, in fact, be linear prior to yielding. The point being that it is not possible to determine if a material is linear just by looking at the shape of an experimentally determined response to a constant strain rate test as generally conducted in the laboratory. Linearity can only be assessed by carefully determining if the material response is independent of stress regardless of loading type, e.g., by the isochronous stress-strain diagrams described earlier. The importance of this principle cannot be overstated.

Using **Eq. 3.35** it is also possible to show that constant strain rate properties vary with temperature for a Maxwell model and would be similar to the results described earlier in **Fig. 3.6** (see problem 3.6).

A constant strain rate test may be used to determine the relaxation modulus and a constant stress-rate test may be used to find the creep compliance. Steady state oscillation tests may also be used to determine the viscoelastic properties of polymers. These details and the interrelation between various test approaches are given in Chapter 5.

3.5.2. Differential Stress-Strain Relations and Solutions for a Kelvin Solid

The Kelvin model is also frequently used to describe the phenomena of creep. Recall the Kelvin solid from **Fig. 3.21**.

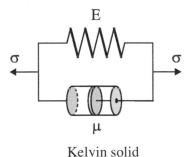

Kelvin solid

The equilibrium equation is,

$$\sigma = \sigma_s + \sigma_d \tag{3.36}$$

and the kinematic condition is,

$$\varepsilon = \varepsilon_s = \varepsilon_d \tag{3.37}$$

The constitutive equations are,

$$\sigma = E\varepsilon_s$$
$$\sigma_d = \mu\dot{\varepsilon}_d$$

(3.38)

and the differential equation becomes,

$$\sigma = E\varepsilon + \mu\dot{\varepsilon}$$

(3.39)

or

$$\sigma = q_0\varepsilon + q_1\dot{\varepsilon}$$

Under creep loading, the solution becomes,

$$\varepsilon(t) = \frac{\sigma_0}{E}\left(1 - e^{-t/\tau}\right)$$

(3.40)

and the creep compliance is,

$$D(t) = \frac{1}{E}\left(1 - e^{-t/\tau}\right)$$

(3.41)

A schematic of the result is given in **Fig. 3.25**.

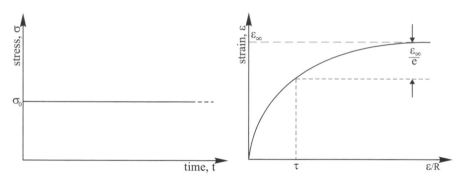

Fig. 3.25 Creep of a Kelvin solid.

There is no initial elasticity as the damper only allows the spring to move slowly with time. Also, the Kelvin model is not useful in understanding the relaxation response of materials because the damper does not allow the spring to move instantaneously.

Note, for a very large time, a constant strain state, ε_∞, is achieved

$$\varepsilon(t = \infty) = \varepsilon_\infty = \sigma_0/E = \sigma_0 D_\infty$$

(3.42)

and **1/E** is the corresponding equilibrium compliance.

The retardation time, τ, is defined as the time required for the strain to come within $1/e$ of its asymptotic value. That is, **Eq. 3.40** becomes,

$$\varepsilon(t = \tau) = \frac{\sigma_0}{E}\left(1 - e^{-1}\right) = \varepsilon_\infty\left(1 - \frac{1}{e}\right) \tag{3.43}$$

Again, the concept of a retardation time can be used as an indication of the intrinsic viscosity of a polymer as the transient strain in a creep test occurs due to viscosity of the assembly of molecular chains. The retardation time of a polymer can be determined from a creep test by considering only the experimental data according to the above definition.

3.5.3. Creep of a Three Parameter Solid and a Four Parameter Fluid

A single Maxwell element is not realistic for characterizing a polymer as no transient response is shown in a creep test, i.e., the creep response is linear with time. A single Kelvin element is also not accurate as no instantaneous elastic response occurs in a creep test. A more realistic result for creep is obtained if a Kevin solid is combined with a Maxwell fluid to obtain the four-parameter fluid as in **Fig. 3.22**.

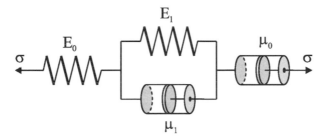

Four parameter fluid

The differential equation can be derived by following similar procedures as previously given for the Maxwell and Kelvin elements (see problem 3.4). The resulting equation can then be solved for the case of creep. However, the creep response can also be obtained by superposition by adding the creep response of Kelvin and Maxwell elements to obtain,

$$\varepsilon(t) = \sigma_0 \left[\underbrace{\frac{1}{E_0}}_{\text{elastic}} + \underbrace{\frac{1}{E_1}\left(1 - e^{-t/\tau}\right)}_{\text{delayed elastic}} + \underbrace{\frac{t}{\mu}}_{\text{flow}} \right] \qquad (3.44)$$

The behavior shown here represents the most general type behavior possible for a viscoelastic material, instantaneous elasticity, delayed elasticity and flow. Some texts do not include the flow term as a viscoelastic component, preferring instead to define viscoelastic behavior only for models with no free damper or flow term.

The response of a four parameter fluid in a creep and creep recovery test is given in **Fig. 3.26** and is recognized as the response of a thermoplastic type polymer as given earlier in **Fig. 3.13**.

By eliminating various elements in the four-parameter model the response of a Maxwell fluid, Kelvin solid, three-parameter solid (a Kelvin and a spring in series) can be obtained and the model can be used to represent thermoplastic and/or thermoset response as illustrated in **Fig. 3.13**. For example the creep response of a three-parameter solid is obtained by eliminating the free damper in **Eq. 3.44** and gives the creep and creep recovery response shown in **Fig. 3.13** for a crosslinked polymer.

The four-parameter fluid can also be evaluated in relaxation but typically, Maxwell elements in parallel are used for relaxation and Kelvin elements in series are used for creep.

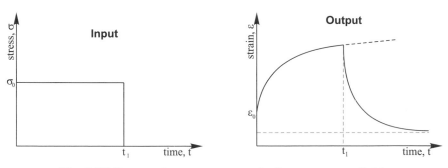

Fig. 3.26 Creep and creep recovery of a four-parameter fluid.

3.6. Review Questions

3.1.   What are some advantages of using polymers as structural materials?

3.2.   What are some disadvantages of using polymers as structural materials?

3.3.   What is a coupling agent?

3.4.   What is a plasticizer?

3.5.   What is the correct name for ABS?

3.6.   Describe vulcanization.

3.7.   Give the complete names for LDPE, LLDPE, HDPE.

3.8.   Sketch creep and creep recovery curves for a VE solid and a VE fluid. Label all significant points. Also identify which curve would be expected to represent a linear polymer. Which would represent a cross-linked polymer.

3.9   Sketch the response of a Maxwell fluid to a creep and a creep recovery tests.

3.10. Sketch relaxation curves for a VE solid and a VE fluid. Label all significant points. Also identify which curve would be expected to represent a linear polymer. Which would represent a cross-linked polymer.

3.11   Describe in detail how the "10 second" modulus is found. Give a sketch of a typical "ten second" modulus curve for an amorphous polymer as a function of temperature and label the five regions of VE response. Show on your sketch curves for amorphous thermoplastic, crystalline thermoplastic and thermosetting polymers. Indicated the location of the T_g and the T_m.

3.12   Give a proper definition for T_g and the T_m and discuss methods for determining these quantities.

3.13   Sketch the variation of the specific volume vs. temperature for an amorphous polymer. Indicate regions of free volume, occupied volume, the T_g and the T_m. Give the correct names for the T_g and the T_m.

3.14   Give an accurate description of how you would determine the linearity of a VE material.

3.15   Indicate on a sketch how stress strain properties of polymers typically depend on a). strain-rate, b). temperature. (Use separate sketches).

3.16 Recalling class discussion and/or class notes, give the proper equation for the creep compliance of a four-parameter fluid. Indicate the instantaneous elasticity term, the delayed elasticity term, and the flow term.

3.7. Problems

3.1 Derive the differential equation for a three parameter solid.

3.2 Derive the differential equation for a four-parameter fluid.

3.3 The DE for a Maxwell model is, $\sigma + \mu/E \, d\sigma/dt = \mu \, d\varepsilon/dt$. Determine the stress output for a relaxation test by solving this DE and sketch the resulting curve.

3.4 Given the relaxation data below. Determine the relaxation time. (The stress remains constant after t = 35 minutes.)

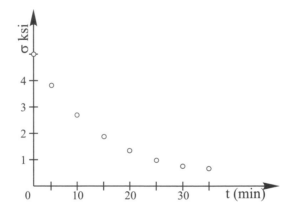

3.5 A schematic of the constant strain-rate response of a Maxwell fluid is shown below. Prove that the constant strain-rate behavior of a Maxwell fluid is linear by constructing an isochronous stress-strain curve. (Note: Use the known form of the analytical solution. Do not attempt to use the schematic curves below as they are not to scale.)

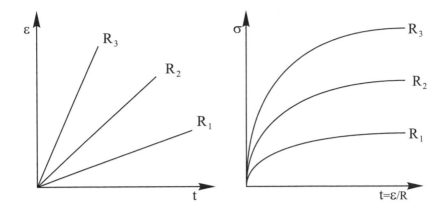

3.6   Prove that the generic (general shape) constant strain rate and temperature properties for a polymer can be phenomenologically explained using a Maxwell model. Hint: think of the parameters in the Maxwell model. Which of these are affected by temperature?

3.7   Given the creep data shown below. Find the necessary parameters to represent the data using a three-parameter solid. Plot your results on the data given. (Clearly indicate your procedures.)

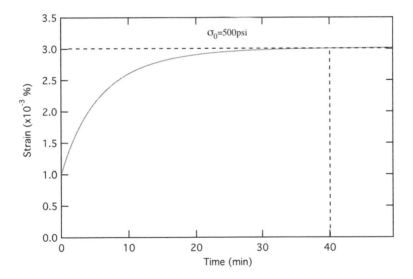

4. Polymerization and Classification

The discussion in previous chapters has provided a glimpse of the relationship between the molecular structure of polymers and their mechanical behavior. In this chapter the intent is to provide more detailed information about the molecular structure of polymers and the relation of such structure to mechanical performance. Typically materials courses taken by engineering students prior to 1980 contained little, if any, information on the structure of polymers that might be useful in the engineering design of polymer based structures. While now most elementary books on materials do include a chapter or two on polymers they are often omitted due to the pressures of schedules and/or time constraints. As a result, engineering students often do not obtain a knowledge base that allows the safe design of polymeric structures. All too frequently, the engineering design of structural polymers is based upon principles that are best used for metals. The purpose of the present chapter is to provide a framework for understanding the structure of polymers and hence the structure-property relationships that give rise to their unique mechanical behavior with time, temperature and other environmental parameters as discussed in subsequent chapters. Due to the prevalence of polymers in industrial uses, a general understanding of the concepts outlined in this chapter are essential for an engineer to be able to make informed design decisions on polymeric components and, importantly, to be able to discuss on common ground with synthesis people the type of polymer needed to be produced for a given application.

4.1. Polymer Bonding

Atomic and molecular bonding of materials are discussed in elementary chemistry, physics and materials science courses. In general, the same bonds are present in polymers but they need to be revisited with an emphasis on the prevalent types found in polymers.

In general there are two types of bonds: (1) primary or chemical bonds and (2) secondary or van der Waals bonds. Primary bonds are metallic, covalent, and ionic. Metallic bonds are unique because all the atoms of a metal give up their valence electrons to share with all other atoms such that the electrons move freely throughout the bulk of the material. Metallic bonds were not generally important in polymers until the recent interest in conducting polymers using metal oxides and the metallacenes. Covalent bonds are when two or more atoms share electrons from their respective valence shells and constitute most of the primary bonds found in polymers. A coordinate bond is a type of covalent bond found in polymers in which the shared electrons come from only one atom (Billmeyer (1984)). Ionic bonds are those in which one atom donates an electron to another atom, e.g. Na^+Cl^-. These bonds are not frequently encountered in polymers but they do occur.

Unlike the case for metals, secondary bonds are of great importance in polymers. These bonds are much weaker than covalent bonds, but for even moderate chain length polymers these bonds have a significant impact on the molecular and bulk properties of these materials. These intermolecular bonds are based on electrostatic interactions and are due to either attractions between permanent dipoles, quadrupoles, and other multipoles, or between a permanent multipole and an induced charge on a second molecule (or moiety, in the case of a polymer), or between transient multipoles. All such secondary bonds can be considered van der Waals forces, but many texts use van der Waals to denote induced and/or transient multipole interactions only. The induced interaction is sometimes referred to as polarization, or sometimes induction bonding. The transient interaction is very weak and is known as dispersion or London dispersion forces, and arises from electrostatic interactions between two molecules due to temporary inhomogeneous electron density distributions in the outermost electron shells of these molecules.

Secondary bonding of the first type, that is, forces between multipoles, are the strongest. This occurs when there is a permanent separation of two atoms with strongly differing electronegativity, such as is found in an oxygen-hydrogen (-OH) bond. Electronegativity can be thought of as the attraction that an atom has for electrons in the outermost shell. Using the OH example, oxygen is strongly electronegative (meaning it has a very strong attraction for an additional electron), whereas hydrogen is very weakly electronegative (meaning that has only a weak attraction for its single electron). This results in the oxygen side of the OH bond having a partial negative charge. Water, consisting of two OH bonds at an angle of 104.45 is strongly polar. Intermolecular forces due to the electrostatic forces between these dipoles give water its special properties. Dipolar van der Waals forces involving hydrogen are referred to as *hydrogen bonds*, and many of the important properties of polymers and polymer side chains are due to hydrogen bonding given the prevalence of hydrogen atoms along most polymer chains.

Using knowledge of the nature of attractive forces and energies between atoms described in Chapter 2, bond lengths and energies of typical covalent bonds found in polymers have been estimated and are shown in **Table 4.1**.

The disassociation energy (kJ/mole or k cal/mole) or cohesive energy density (J/cm^3) is the energy required to move a molecule far enough away from another molecule so that the attractive force or energy between the two is negligible. The cohesive energy densities to break the bond between mer units of a number of linear polymers is shown in **Table 4.2**. In linear or thermoplastic polymers, it is only the secondary bonding forces that hold the polymer together if entanglements are neglected. Therefore the energies in **Table 4.2** give only an estimate of the breaking strength of a highly oriented samples of the various polymers listed.

Table 4.1 Typical covalent bond lengths and energies found in polymers. (Data from Billymeyer (1984))

Bond	Bond Length (Å)	Dissociation Energy (kJ/mole)
C — C	1.54	347
C = C	1.34	611
C — H	1.10	414
C — N	1.47	305
C = N	1.15	891
C — O	1.46	360
C = O	1.21	749
C — F	1.35	473
C — Cl	1.77	339
N — H	1.01	389
O — H	0.96	464
O — O	1.32	146

Of even more interest is a comparison of bond lengths and energies given in **Table 4.3** for primary and secondary bonds which assists in understanding the differences between linear and crosslinked polymers.

Table 4.2 Cohesive energies of linear polymers. (Data from Billmeyer (1984)).

Polymer	Repeat Unit	Cohesive Energy Density (J/cm^3)
Polyethylene	$— CH_2CH_2 —$	259
Polyisobutylene	$— CH_2C(CH_3)_2 —$	272
Polyisoprene	$— CH_2C(CH_3) = CHCH_2 —$	280
Polystyrene	$— CH_2C(C_6H_5) —$	310
PMMA	$— CH_2C(CH_3)COOCH_3 —$	348
PVC	$— CH_2CHCL —$	381
PET	$— CH_2CH_2OCOC_6H_4COO —$	477
PAN	$— CH_2CHCN —$	992

Table 4.3 Comparsion of primary and secondary bond distances and energies. (Data from Rosen (1993))

Bond Type	Interatomic Distance (nm)	Dissociation Energy (kcal/mole)
Primary covalent	0.1-0.2	50-200
Ionic	0.2-0.3	10-20
Hydrogen	0.2-0.3	3-7
Dipole	0.2-0.3	1.5-3
van der Waals	0.3-0.5	0.5-2

Except for the dispersion bond, all bonds are functions of temperature. As a result, variations in temperature for the same polymer lead to different physical states as represented by **Fig. 4.1**. The relation of these states to mechanical properties will be discussed further in later sections and chapters. Notice that both linear and cross-linked polymers are indicated and temperature can be used to alter the state and or the chemistry of a polymer.

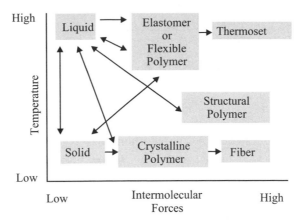

Fig. 4.1 The interrelation of states in a bulk polymer. (After Billmeyer (1984)).

4.2. Polymerization

The polymerization process can be illustrated by the conversion of ethylene into polyethylene which is one of the most widely produced polymers in the world. The unsaturated ethylene molecule or monomer is shown in **Fig. 4.2** below. (In general, the term unsaturated refers to molecules with double or triple bonds while those with only single bonds are termed saturated.)

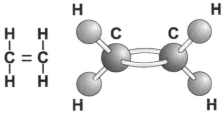

Fig. 4.2 The ethylene molecule.

Under appropriate conditions of heat and pressure in the presence of a catalyst, the double bond between the two carbon atoms can be "opened" or broken and replaced by a single saturated bond with other similarly opened monomeric units on either side to form a long replicated strand of mer units as illustrated in **Fig. 4.3**.

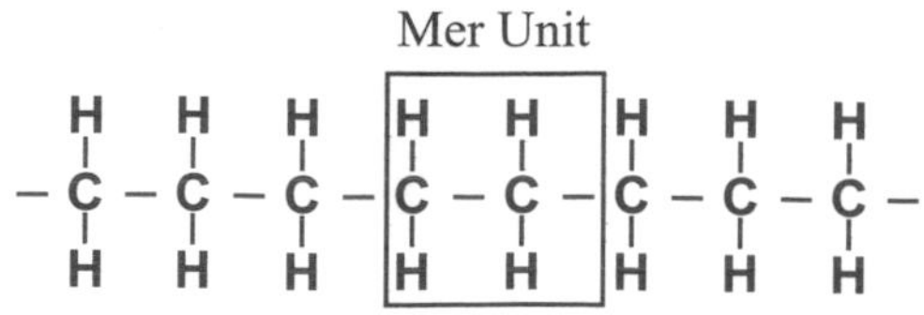

Mer Unit

$$-\overset{\overset{\displaystyle H}{|}}{\underset{\underset{\displaystyle H}{|}}{C}}-\overset{\overset{\displaystyle H}{|}}{\underset{\underset{\displaystyle H}{|}}{C}}-\overset{\overset{\displaystyle H}{|}}{\underset{\underset{\displaystyle H}{|}}{C}}-\boxed{\overset{\overset{\displaystyle H}{|}}{\underset{\underset{\displaystyle H}{|}}{C}}-\overset{\overset{\displaystyle H}{|}}{\underset{\underset{\displaystyle H}{|}}{C}}-}\overset{\overset{\displaystyle H}{|}}{\underset{\underset{\displaystyle H}{|}}{C}}-\overset{\overset{\displaystyle H}{|}}{\underset{\underset{\displaystyle H}{|}}{C}}-\overset{\overset{\displaystyle H}{|}}{\underset{\underset{\displaystyle H}{|}}{C}}-$$

Fig. 4.3 Repeating mer units of polyethylene.

In an actual polymer each individual chain may contain from several thousand to hundreds of thousand repeating mers or units.

The resulting solid polyethylene will contain a great many chains but each chain will vary in length. This leads to the need to have special methods to quantify the molecular weight of polymers and these will be discussed in a subsequent section. In the case of polyethylene, the molecular bonds between carbon atoms along the length of the chain are all primary or covalent. However, the bonds between individual chains are secondary. For this reason, under a sufficient increase in temperature, the secondary bonds become ineffective or broken and the various long chains can move or flow past each other with relative ease. Therefore, polyethylene is called a thermoplastic polymer as it can be melted and molded or reformed. Polyethylene and other polymers with similar characteristics are also called linear polymers because the backbone chain as shown in **Fig. 4.3** appears to be one-dimensional or like a long string.

It is important to note here that the use of the term linear to describe a type of polymer refers only to the geometry of the chain and/or the bonding state between chains and should not be confused with the term linear used to describe the relation between stress and strain in earlier sections.

The term linear is also somewhat misleading with respect to chain geometry as even a fully extended PE chain has more of a "zig-zag" shape as shown in **Fig. 4.4** because the equilibrium angle between alternate carbon atoms is 109° 28'. And in reality, the chains are neither linear or of a zig-zag shape as, depending on the temperature, carbon atoms can rotate relatively easily about adjacent carbon atoms as shown in **Fig. 4.5**. As a result, an individual chain within a polymer will form in a random manner during polymerization and the final shape of a chain will appear as given in **Fig. 4.6**. Each long chain molecule will exist together with many, many other chains in a tangled mass which has often been said to resemble a tangled ball of many pieces of individual strings of different length. A more precise description of a tangle ball of very long worms has been used as each atom is in a state of constant motion or vibration. Indeed, in recent years this analogy has been used to develop a reptation model to explain the manner in which one long molecule can move through a seemingly continuous mass of other chains (Aklonis and McKnight (1983)). The method of displaying the atoms as bar-mass linkages in **Figs. 4.3** and **4.4** is traditional. It is often used to visualize bonding arrangements and many types of computations associated with molecular geometry and motion.

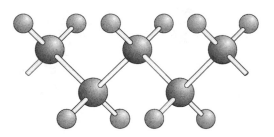

Fig. 4.4 Zigzag shape of polyethylene molecule.

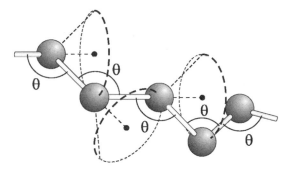

Fig. 4.5 Random nature due to rotation of carbon molecules.

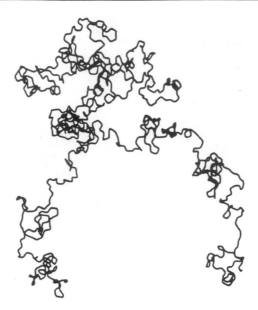

Fig. 4.6 Shape of a 1,000 link polyethylene chain (Treolar (1975), reprinted by permission of Oxford University Press).

The mer units of a number of frequently used thermoplastic polymers are given in **Fig. 4.7**.

Polymer	Repeating (Mer) structure
Polyethylene (PE)	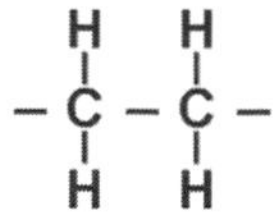
Polyvinyl Chloride (PVC)	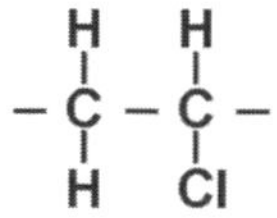
Polytetrafluoroethylene (PTFE)	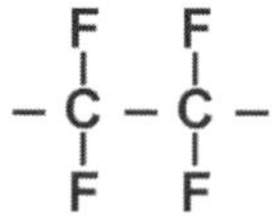
Polypropylene (PP)	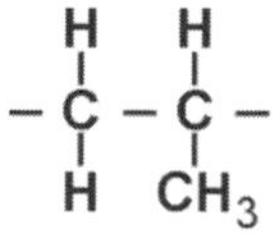
Polystyrene (PS)	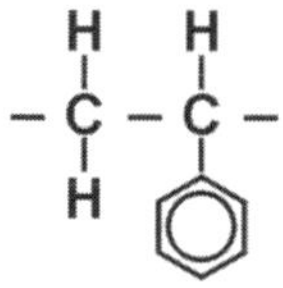
Polymethylmethacrylate (PMMA)	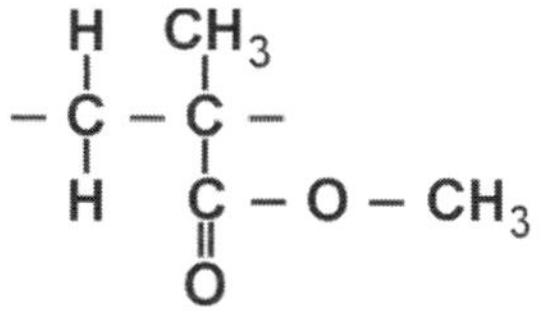
Polycarbonate	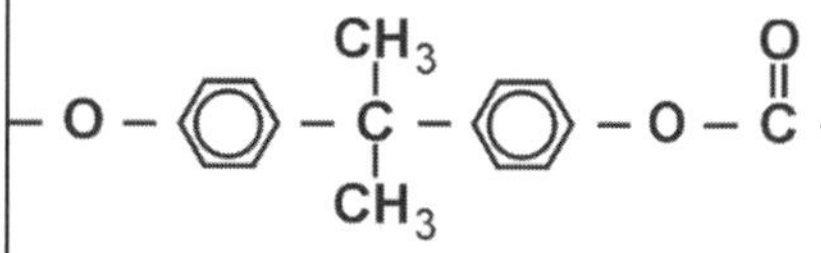

Fig. 4.7 Mer units of selected thermoplastic polymers.

Thermosetting or "cross-linked" polymers are also formed under catalytic conditions of heat and pressure (often pressure is not needed). However, in this case covalent bonds do exist between individual chains. This "cross-linking" may vary considerably from polymer to polymer but generally leads to a solid material which cannot be melted. Examples of several chemical units that lead to cross-linked polymers are shown **Fig. 4.8.**

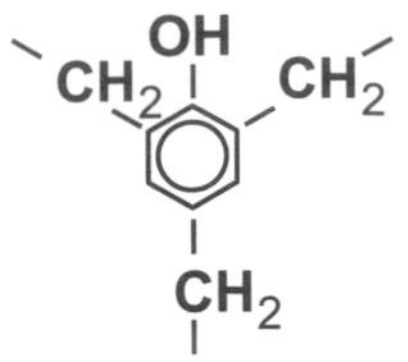

Phenol-formaldehyde (bakelite)

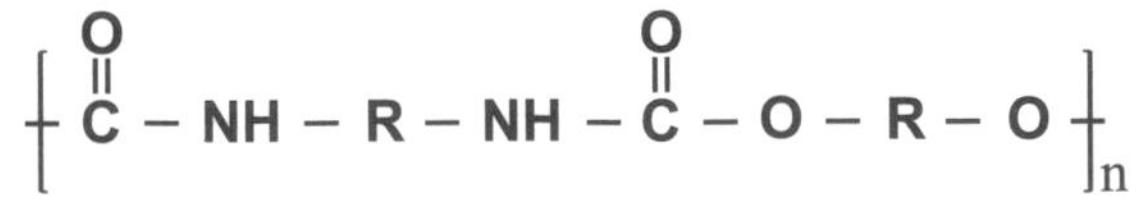

Polyurethane

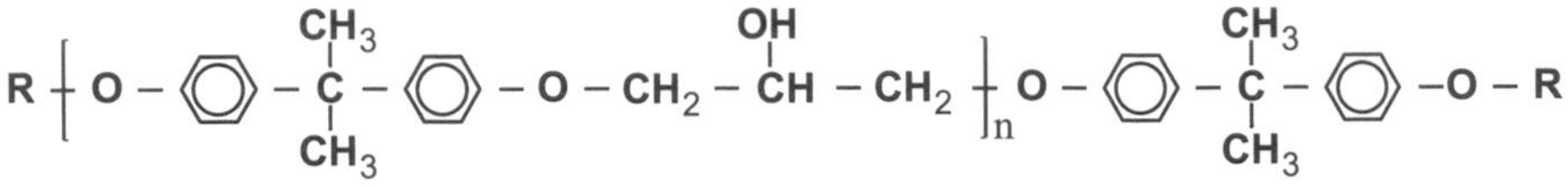

Bisphenol-A epoxy based polymer

Fig. 4.8 Mer units of selected thermoset polymers.

Pheno-formaldehyde or Bakelite was one of the first polymers introduced in the US by Leo Bakeland in 1907. Polyurethane can be polymerized with other elements to give either elastomeric or rigid polymers. The epoxy precursor shown can be reacted with several other compounds to give well known epoxy resins.

4.3. Classification by Bonding Structure Between Chains and Morphology of Chains

One simple classification scheme according to bonding structure is shown in **Fig. 4.9**. Here it is appropriate to emphasize the distinction between thermoplastic and thermosetting polymers,

> **Linear or Thermoplastic Polymers:** Intrachain bonds are primary (covalent). Interchain bonds are secondary (hydrogen, induction, dipole, etc.).

Crosslinked or Thermosetting Polymers: Intrachain bonds are primary. Interchain bonds are both secondary and covalent. Very heavily crosslinked polymers are often called network polymers.

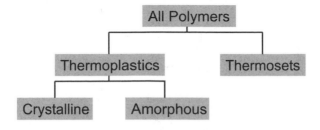

Fig. 4.9 A simple classification scheme for polymers.

It is noted that there are variations in each type and schematically these may be represented as given in **Fig. 4.10**. The branches in branched polymers may vary from very short to very long. Long branches may be further classified as comb-like, random or star shaped as shown in **Fig. 4.11**.

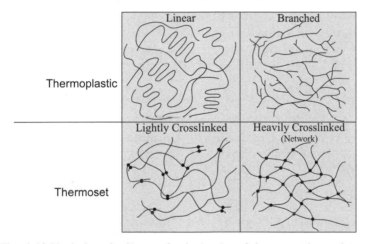

Fig. 4.10 Variations in thermoplastic (top) and thermosetting polymers (bottom).

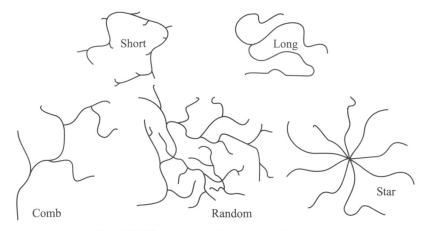

Fig. 4.11 Variations in branched polymers.

Crystalline regions of a linear polymer in **Fig. 4.10** are shown schematically as small parallel segments within a chain. This ordering of the structure will be discussed in a later section on morphology. However, it is important to note here the relative amount of ordering (crystallinity) for polyethylene and the effect on density and mechanical properties. This information is given in the **Table 4.4**.

Characteristics and applications of several linear polymers are given in **Table 4.5.**

Table 4.4   Effect of crystallinity on density and strength of polyethylene. (Data from Hertzberg (1989))

Density (g/cm³)	% Crystallinity	Ultimate Tensile Strength	
		MPa	ksi
0.920	65	13.8	2.0
0.935	75	17.8	2.5
0.950	85	27.6	4.0
0.960	87	31.0	4.5
0.965	95	37.9	5.5

Table 4.5 Characteristics of several linear polymers.

Material	Characteristics	Applications
Low Density Polyethylene	Branched crystalline, inexpensive, good insulator	Films, moldings, squeeze bottles, cold water plumbing
Polypropylene	Crystalline, corrosion and fatigue resistance	Fibers, pipe, wire covering
Nylon 66	Crystalline, tough, resistance to wear, high strength	Gears, bearings, rollers, pulleys, fibers
PTFE (Teflon)	Crystalline, corrosion resistance, very low friction, non-sticking.	Coatings, cookware, bearings, gaskets, insulation tape, non-stick linings
PVC	Amorphous, inexpensive, good processability	Film, water pipes, insulation
PMMA	Amorphous, high transparency	Signs, windows, decorative products

4.4. Molecular Configurations

The terms configuration and conformations are often used to describe the arrangement of atoms in a polymer and sometimes it seems as if they can be used interchangeably. However, herein the description for each given by Billmeyer (1962) will be used. Configurations describe those arrangements of atoms that cannot be altered except by breaking or reforming chemical bonds. Conformations are arrangements of atoms that can be altered by rotating groups of atoms about a single bond. Each will be discussed in the subsections below.

4.4.1. Isomers

Polymers that have the same composition but with different atomic arrangements are called isomers. Two basic types are: stereoisomers and geometrical isomers. Isomers occur because polymers may have more than one type of side atom or side group bonded to the main chain (e.g. PVC, see **Fig. 4.7**) such that a mer unit would appear as in **Fig. 4.12(a)** in which R represents an atom or side group other than hydrogen. Polymers with only one extra side group are called "vinyl" polymers. A "head-to-head" arrangement of mers occurs when the R groups are adjacent to each other, and a "head-to-tail" arrangement occurs when the R groups bond to alternate carbon atoms in the chain as shown in **Fig. 4.12**. The head-to-tail configuration predominates as polar repulsion occurs between R groups in head-to-head configurations.

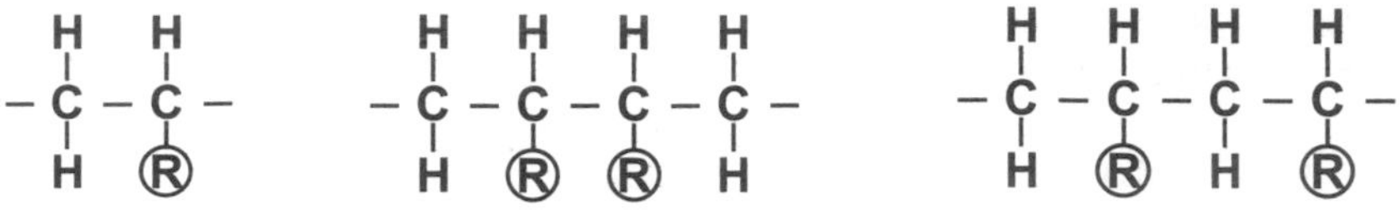

(a) Basic mer unit　　　(b) Head to head config.　　　(c) Head to tail config.

Fig. 4.12 Sequences for isomers.

For a polymer chain with a given sequence of mer groups, stereoisomers or geometrical isomers can then be distinguished. The three types of stereoisomers (isotactic, syndiotactic and atactic) for a head-to-tail sequence are shown in **Fig. 4.13** and the two types of geometric isomers (trans and cis) for a mer unit containing a double bond are shown in **Fig. 4.15**. In steroisomerism, the atoms are linked together in the same order (e.g., head-to-tail) but their spatial arrangement is different. The isotactic configuration (a) is when the R groups are all on the same side of the chain. The syndiotactic configuration (b) is when the R group is on alternate sides of the chain and the atactic configuration (c) is when the R group alternates from one side to the other in a random pattern. Examples of stereoisomers for polypropylene are given in **Fig. 4.14**.

Conversion from one type to another is only possible by breaking a carbon to carbon bond, rotating and reattaching. This constraint can be seen best by use of molecular models or three dimensional chain representations (eg, **Fig. 4.20**). A specific polymer may contain more than one type of steroisomer but one may predominate depending only on the synthesis procedure used.

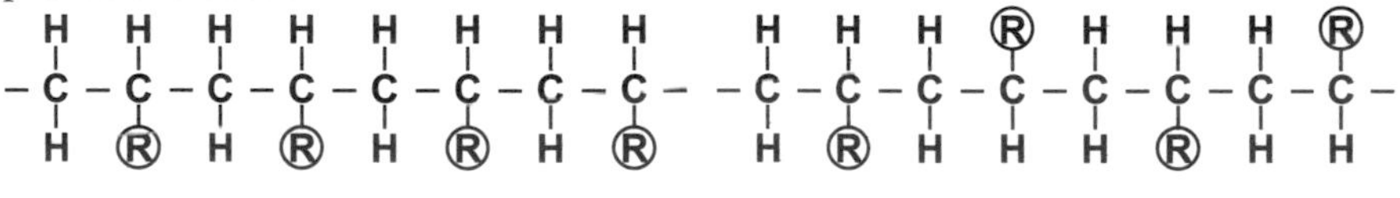

(a) Isotactic　　　　　　　　　　(b) Syndiotactic

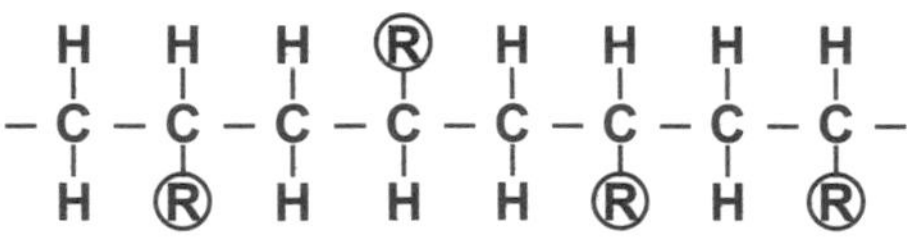

(c) Atactic

Fig. 4.13 Stereoisomers

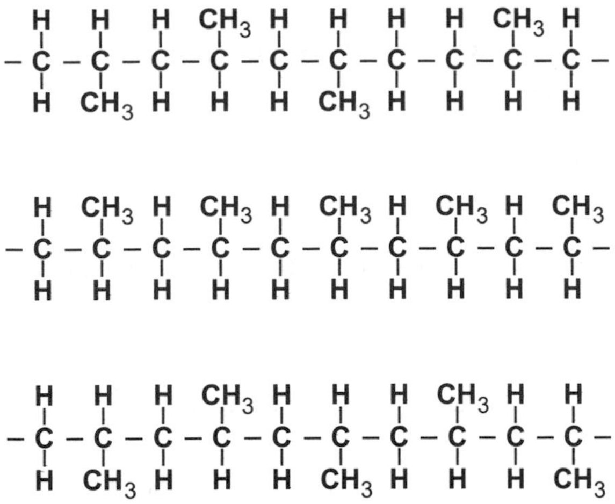

H   H   H   CH₃ H   H   H   H   CH₃ H
- C - C - C - C - C - C - C - C - C - C -   (a)
H   CH₃ H   H   H   CH₃ H   H   H   H

H   CH₃ H   CH₃ H   CH₃ H   CH₃ H   CH₃
- C - C - C - C - C - C - C - C - C - C -   (b)
H   H   H   H   H   H   H   H   H   H

H   H   H   CH₃ H   H   H   CH₃ H   H
- C - C - C - C - C - C - C - C - C - C -   (c)
H   CH₃ H   H   H   CH₃ H   H   H   CH₃

Fig. 4.14 Atactic (a), Isotactic (b) and Syndiotactic (c) polypropylene.

An example of geometrical isomerism is given by the isoprene mer and is shown in **Fig. 4.15**. The cis-isoprene is when the structure is such that the CH_2 groups are on the same side of the carbon to carbon double bond and the trans-isoprene is when the CH_2 groups are on the opposite side of the carbon to carbon double bond. Conversion between the two configurations is not possible by a simple rotation as the double bond is rotationally rigid.

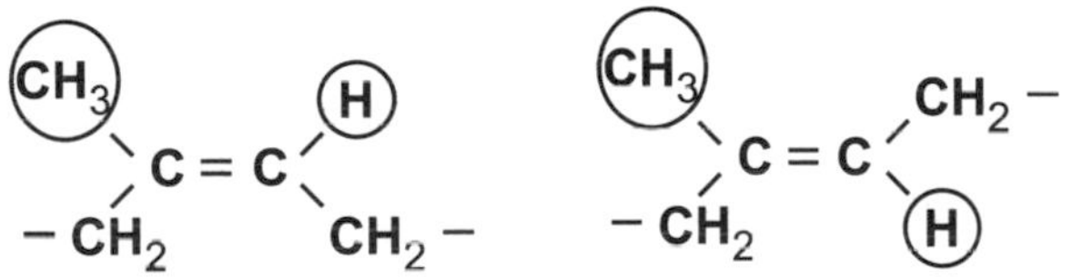

(a) Cis-isoprene   (b) Trans-isoprene

Fig. 4.15 Geometrical isomers.

With these various molecular characteristics, it is now possible to have a more precise classification scheme of polymers as is illustrated in **Fig. 4.16**.

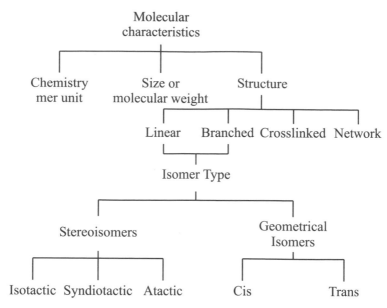

Fig. 4.16 Classification of polymers by molecular characteristics.

4.4.2. Copolymers

The polymers described previously are generally referred to as homopolymers because the mer units along the backbone chains are identical. However, it is possible to form copolymers such that the mer units along the backbone chain may vary. Depending on the process of polymerization, various sequences of mers may occur along the backbone chain in random, alternating, block or graft arrangement as shown in **Fig. 4.17**.

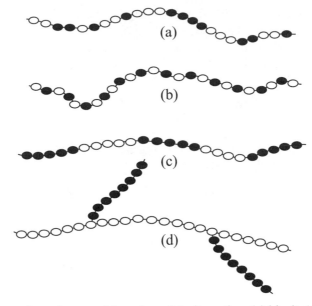

Fig. 4.17 Types of copolymers: (a) random, (b) alternating, (c) block, (d) graft.

4.4.3. Molecular Conformations

In an earlier section, it was suggested that the shape of a polymer molecule could change because of a rotation about the bond between carbon atoms. An example of a possible rotation is given in **Fig. 4.18** where two possible positions, staggered and eclipsed, of hydrogen atoms attached to two adjacent carbon atoms are shown for the ethane molecule.

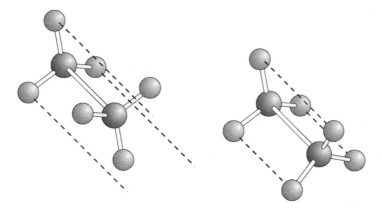

Fig. 4.18 Staggered and eclipsed conformations of ethane.

A better understanding of the geometry is possible by looking along the carbon to carbon bond as shown below in **Fig. 4.19**. The potential energy of the staggered position is slightly less than the potential energy of the eclipsed position as the hydrogen atoms are slightly further apart. For this reason, the staggered position is more favored or more stable than the eclipsed position. The energy varies with position as shown.

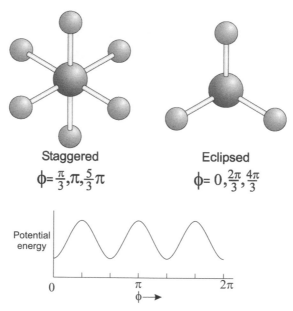

Staggered
$$\phi = \frac{\pi}{3}, \pi, \frac{5}{3}\pi$$

Eclipsed
$$\phi = 0, \frac{2\pi}{3}, \frac{4\pi}{3}$$

Potential energy

0 $\qquad$ π $\qquad$ 2π

$\phi \longrightarrow$

Fig. 4.19 Potential energy vs, angle for ethane.
(Eclipsed is maximum and staggered is minimum.)

Rotation may be quite restricted in a molecule with larger side groups. However, many jumps between staggered positions will occur per second but the amount of time spent in the unstable eclipsed position is small (Alfrey and Gurne (1967)).

Recalling the zig-zag shape of a polyethylene chain from **Fig. 4.4** and that the shape of the chain can change dramatically by rotation about the C-C bonds as described in **Fig. 4.5**, it is easy to see that the chain can take on many conformations. Again, as rotation about the C-C bond occurs, the energy state between atoms changes because the distance between atoms changes slightly. The extended "trans" conformation of the chain is shown in **Fig. 4.20**, where here the term "trans" indicates that the bonds are rotated such that the hydrogens on neighboring carbon atoms are in the staggered position. **Fig. 4.21** shows a kinked conformation including both the trans and the gauche positions, where the gauche configuration is such that the hydrogens on neighboring carbon atoms are in a position intermediate to the staggered and eclipsed positions. The variation of the energy state from the maximum state (where hydrogen atoms on adjacent carbons are in the eclipsed position, not shown) to the gauche then trans and back to the maximum is shown in **Fig. 4.22.**

Fig. 4.20 Zig-zag shape of polyethyline molecule in the extended trans position.

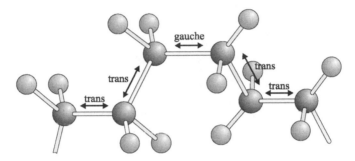

Fig. 4.21 Kinked polyethylene molecule in the trans and gauche positions.

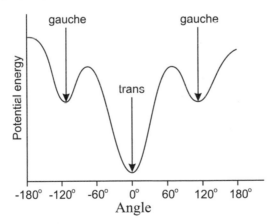

Fig. 4.22 Energy level in polyethylene chain with rotation about the C-C Bonds.

Clearly the trans state is still preferred but the gauche can be relatively stable state as well. It is to be noted that two gauche states occur. One is obtained by rotating a single bond 120° CW while the other is obtained by rotating 120° CCW.

The fully extended polyethylene chain is shown in **Fig. 4.23** along with a chain with several bond rotations and a convoluted chain that might result from many rotations. Notice the similarity between the convoluted chain and that given in **Fig. 4.6**, which was calculated using the general procedures in the next section. It is easy to now visualize many intermingled chains giving rise to the analogy of a tangled ball of string, which is in a constant state of agitation.

Again, depending on the temperature, many changes from one state to another may occur per second. In the glassy or solid state few will take place, while in the liquid state many rotations will occur. Further, which state is preferred will depend upon whether the molecule is in a crystalline close packed state or in the more loosely packed amorphous state. As a result, it is clear that many factors tend to determine the conformations of a polymer molecule. Effects of orientation and temperature will be discussed in later Sections and Chapters.

Fully Extended Chain

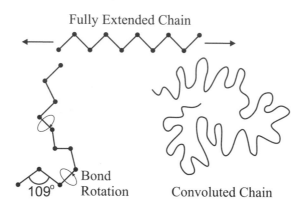

109° Bond
Rotation   Convoluted Chain

Fig. 4.23 Various stages of conformations.

4.5. Random Walk Analysis of Chain End-to-End Distance

From the preceding sections it is seen that a single polyethylene chain could virtually have any shape depending only upon rotations about each C-C bond. As shown in **Fig. 4.23**, the chain could be fully extended or a tightly coiled ball. If fully extended, the end-to-end distance would be the length of the chain while in a tightly coiled arrangement, the end-to-end distance would be nearly zero. Assuming a polyethylene chain with 10,000 mer units, that the C-C bond angle is fixed at approximately 109° and that each bond is restricted to the three positions of one trans and two gauche gives rise to $10^{4.771}$ possible conformations (Painter and Coleman, 1994). Due to thermal agitation, a single chain, if it were not confined by other chains, might see many conformations from fully extended to tightly coiled over a long period of time. However, very little time would be spent in the extreme positions and most of the time would be spent in an average convoluted state. Obviously, if one could apply a force to the opposite ends of a single chain, the fully extended chain would be more difficult to deform than the highly convoluted one. Thus, it is possible to begin to see a relation between the shape of a molecule and its mechanical properties. More of this will be discussed in following sections but for now, it is important to note the relationship between end-to-end distance and mechanical properties. In a solid polymer each chain will interact physically through entanglements with other chains and there will be additional parameters associated with the interaction. The purpose here, however, is to give an introduction to methods to estimate the end-to-end distance or the shape of a chain.

Each chain will have its own length (number of mer units or molecular weight), shape and end-to-end distance. The ability to calculate the average shape or end-to-end distance of each chain and the average for all chains in a polymer can give insight to the relation between structure (conformations in this case) and properties. Because of the large number of possible arrangements of atoms in a chain, a statistical approach is necessary. A simple random walk or random flight method gives the correct form for end-to-end distance of a polymer chain. The only objective here is to show that this can be done and a more in depth study of the required statistical thermodynamics can be found in Painter and Coleman, (1994); Billmeyer, (1984) and Flory, (1953).

In a random walk, a person starts from an initial position and walks x distance in a straight line. The person then turns an arbitrary angle and walks another x distance in a different direction. After n such operations, the objective is to compute the probability that he or she is a distance between **R** and **R+dR** from the starting point. In a random flight, the same procedure is used except the process is accomplished in three dimensions instead of two.

To utilize this procedure to find the correct form of the end-to-end distance of a single polymer chain, a number of assumptions must be made. In addition to free rotation about the C-C bonds, it is assumed that the chain valance angle is free, i.e. for polyethylene the angle of ~109° is no longer fixed but may be any value. Further, it is assumed that the chain can move through itself, i.e. no entanglements result.

Consider the distance between two carbon atoms to be a vector and that a chain can be represented by a series of vectors as shown in **Fig. 4.24**,

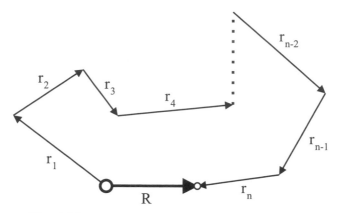

Fig. 4.24 Vector representation of polymer chain.

After $\mathbf{r_i}$ steps in three dimensions, the distance between the starting and ending point will be the sum of the vectors,

$$\mathbf{R} = \sum_{i=1}^{n} \mathbf{r_i} \tag{4.1}$$

where both $\mathbf{R}$ and $\mathbf{r_i}$ are vectors. To find the scalar distance between chain ends, the dot product is used,

$$R = \left[\mathbf{R} \cdot \mathbf{R}\right]^{1/2} = \left[\left(\sum_{i=1}^{n} \mathbf{r_i}\right) \cdot \left(\sum_{j=1}^{n} \mathbf{r_j}\right)\right]^{1/2} \tag{4.2}$$

The dot product of the vector sums can be expanded as

$$\begin{aligned}\mathbf{R} \cdot \mathbf{R} &= \left(\mathbf{r_1} + \mathbf{r_2} + \mathbf{r_3} + \cdots\right) \cdot \left(\mathbf{r_1} + \mathbf{r_2} + \mathbf{r_3} + \cdots\right) \\ &= \left(r_1^2 + r_2^2 + r_3^2 + \cdots\right) + 2\left(r_1 r_2 \cos\vartheta_{12} + r_1 r_3 \cos\vartheta_{13} + \cdots\right) \\ &= \sum_{i=1}^{n} r_i^2 + 2r^2\left(\cos\vartheta_{12} + \cos\vartheta_{13} + \cdots\right)\end{aligned} \tag{4.3}$$

Since the vectors are all the same length, $r_1 = r_2 = \ldots = r_n = r$ (r is the length of the C-C bond), and since the chain is freely jointed, all angles are equally probable and the average of $\cos\theta_{12} + \cos\theta_{13}$, etc. will be zero. As a result, the average end to end distance is,

$$R = \left(\sum_{i=1}^{n} r_i^2 \right)^{1/2} \tag{4.4a}$$

and since all links are the same length and there are n links,

$$R = r(n)^{1/2} \tag{4.4b}$$

Thus, the end-to-end distance of the idealized molecule, using the rather restrictive assumptions, is proportional to the number of mer units. For example, the end-to-end distance for a chain with 10,000 units would be 100 bond lengths. This procedure is the method used by Treloar to obtain the estimated convolution of a polyethylene chain shown in **Fig. 4.6** which is reproduced again in **Fig. 4.25** for emphasis.

Eq. 4.4, while not exact for a real chain in a solid polymer, is of the correct form as the end-to-end distance found using more sophisticated procedures which is also proportional to the number of links or mer units (Painter and Coleman, (1994)).

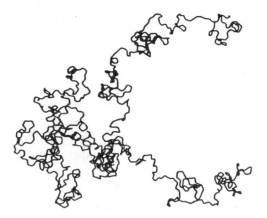

Fig. 4.25 Shape of a 1,000 link polyethylene chain (Treloar (1975), reprinted by permission of Oxford University Press).

4.6. Morphology

At one time it was thought that polymers could not be crystalline because of the supposed tangled nature of the many long chains composing a bulk polymer. That is, the concept of a tangled ball of strings seemed to preclude long range order. However, it was found that some polymers do cause diffraction of x-rays and exhibit diffraction patterns indicative of short range order. See (Painter and Coleman, 1994) for a good discussion of the x-ray technique as applied to crystals and to polymers. **Fig. 4.26** gives an example of an x-ray diffraction pattern for unoriented and oriented Polyoxymethylene (from Billmeyer, (1984)).

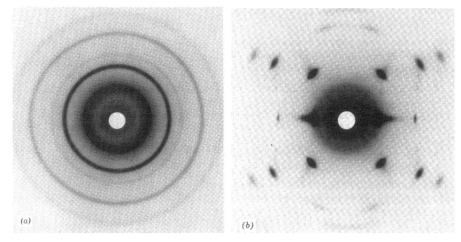

Fig. 4.26   X-ray diffraction pattern for unoriented (a) and oriented (b). (Billmeyer (1984), p. 294, reprinted with permission of John Wiley and Sons, Inc.)

Now it is well accepted that some but not all polymers can be crystalline. The amount of crystallinity may vary anywhere from a few percent to as high as 98% (Rosen, 1993). Normally, however, polymer crystallinity is much less than 98 % and is most often less than 50%. Polymers containing chains with bulky side groups or branches do not generally crystallize and cross-links prohibit crystallization. In general, transparent polymers are completely amorphous while opaque or translucent homopolymers are generally crystalline. On the other hand polymers with fillers or a second phase may be opaque due to the added constituents and not due to their crystallinity.

Probably the best method to evaluate crystallinity is through density measurements. If, for example, specific volume (the inverse of density or vol./g) is measured as temperature is decreased, a sudden and nearly discontinuous change occurs at the melting point (due to a phase change from a semi-solid to a very viscous fluid) for a crystalline thermoplastic polymer as shown in **Fig. 4.27**,

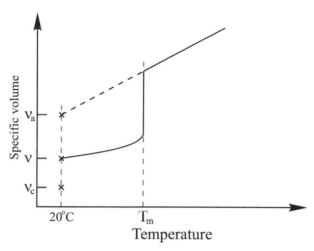

Fig. 4.27 Specific volume vs. temperature for semicrystalline polyethylene.

The degree of crystallinity, c, in percent can be obtained from,

$$c = \frac{v_a - v}{v_a - v_c} \cdot 100 \qquad (4.5)$$

where v_a is the specific volume of the amorphous phase, v_c is the specific volume of the crystalline phase and v is the specific volume of the total sample. In **Fig. 4.27**, v_a is found by extrapolating the v-T curve from above the melt temperature, T_m, to 20° C. In order to find the degree of crystallinity, c, a measure of v_c is needed which is normally obtained from x-ray diffraction measurements (McCrum et al (1997)).

While much has been learned about the crystalline structure of polymers, the exact shape and structure of crystalline regions is still under intense study as increasing the degree of crystallinity leads to improved thermo-mechanical properties (**Table 4.4**). Relations between crystalline structure and mechanical response will be discussed in more detail later. The first interpretation of crystalline structure was suggested by x-ray diffraction studies and is known as the "fringed micelle model". The Bragg diffraction patterns for polymers are broad and diffuse as compared to those from the more perfect forms of metals and other crystalline materials. As a result, it was inferred that the size of the crystallites were very small, being on the order of few hundred Angstroms (Billmeyer, 1984). In this model, a schematic of which is shown in **Fig. 4.28**, a single molecule would traverse a number of amorphous and crystalline regions because a polymer chain is much longer than a few hundred Angstroms. The regions at the end of the crystallite would be the "fringe" and the crystallite would be the "micelle".

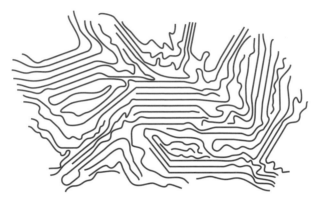

Fig. 4.28 Fringed micelle model.

Polymer crystallinity was later observed experimentally by growing single crystals from a dilute solution by either cooling or evaporating the solvent. A single crystal grown by such a procedure is given in **Fig. 4.29**. In this manner, thin plate like structures can be obtained that are about 10^5 Å long and about 10^2 Å thick as shown in **Fig. 4.30**. X-ray measurements indicated that the chains were perpendicular to the face of the lamellae and the only way a long polymer chain could fit in such a small space was to be folded. It was not clear if a chain was completely contained in the lamellae or if it exited and reentered.

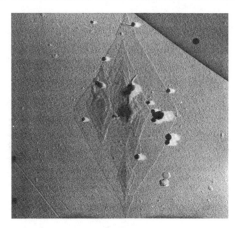

Fig. 4.29 Electron micrograph of a nylon 6 single crystal. The lamellae thickness are 50-100 Å. (Geil, (1960), reprinted with permission of John Wiley and Sons, Inc.)

One school of thought was that the chain exited the lamellae smoothly and reentered at adjacent lattice sites as shown on the left (regular reentry model) and others (especially Flory) thought the portion outside of the lamellae was quite chaotic as shown on the right (switchboard reentry model). The controversy served a good cause as, to prove his point, Flory is reported to have returned from a conference where the subject was intensely debated and began a research program to understand the reentry model. His effort was, in fact, successful and in the process he was a forerunner in the development of a major new field of study of polymers based upon statistical thermodynamics (for a more complete discussion, see Painter and Coleman, 1994). For single crystals, it has subsequently been shown that an intermediate model is more correct, with virtually all of the chains reentering the crystal within 3 lattice sites from their exit point. For highly flexible polymers the number of adjacent reentry points can be as high as 80%.

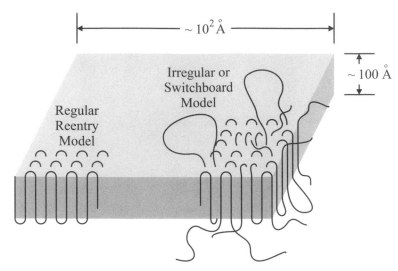

Fig. 4.30 Folded chain model for a crystalline lamellae in polymers.

The folded chain model is now well accepted as also occurring in bulk polymers crystallized from the melt but the lamellae may be as large as one micron thickness. In addition, for bulk crystallization amorphous regions are interspersed between crystalline lamellae and the degree of regular reentry of chains into a given lamellae is small. A more accurate picture is given in **Fig. 4.31**, where a significant number of "tie molecules" are shown connecting the crystalline regions; these molecules are important in the improved mechanical properties of crystalline polymers. According to Rosen (1994), recent data indicates the existence of a third interfacial phase of significant volume fraction between the lamellae and amorphous regions, but little is understood about this interphase region at present.

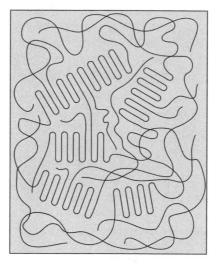

Fig. 4.31 Compromise model: Folded chains tied together by amorphous regions as in the Fringed micelle model.

As a polymer is super-cooled below the melt temperature, T_m, the crystalline regions nucleate at minute impurity sites, growing to form spherical domains called spherulites. These spherulites grow radially until another spherulite is encountered as shown in **Fig. 4.32**. The rate of cooling determines the degree of crystallinity of the solid polymer and for many materials a totally amorphous glass is possible by very rapid cooling rates. Lower cooling rates allow formation of spherulitic crystals and the number and size of spherulites can be modified by choosing cooling rates and temperatures.

Examples of spherulites obtained using microscopy are shown in **Figs. 4.33-4.34**. Shown in **Fig. 4.33(a)** is a branched spherulite in polypropylene observed via AFM while **Fig. 4.33(b)** and **Figs. 4.34(a)** and **(b)** show spherulites in polystyrene, polyethylene and poly(hydroxybutyrate) respectively. The latter figures are optical micrographs taken of thin sections of polymers as seen under polarized light (crossed polarizers). The dark areas are the characteristic "maltese cross" created due to the birefringent properties of polymers indicating a crystalline structure.

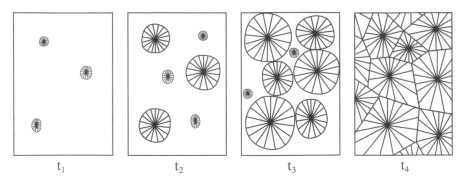

t_1 $\qquad$ t_2 $\qquad$ t_3 $\qquad$ t_4

Fig. 4.32 Stages in the growth of a spherulite.

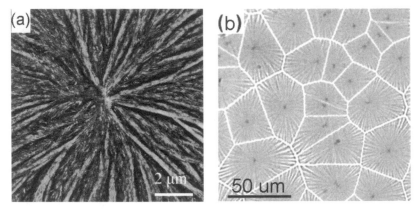

Fig. 4.33 Examples of spherulites: (a) Branched spherulite in polypropylene from AFM (Zhou et al. (2005), reprinted by permission from Elsevier.) (b) Field of growing spherulites in polystyrene (Reprinted by permission from Beers et al. Copyright 2003, American Chemical Society)

Birefringence occurs because polarized light passing through a crystal is broken up into two components propagating along a plane perpendicular to the principal axis of the crystal. Each component travels at a different velocity and therefore one is retarded relative to the other. On emerging and passing through a second polarizer, interference between the two waves gives rise to the dark "fringes" and hence the maltese cross. Actually the maltese cross is indicative of the directions of principal stress which are perpendicular to each other in the plane of the section. The nature of birefringence in crystals has been known for well over a century (as discussed in the Introduction) and is the basis for the well known photoelastic stress analysis method. In this procedure, initially amorphous polymers become optically anisotropic due to the application of external forces. That is, the external forces cause a slight realignment of the molecular structure such that the polymer reacts to light as if it were a crystal. As a result, the stress inside the material can be visualized and analyzed using the birefringence effect. The isochromatic fringes (lines of equal shear stress) in a sample of polycarbonate containing a crack are shown in **Fig. 4.35**.

Cross-linked or thermosetting polymers are typically used for photoelastic stress analysis. Thus, it is clear that a certain amount of crystallinity can be induced by stresses in network polymers but the degree of crystallinity is necessarily very small.

A schematic visualization of a spherulite is given in **Fig. 4.36**. Here the spherical nature is apparent and it is to be noted that the individual fibrils/lamellae grow radially. The individual fibrils have a folded chain structure and the chain traverses both crystalline regions and amorphous regions as illustrated in **Fig. 4.31** of the folded chain model.

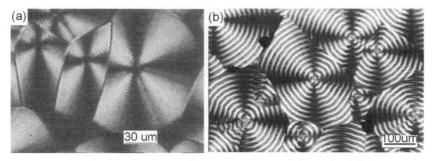

Fig. 4.34 Examples of spherulites: (a) Spherulites in polyethylene (Armistead et al. (Reprinted by permission from Armistead et al. Copyright 2003, American Chemical Society). (b) Ringed spherulites of poly(hydroxybutyrate), (Hobbs et al. (2000)) Reprinted by permission of John Wiley and Sons, Inc.

Fig. 4.35 Birefringence photograph of polycarbonate showing isochromatic fringes surrounding a crack.

The application of large external loads to linear or thermoplastic polymers can cause the material to yield and for plastic flow to occur. An example for the plastic flow or the creation of a necked region in polycarbonate was given in **Fig. 3.7** in Chapter 3. Further application of the load can produce a severely drawn material in which the molecular chains have become oriented due to the external load. A schematic illustration of the progression of the drawn material is given in **Fig. 4.37** together with a description of how the fold chains move in order to create the oriented structure in the drawn material. Orientation of the lamellae in the direction of drawing along with deformation induced crystallinity in the amorphous regions leads to an overall increase in crystallinity with drawing.

The cold drawing of thermoplastic polymers can drastically improve mechanical properties and is often accomplished to create favorable properties for certain applications. A case in point is the biaxial stretching of polycarbonate for use in aircraft canopies. The ability to be drawn (either cold or hot) is of great use commercially. For example, PET (polyethylene terephthalate) which is often used for soft drink bottles is first produced by injection molding as a small test tube size object. Before filling with liquid, the material is heated and "blown" as large as the standard 2 liter soft drink container.

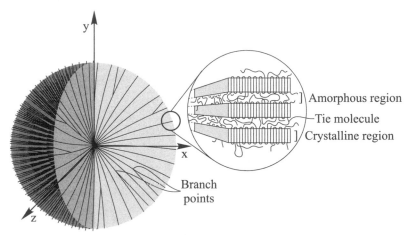

Fig. 4.36   Schematic diagram of a spherulite. Inset detail after Callister (1991).

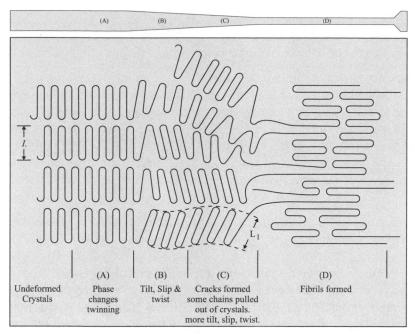

Fig. 4.37   Illustration of the transformation from a lamellar to a fibrillar structure by drawing. (After Painter and Coleman (1994), Original from Peterlin (1965))

4.7. Molecular Weight

The atoms in polymer chains, as in metals and all other materials, consist of electrons orbiting a nucleus containing protons and neutrons. The atomic mass (weight) of an element is the sum of the masses of the protons and neutrons in its nucleus, since the mass of the electrons is several orders of magnitude smaller and therefore negligible. Note that although atomic mass is the more appropriate (and ISO standard) term, by common usage atomic weight is most often found in polymer literature. The number of protons defines the element, but for some elements several isotopes are possible, all having the same number of protons but different numbers of neutrons. The atomic mass of such atoms is given in the periodic table as the weighted average (according to abundance in nature) of the atomic masses of the naturally occurring isotopes. A proton and a neutron have the same mass to three significant digits and the **atomic mass unit** (amu) is defined on the basis of Carbon-12, the most common isotope of Carbon containing 6 protons and 6 neutrons, with an atomic mass of exactly 12.000 amu. The atomic mass shown in the periodic table for Carbon is slightly higher (12.011) as it accounts for small amounts of the isotope ^{13}C.

Since one does not typically work with single atoms or molecules, quantities of chemical substances are given in **moles**. A mole of an element is defined as 6.02214×10^{23} (**Avogadro's number**) atoms; a mole of a given type of molecule is 6.02214×10^{23} molecules. Avogadro's number is defined to provide a simple conversion to grams: 6.022×10^{23} atoms (or molecules) have the mass in grams of the atomic mass of a single atom (molecule). For example, 1 mole (6.022×10^{23} atoms) of ^{12}C has a mass of exactly 12.0g. The conversion is therefore

$$6.02214 \times 10^{23} \text{ amu} = 1 \text{ gram}$$
$$\text{or } 1 \text{ amu} = 1.66054 \times 10^{-24} \text{ g}$$

As an example, consider a mole of water molecules (H_2O) which contains 6.022×10^{23} atoms of oxygen and $2 \times (6.022 \times 10^{23})$ atoms of hydrogen. The atomic masses of oxygen and hydrogen are 15.9994 amu and 1.0079 amu respectively. Therefore a mole of water has a mass of $2 \times (1.0079g) + 15.9994g = 18.015$ grams. This example also emphasizes that moles are the necessary units to use for chemical reactions as the proper number of atoms must be tracked: e.g. one mole of oxygen and two moles of hydrogen can be combined to form 1 mole of water; 1 gram of oxygen and 2 grams of hydrogen are not in the proper ratio to form a gram of water owing to the differing masses of the elements.

Historically, the terms gram-atom and gram-molecule were originally coined to refer to the mass in grams of Avogadro's number of atoms or molecules, respectively; with the introduction of the term mole, these early terms are less used but can still be found in the literature. The term "molecular weight" is by far the most common expression used to refer to the mass of a molecule. A 1992 ISO standard dictates that the term "relative molecular mass" should replace "molecular weight" in all publications, but in practice adoption of the terminology has been slow. The word "relative" is used in the expression to convey that the mass is given relative to 1/12 the mass of an atom of Carbon-12. Since the mass of 12C is exactly 12.00amu, the relative molecular mass provides the mass of a molecule in amu although technically the quantity is unitless. Another term sometimes seen is "molar mass" or "relative molar mass". Both of these latter terms refer to the mass per mole of a substance and are expressed in grams/mole.

To give an example for a polymer, a single polyethylene chain with a degree of polymerization of 10^4 (or 10^4 mer units) has a relative molecular mass (molecular weight) of

Mass of 1 PE chain: 10^4 (2 x 12 + 4 x1) = 280,000 amu

A mole of polyethylene chains, where each chain is 10^4 mer units long, has a molar mass of

Mass of 1 mole of PE chains: 280,000 grams (or grams/mole)

neglecting chain end effects. Note that the molecular mass of a chain end (or at a branch point) is not the same as the molecular mass of a mer unit but the difference is neglected because the effect is small in terms of the total molecular mass of a chain.

While "relative molecular mass" is the official and more correct terminology for polymers (as used in McCrum, 1997), in the following the term molecular weight will be most often used as is common in many polymer texts.

A useful term to describe the extent of polymerization in polymers is the "degree of polymerization" (**DP**) which is defined as the number of mer units per chain or,

$$n = \frac{M}{M_r} = DP \qquad (4.6)$$

where **M** is the molar mass (weight) of a chain and $\mathbf{M_r}$ is the molar mass (weight) of a mer or repeat unit. (Number average and weight average degrees of polymerization are also used as will be evident directly.)

The degree of polymerization or the length of a polymer chain is an indicator of the nature and mechanical characteristics of a polymer composed of similar length chains. The following table illustrates the relationship between chain length and the character of a polymer at 25 °C and a pressure of one atmosphere.

Table 4.6   Degree of polymerization – phase relationship (data from Clegg and Collyer (1993), p. 11)

Number of $-CH_2-CH_2$ Repeat units per chain (Degree of Polymerization	Molar Mass kg mol^{-1}	Softening Temperature °C	Character at 25°C and 1 at.
1	28	-169	Gas
6	170	-12	Liquid
35	1000	37	Grease
140	4000	93	Wax
430	12000	104	Resin
1350	38000	112	Hard Resin

It is now clear how to calculate the molecular weight of a single chain or of a mole of polymer chains of identical lengths. Unfortunately, however, the lengths of chains in a polymer vary greatly and depend to a large degree on the circumstances and the manner in which the polymerization reaction proceeds. That is, a wide distribution of chain lengths (DP's or chain molecular weights) exist in a typical polymer as shown in **Fig. 4.38**. The distribution is seldom symmetrical and the breath of distribution varies with the type of reaction. For example, the distribution is often quite broad for polyethylene while the distribution for polystyrene may be quite narrow (Fried, 1995).

Because of the distributed nature of the lengths of chains in a polymer it is necessary to define the molecular weight using an averaging process. The most common averaging processes used are the number average, the weight average and the z-average. Only the number and weight average methods will be described here. Both discrete and continuous distributions are possible. For

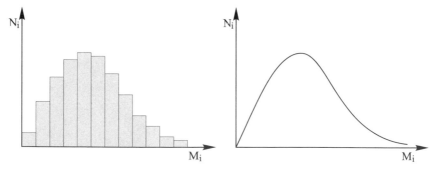

Fig. 4.38 Typical molecular weight distribution of the number of chains in a polymer.

example the continuous distribution in **Fig. 4.38(b)** is obtained by drawing a smooth curve through the discrete distribution shown in **Fig. 4.38(a).**

For a discrete distribution the *number average molecular weight* is defined as,

$$\overline{M}_n = \frac{\sum_{i=1}^{k} N_i M_i}{\sum_{i=1}^{k} N_i} = \frac{\sum_{i=1}^{k} N_i M_i}{N} \tag{4.7}$$

where N_i is the number of chains within an interval, M_i is the median (middle) molecular weight in an interval, k is the total number of intervals and N is the total number of chains. If a continuous curve is fit to the discrete data such that N is given as a function of M, i.e., $N=N(M)$, the summations can be replaced by an integration to obtain (Kumar and Gupta, (1998)),

$$\overline{M}_n = \frac{\int_0^M N(M)\, dM}{\int dN} = \frac{\int_0^M N(M)\, dM}{N} \tag{4.8}$$

The product of the number of chains in an interval, N_i, and the molecular weight of an interval, M_i, equals the total weight of an interval. Exchanging N_i in **Eq. 4.10** by $N_i M_i$ defines the *weight average molecular weight* and can be written as,

$$\overline{M}_w = \frac{\sum_{i=1}^{k} N_i M_i^2}{\sum_{i=1}^{k} N_i M_i} = \frac{\sum_{i=1}^{k} N_i M_i^2}{M} \tag{4.9}$$

where M is the total molar mass of the sample. Some have likened the number average and weight average molecular weights to the first and second moments of masses (or areas) in elementary mechanics courses. Such an analogy is appropriate if the number of chains, N_i, is replaced by a lever arm d_i with units of length. One text incorrectly relates the weight average molecular weight to a radius of gyration.

Consider the example where,

i	M_i	N_i
Interval No.	g/mole of chains in interval	No. of chains in interval
1	5,000	2
2	15,000	4
3	30,000	5
4	50,000	1

The number average and the weight average molecular weight from **Eqs. 4.10** and **4.11** will be,

$$\overline{M}_n = \frac{\sum\limits_{i=1}^{k} N_i M_i}{\sum\limits_{i=1}^{k} N_i} \qquad (4.10a)$$

$$\overline{M}_n = \frac{2(5,000) + 4(15,000) + 5(30,000) + 1(50,000)}{12} = 22,500 \text{ g/mole} \qquad (4.10b)$$

Some experimental approaches separate the chains in a polymer into discrete number or weight fractions. The number fraction, x_i, is defined as the ratio of the number of chains in an interval to the total number of chains in the sample,

$$x_i = \frac{N_i}{N} \qquad (4.11)$$

and the weight fraction, w_i, is the ratio of the total weight of the chains in an interval to the total weight of the sample,

$$w_i = \frac{M_i}{M} \qquad (4.12)$$

An example illustrating this approach is shown in the hypothetical distribution given below,

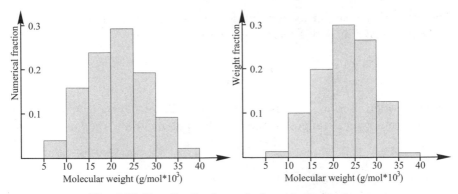

Fig. 4.39 Size distributions of a hypothetical polymer

Using this definition, the number average molecular weight or weight average molecular weight can be written as,

$$\overline{M}_n = \sum_{i=1}^{k} x_i M_i \tag{4.13}$$

$$\overline{M}_w = \sum_{i=1}^{k} w_i M_i \tag{4.14}$$

where all quantities are as previously defined and (4.13) and (4.14) yield identical results to (4.10) and (4.12) respectively.

The number average emphasizes the importance of the smaller molecular weight chains while the weight average emphasizes the higher molecular weight chains. This is demonstrated in **Fig. 4.40**.

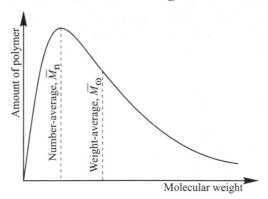

Fig. 4.40   Distributions of molecular weight in a typical polymer.

The ratio of the weight average molecular weight to the number average molecular weight is defined as the polydispersity index,

$$PDI = \frac{\overline{M}_w}{\overline{M}_n} \qquad (4.15)$$

which is often used as a measure of the breadth of the molecular weight distribution. Typical ranges in the PDI for polymers are shown in **Table 4.7**.

Table 4.7 Typical ranges of $\overline{M}_w / \overline{M}_n$ in synthetic polymers. (Data from Billmeyer (1984), p. 18)

Polymer	Range
Hypothetical monodisperse polymer	1.0
Actual monodisperse living polymers	1.01 – 1.05
Addition polymer, termination by coupling	1.5
Addition polymer termination by disproportionation	2.0
High conversion vinyl polymers	2-5
Polymers made with autoacceleration	5-10
Addition polymers made by coordination polymerization	8-30
Branched polymers	20-50

When $\overline{M}_n$ is high and **PDI** is low there are more chance for entanglements which in turn increases strength and rigidity because the strain is lower for a given stress. When $\overline{M}_w$ or **PDI** is high, chains are likely longer and the temperature resistance is increased.

Molecular weight is an important indicator of mechanical properties. For example the variation of tensile strength of a lightly crosslinked rubber is shown in **Fig. 4.41** and the variation of the elastic modulus above the glass transition temperature is shown in **Fig. 4.42**. As may be observed, above the Tg the modulus becomes very small when the molecular weight is low but increases to a plateau when the molecular weight is very high. This plateau extends to relatively high temperatures until sufficient energy is input to begin to degrade cross-links and the backbone chain. This is often indicated by a change in color of the polymer due to charring. The reason for the different behavior as a function of molecular weight is due to increased entanglements for higher molecular weights (Clegg and Collyer (1993)).

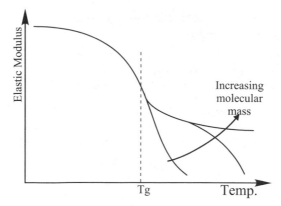

Fig. 4.41 The effect of molecular weight on the elastic modulus of an amorphous thermoplastic polymer above the T_g.

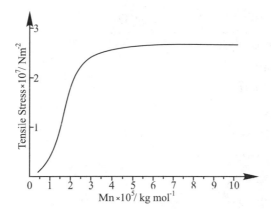

Fig. 4.42 Approximate tensile strength of a lightly crosslinked rubber as a function of number average molecular weight. (Data from Clegg and Collyer (1993))

4.8. Methods for the Measurement of Molecular Weight

The preceding section illustrates the importance of molecular weight on mechanical properties. Molecular weight also has a large influence on manufacturing and processing. For example, a resin suitable for extrusion must have a high viscosity at low shear rates while a resin suitable for injection molding must have a low viscosity at high shear rates. Both of these requirements can be met for a polymer by the adjustment of the molecular weight distribution. Molecular weight distribution also influences the extent of chain entanglement and the amount of melt elasticity. For these and many other reasons, it is necessary to measure the molecular weight and molecular weight distribution. Indeed, as mentioned in the introduction, the lack of accurate methods to measure high molecular weights impeded the initial development and understanding of polymers.

Many of the methods used to measure molecular weight are listed in the **Table 4.8**. The usual range of weights that can be found by each method is also given. Note that the end group analysis and colligative property methods give number average molecular weight while the light scattering method gives the weight average molecular weight. The other methods give only a relative measure of the molecular weight and one of the former methods must be used to provide a calibration of the method and therefore it is possible to obtain either quantity.

Table 4.8 Average molecular weight measurement.

End Group Analysis	$\overline{M}_n < 10,000$
Colligative Properties	
Ebulliometry	$\overline{M}_n < 100,000$
Cryoscopy	$\overline{M}_n < 50,000$
Vapor Pressure Osmometry	$40,000 < \overline{M}_n < 50,000$
Membrane Osmometry	$50,000 < \overline{M}_n < 1,000,000$
Intrinsic Viscosity	$- < \overline{M}_n, \overline{M}_w < -$
Light Scattering	$10,000 < \overline{M}_w < 10,000,000$
Size Exclusion Chromatography (SEC)	$\overline{M}_w, \overline{M}_n < 10,000,000$
Ultracentrifuge	$\overline{M}_w, \overline{M}_n < 40,000,000$

The intent here is not to give a complete description of each method including the necessary equations needed to convert a particular measurement into a molecular weight. Rather, the essential features of each technique will be discussed briefly.

The end group analysis method relies on a knowledge of the nature and types of end groups present. In this method the number of molecules are simply counted. This is accomplished by using standard analytical techniques to determine the concentration of the end groups and thereby the number of polymer molecules. See Rosen, (1993) for a more complete description of this procedure.

When a material (solute) is dissolved in a liquid (solvent) the boiling point, freezing point and vapor pressure are changed. As a result, if a small amount of a solute (polymer) is dissolved in a solvent it is possible to use the thermodynamics of solutions to calculate the change in the temperature at the boiling point (ebulliometry) and freezing point (cryoscopy) which in turn can be related to the number average molecular weight. It should be noted, however, that only small changes in temperature occur and the precision of the method depends on the accuracy of temperature measurement. See Billmeyer, (1984) for a more complete description of these methods.

The vapor pressure method uses two thermister probes to measure the temperature difference between a drop of solvent placed on one probe and a drop of a solution of solute and solvent on the other probe. The difference in rates of vaporization at the two probes leads to a difference in temperature at the two probes. This difference in temperature can be related to the number average molecular weight. For more insight into the method and the magnitude of the temperature differences encountered (see Kumar and Gupta (1998)).

The membrane osmometry method depends upon finding a suitable membrane which will allow solvent to move through the membrane but not allow motion of the solute in the reverse direction. That is, if a solute and solvent are separated by a semipermiable membrane as shown below in **Fig. 4.43**, the motion of the solvent will create an increase in pressure in the solute which can be measured by the relative difference of the height of the two fluids in their respective capillaries. This pressure differential can be related to the number average molecular weight.

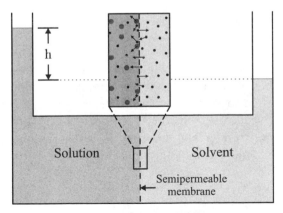

Fig. 4.43 Osmosis through a semipermiable membrane.

The viscosity of a dilute solution of a polymer and a solvent is obviously larger than the viscosity of the solvent alone. As a result, measurement of the viscosity of the two fluids will give a relative measure of the molecular weight of the mixture. If varying concentrations of polymer are placed in solution, the relative viscosity will vary. If the molecular weight fraction(s) of the same polymer has been made by one of the other methods, then the molecular weight through a relative viscosity measurement can be obtained by comparison. An illustration of the relation between viscosity and molecular weight of polyisobutylene in two solutions of cyclohexane and diisobutylene are shown in **Fig. 4.44.**

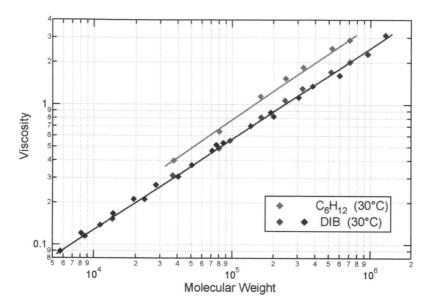

Fig. 4.44 Instrinsic viscosity molecular weight distributions in two solvents. (Data from Flory (1953)).

Typical glassware viscometers which are used for viscosity measurement are shown in **Fig. 4.45**. The flow times through the capillaries are measured and converted to viscosity measurements using the concepts of Newtonian flow. It should be remarked, however, that polymer solutions are normally non-Newtonian but the error is small with properly designed equipment. For a more complete discussion of this technique, see Rosen, (1993).

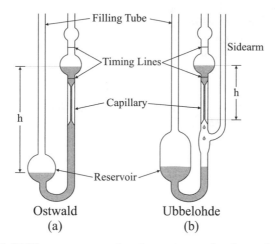

Fig. 4.45 Viscometers used to determine molecular weight.

One of the most popular techniques to determine molecular weight is through gel permeation chromatography (GPC). Because a gel is no longer used some prefer to call the technique size exclusion chromatography (SEC). A schematic of a GPC is shown in **Fig. 4.46**.

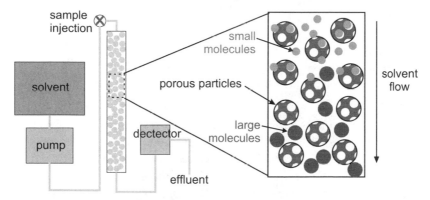

Fig. 4.46   Schematic of a gel permeation chromatograph (GPC). Basic instrument design (left) and separation column detail (right).

In this procedure a column is packed with small porous beads. The beads may be a porous gel (a low molecular weight polymer) or porous beads made of polystyrene or glass. A solvent containing a polymer sample is pumped through the column of beads at a very low rate. As shown in the schematic of the column, the large molecular fractions cannot penetrate the beads and are retained in the solvent which moves through the column at a faster rate than the lower molecular weight fractions. The smaller molecular weight fractions pass through the porous bead microstructure and therefore move through the column at a slower rate. The mass concentration leaving the column, the effluent, passes through a detector (a refractometer) which measures the refractive index of the emerging volume. The refractive index will vary with molecular weight but again another absolute molecular weight method must be used to calibrate the system. A typical calibration curve is shown in **Fig. 4.47.**

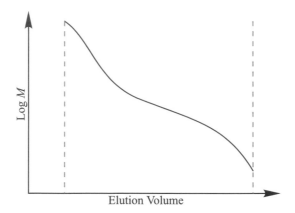

Fig. 4.47 Calibration curve for a GPC.

The GPC method gives a continuous variation of the molecular fractions with volume (or mass) and thus is well suited to yield a continuous curve such as the one shown in **Fig. 4.38(b)**. For more details of the procedure, see Rodriguez, (1996).

Light passing through a medium other than a vacuum will interact with the molecules of that medium. Light energy interacts with the normal oscillatory nature of a molecule and induces additional oscillations. The amount of additional oscillation depends on the atomic nature and size of the molecule and is measured by the polarizability of the molecule. The interaction also causes the molecule to become a source of radiation. As a result, the light will be scattered by the molecule (or will radiate from the molecule) such that it can be seen from all directions. That is, consider light from a point source (such as a laser) being directed at a container of a dilute polymer solution. It will be possible to see the light from different sides of the container. It is possible to relate the intensity of emanating light and the appropriate geometry of observation to the molecular weight of the solution in the container. The molecular weight of a dilute polymer can be obtained by measuring the intensity of light scattered from various concentrations of the polymer solution and comparing with the intensity of light scattered only by the solvent. It should be noted that the weight average molecular weight is determined with this approach. For an excellent description of this method, see Painter and Coleman, (1994). Mechanical engineers who are interested in the relation between the polarization of light and the relative retardation of light in polymers will find the discussion provided by Painter and Coleman very interesting. Further, this source will assist in understanding the relationship between molecular parameters and birefringence parameters.

The invention of the ultracentrifuge was one of the milestones in development of polymer science and technology. The basic principle of this device is illustrated by observing a small sphere moving through a Newtonian liquid under the action of gravity as shown in the **Fig. 4.48.**

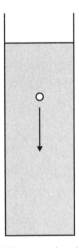

Fig. 4.48 Small sphere falling gravity through a viscous fluid.

The sphere will reach a terminal velocity and, assuming the cylinder radius is large compared to the sphere, equilibrium of the forces on the sphere (gravity, buoyancy and drag) will give,

$$\rho_s Vg - \rho Vg - \frac{kTv}{D} = 0 \qquad (4.16)$$

where ρ_s is the density of the sphere, ρ is the density of the fluid, V is the volume of the sphere, g is the acceleration of gravity, k is Boltzman's constant, T is the absolute temperature, v is the terminal velocity and D is the sphere diffusion coefficient. Solving for sphere volume and multiplying the result by the sphere density gives the sphere mass,

$$m = \frac{kTv}{Dg\left[1 - \left(\dfrac{\rho}{\rho_s}\right)\right]} \qquad (4.17)$$

If it is assumed that the sphere is a molecule, the molecular weight can be found by multiplying both sides of the equation by Avogradro's number and is given as,

$$M = \frac{RTv}{Dg\left[1 - \left(\dfrac{\rho}{\rho_s}\right)\right]} \qquad (4.18)$$

where R is the universal gas constant.

In an ultracentrifuge a polymer sample (solute) in a pie shaped container is rotated at a very high speed (70,000 rpm) in the horizontal plane. At such high speeds the higher molecular weight polymer fractions will separate from the solvent and will be forced to the outer wall of the container. **Eq. 4.18** can be modified by replacing the acceleration of gravity by the angular acceleration, $r\omega^2$. The sedimentation rates are measured by Schlieren optics or by UV absorption. The procedure is used most often for biological materials and is not used so much with synthetic polymers because of experimental difficulties and other approaches give more reliable results. For more details on this procedure, see Kumar and Gupta, (1998) and Rodriguez, (1996, p. 202).

4.9. Polymer Synthesis Methods

The two fundamental approaches to obtaining polymers used by Carothers in his pioneering synthesis efforts in the 1920's were condensation and addition polymerization. In a condensation reaction water, ammonia or some other substance is a byproduct and generally must be removed from the final polymer. An example of a condensation reaction is given by the formation of nylon 6,6 from the combination of hexamethylene and adipic acid in **Fig. 4.49**.

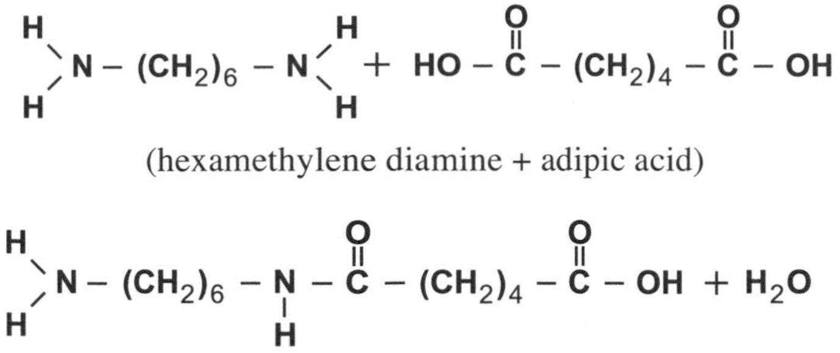

$$
\begin{array}{c}
H \\
\diagdown N - (CH_2)_6 - N \diagup \quad + \quad HO - \overset{\displaystyle O}{\overset{\|}{C}} - (CH_2)_4 - \overset{\displaystyle O}{\overset{\|}{C}} - OH \\
H \qquad\qquad\qquad H
\end{array}
$$

(hexamethylene diamine + adipic acid)

$$
\begin{array}{c}
H \\
\diagdown N - (CH_2)_6 - \underset{\underset{H}{|}}{N} - \overset{\displaystyle O}{\overset{\|}{C}} - (CH_2)_4 - \overset{\displaystyle O}{\overset{\|}{C}} - OH \ + \ H_2O \\
H
\end{array}
$$

Fig. 4.49 Example of a condensation reaction.

The number of reactive sites in the monomer is known as the functionality of the unit and will determine if a polymer can be formed and if the resulting polymer will be a thermoplastic or a thermoset. A bifunctional monomer leads to a thermoplastic while a trifunctional monomer is needed to produce a thermoset. Generally speaking the N—H bond of an amine, the O—H bond of an alcohol, and the C—OH bond of an acid can be split to form another bond. Also, unsaturated bonds such as those that exist between the two carbon elements in the ethylene molecule shown in **Fig. 4.2** can be broken to from bonds with other elements and ring structures such as those shown in **Fig. 4.50** can split to form other bonds.

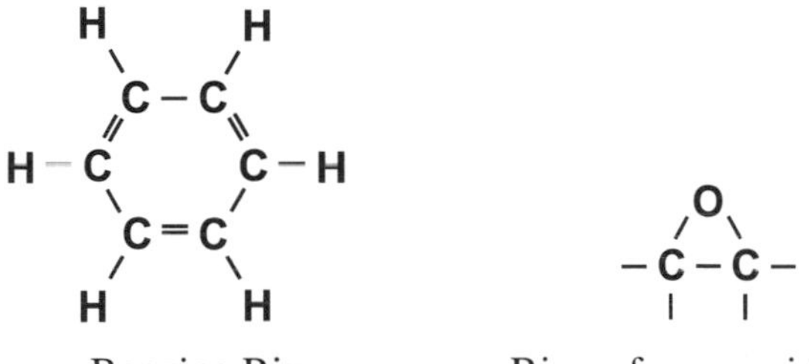

Benzine Ring   Ring of an epoxide group

Fig. 4.50 Ring structures.

It will be noted that both hexamethylene diamine and the adipic acid in the above example are each bifunctional and the resulting molecule after combination of the two is bifunctional as well. That is, an active site exists on each end of the original molecules and on each end of the product molecule. Therefore, the reaction can continue by additional linking of the diamine and the acid with the product molecule. The mer unit for nylon 6,6 can now be identified as given in **Fig. 4.51**. Nylon is called a polyamide because it is formed by splitting the NH$_2$ group and is given the notation

6,6 due to the number of carbon elements in each mer. Other nylons can be made such that the notation would

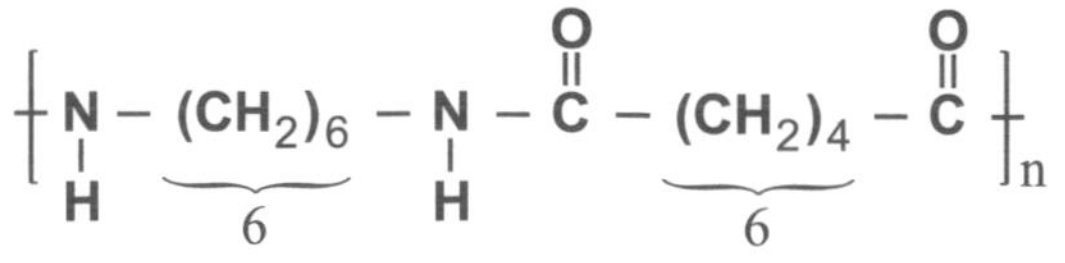

$$\left[\begin{array}{c} N - (CH_2)_6 - N - \overset{\displaystyle O}{\overset{\|}{C}} - (CH_2)_4 - \overset{\displaystyle O}{\overset{\|}{C}} \\ H \hspace{1.5cm} H \end{array} \right]_n$$

$\underbrace{}_{6}$ $\underbrace{}_{6}$

Fig. 4.51 Nylon 6.6 monomer.

be 5,10, 6,10, etc. An interesting procedure to form nylon 6,6 is by the so-called "nylon rope trick" illustrated in **Fig. 4.52**. This is called interfacial condensation and occurs, in this case, when solutions of acid chloride in chloroform and hexamethylene diamine in water are combined. The two solutions do not mix and a skin is formed at the interface between the two. It is possible to carefully withdraw the skin from the interface and to form a thread or film as shown.

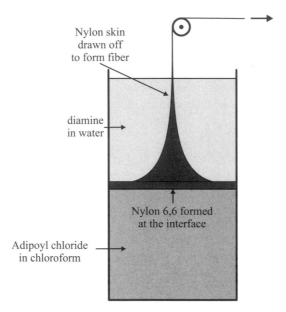

Nylon skin drawn off to form fiber

diamine in water

Nylon 6,6 formed at the interface

Adipoyl chloride in chloroform

Fig. 4.52 Nylon rope trick.

Examples of polymers formed by a condensation reaction are,

Polyamides (nylon)
Polyurethanes
Polycarbonates

Addition reactions take place by the combination of monomers with two reaction sites such as the case for ethylene given in **Fig. 4.3** and repeated below **Fig. 4.53** for emphasis. The unsaturated double bond of carbon is another example of a bond that can be broken to from bonds with other elements.

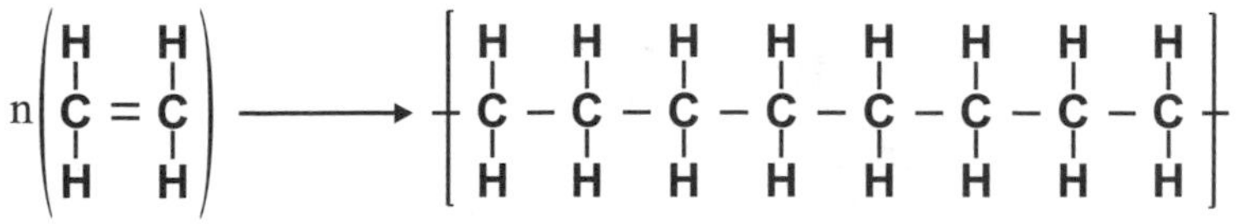

$$n\begin{pmatrix} H & H \\ | & | \\ C = C \\ | & | \\ H & H \end{pmatrix} \longrightarrow \begin{bmatrix} H & H & H & H & H & H & H & H \\ | & | & | & | & | & | & | & | \\ C - C - C - C - C - C - C - C \\ | & | & | & | & | & | & | & | \\ H & H & H & H & H & H & H & H \end{bmatrix}$$

Fig. 4.53 Example of an addition reaction.

With this process, chains grow in a sequential manner. That is, one monomer unit reacts with another monomer unit to produce a sequence of two mer units or "dimer". The resulting dimer reacts with a monomer to produce a sequence of three mer units or "trimer". Trimers, dimers and monomers can react to produce an "oligomer" or a chain composed of a small number of mer units (often considered to be less than ten). In this way, the reaction continues (or propagates) until eventually a chain stops growing or terminates. As a result, an addition reaction is usually characterized by three stages,

> Initiation
> Propagation
> Termination

Polyethylene and polypropylene are in a group of polymers called polyolefins and their production constitutes one of the largest polymer markets in the world.

Several important vinyl polymers (those in which a single hydrogen element the monomer in **Fig. 4.53** is replaced another element) formed by addition polymerization are,

> Polyethylene
> Polyvinylchloride
> Polystyrene
> Polymethylmethacrylate

There are a number of different types of addition polymerization methods. Several of these are,

> Free radical
> Ionic
> Coordination

Free radicals are intermediate compounds containing a free (unpaired) electron and are highly reactive. To initiate free radical polymerization, the unpaired electron of a free radical steals an electron from a vulnerable bond in the monomer (such as a double bond), leaving the monomer with an unpaired electron to propagate the reaction. The most common free radicals used as initiators are peroxides, which are easily broken down as

$$R - O - O - R \quad \rightarrow \quad 2RO^{\bullet}$$

where the dot indicates oxygen to have a free (unpaired) electron. This type of polymerization has a natural chain termination step which occurs when two free radicals collide.

Ionic polymerization follows a similar process for initiation and propagation of the reaction, where instead of a free radical the reactive unit on the end of a chain is either positively (cationic) or negatively (anionic) charged. Unlike free radical polymerization, termination occurs when the monomer is depleted. This type of polymerization is often used to produce block copolymers.

Coordination polymerization is a type of addition reaction in which a fragment of the catalyst is said to be inserted into a growing chain. Much of this type of reaction is based upon the Noble prize winning efforts of Giullio Natta and Karl Ziegler after whom the Ziegler-Natta catalyst is named.

It is to be noted that not all polymers made by the condensation method form a condensate during the reaction. Polyurethanes which are formed by a reaction of isocyanates and alcohols are such an example. Also, ring opening polymerization reactions are considered to be of the addition type even though they form polymers which can also be formed by a condensation reaction, e.g., the polymerization of caprolactam to form nylon 6,6 (see Painter and Coleman, (1994)). As a result, most modern texts do not use the polymerization descriptions, condensation and addition. Rather, the terms "step growth" and "chain" are used in place of condensation and addition respectively.

Thermosetting polymers can be made using either step growth or chain polymerization procedures. To obtain a crosslinked polymer, at least one of the monomers used must be trifunctional. A condensation production process is used to produce phenolic polymers which have the highest volume usage of all thermosets. The reaction between phenol and formaldehyde to form a thermoset phenolic polymer is shown in **Fig. 4.54.** Three

CH_2 bonds can be made per phenol group, allowing formation of a crosslinked network structure.

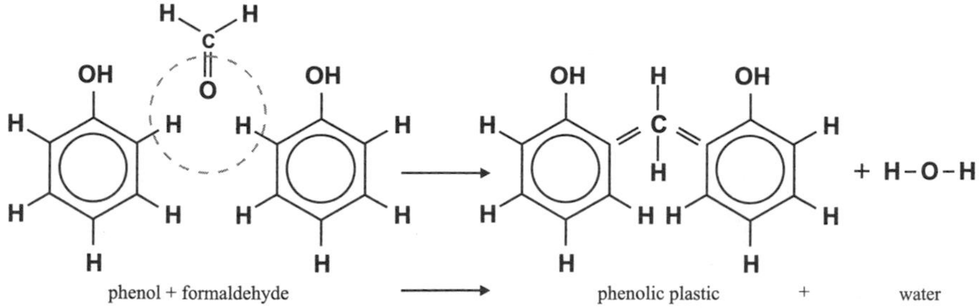

Fig. 4.54 First step in formation of a phenolic network polymer.

Examples of important crosslinked polymers are,

Phenolics
Polyesters
Epoxies
Urethanes
Silicones

Polymerization processes are important in determining the molecular weight, thermal and mechanical properties of a polymer. Usually either batch or continuous processes are used. The former is normal for research laboratory operations but to produce large quantities of polymer, the latter is preferred. However, step growth or condensation reactions are often very slow and a batch process is normally used. There are also single and multiple phase processes. Generally, chain polymerization is not often performed in single (bulk) phase process because of difficulties controlling the reaction. Reactions are often either endothermic or exothermic. If the latter is the case, it is sometimes difficult to control the temperature of the reaction and a "run away reaction" may occur.

There are other important polymerization processes such as those of the suspension or emulsion type. Examples of these two types are given in **Figs. 4.55** and **4.56**. For an excellent description of the synthesis and kinetics of polymerization methods and processes the reader is referred to more detailed texts focusing on polymer science, eg. Painter and Coleman, (1994) and Billmeyer, (1984).

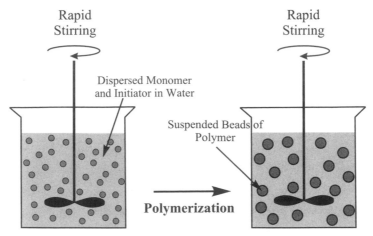

Fig. 4.55 Schematic representation of suspension polymerization. Polymerization occurs by chain growth and the aqueous media serves to disperse the heat of reaction. (After Painter and Coleman (1994), p. 54)

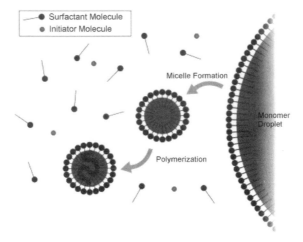

Fig. 4.56 Schematic representation of emulsion polymerization. Large monomer droplets are stabilized by surfactant molecules in water. Excess surfactant forms micelles into which monomer molecules diffuse. Initiator molecules interact predominantly with the numerous small micelles (larger surface area) where the monomer polymerizes, resulting in a suspension of polymer beads in water.

4.10. Spectrography

One might ask the question, "how is it possible to know the structure of a new polymer that has been synthesized for the first time"? One answer to the question, is "through the use of spectrographic analysis of the resulting polymer." For this reason, many people involved in research and development of polymers are specialists in spectrographic analysis of one type or the other. A list of many of the various types of spectroscopy are given below,

UPS	UV (ultraviolet) Photoelectron Spectroscopy
XPS	X-ray Photoelectron Spectroscopy
ESCA	Electron Spectroscopy for Chemical Analysis
(S)AES	(Scanning) Auger Photoelectron Spectroscopy
ISS	Ion Scattering Spectroscopy
LEIS	Low Energy Ion Scattering (Spectroscopy)
SIMS	Secondary Ion Mass Spectrometry
SNMS	Secondary Neutral Mass Spectrometry
SSMS	Scanning Secondary Ion Mass Spectrometry
FAB	Fast Atom Bombardment (Spectroscopy)
(S)EXAFS	(Surface) Extended X-ray Absorption Fine Structure (Spectroscopy)
RBS, HEIS	Rutherford Back Scattering, High Energy Ion Scattering (Spectroscopy)
LAMMA	Laser Micro Mass Analysis (Spectroscopy)
IETS	Inelastic Electron Tunneling Spectroscopy
LEELS	Low Energy Electron Tunneling Spectroscopy
ESD	Electron Stimulated Desorption (Spectroscopy)
NMR	Nuclear Magnetic Resonance Spectroscopy
IR	Infrared Spectroscopy
Raman	Raman Spectroscopy

Most of these procedures are limited to analyzing the surface of a material and the area of specialization is often called "surface chemistry". Persons with this capability are often trained in chemistry, physics, or materials engineering (science) departments. The equipment used is usually very expensive and is often designed through a collaboration of one of the above groups with mechanical engineers.

While these procedures are usually limited to the surface of a material, some of the methods can be used on the interior of the material by actually atomically drilling (sputtering) into the interior. Also, a newer procedure involving infrared spectroscopy and FFT (fast Fourier transform) techniques can examine fluorescing phenomenon of molecules on the interior of materials.

Generally spectroscopy is the study of the interaction of electromagnetic radiation with matter. The wavelength (frequency) range for various types of electromagnetic radiation is given in **Fig. 4.57**.

The essential features to measure the spectrum of a particular polymer is given in **Fig. 4.58**. If, for example, a beam of light is focused on a material, it can be reflected or transmitted. In either case, some of the energy may be absorbed or scattered. The amount of absorbtion or scattering is related to the type of molecules or atoms encountered by the radiation. With a proper detector, the radiation transmitted or reflected from a material can analyzed to determine the amount absorbed. A schematic of the readout of absorbed light is given in **Fig. 4.58** and a typical spectra is shown in **Fig. 4.59** and is an example of an infrared absorption spectrum for isotactic polystyrene either annealed or quenched from the melt. Differences between the spectra infer changes in the molecular structure which occur during annealing.

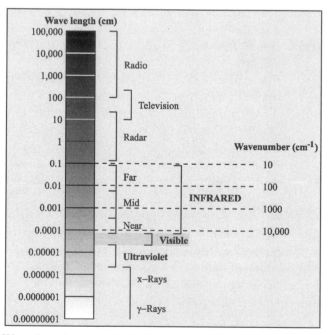

Fig. 4.57 Wave length ranges for various types of eletromagnetic radiation.

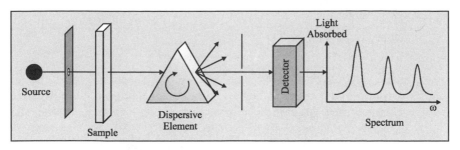

Fig. 4.58 Basic elements of a spectrometer.

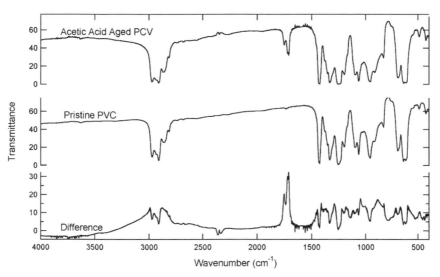

Fig. 4.59 Infrared spectrum of polyvinyl chloride (PVC) showing pristine PVC, after aging in acetic acid, and the difference between the two spectra.

4.11. Review Questions

4.1. Give the chemical formula for the mer units of PVC, PTFE, PP, PS and PMMA.

4.2. Explain the meaning of configuration and conformation when applied to a polymer chain.

4.3. Give an examples of different types of configurations.

4.4. What are chemical bonds?

4.5. Name three primary bonds and describe each.

4.6.   Name four types of secondary bonds. By what other name are secondary bonds known?

4.7.   Give approximate energies (kcal/mole) for covalent bonds. Hydrogen bond. Secondary bonds.

4.8.   Outline (i.e., give a schematic diagram) a simple classification scheme based on bonds within chains and between chains and based on their morphology.

4.9.   Describe a branched polymer. Name different types of branched polymers.

4.10.   Name three types of isomers and describe each.

4.11.   Describe (i.e., give a schematic diagram) a classification system based on molecular characteristics, structure and isomeric states.

4.12.   Name four types of copolymers and describe each.

4.13.   What are two possible conformations of the ethane molecule? Draw the variation of energy with angular rotation and explain the reason for the variation.

4.14.   Explain the trans and gauche positions along the backbone chain of a polyethylene molecule.

4.15.   Derive (and explain all assumptions) an approximate relationship for the end-to-end distance of a single molecular chain.

4.16.   Explain the folded chain model for crystallinity. The fringed micelle model.

4.17.   Describe a spherulite. How might they be formed.

4.18.   What method (or instrument) is generally used to provide evidence of a spherulite?

4.19.   What method(s) is used to evaluate crystallinity? Explain.

4.20.   Give approximate characteristics of a molecular chain containing the number of mer units shown in the following table.

No.	Softening Temperature	Character at 25°C
1		
6		
35		
140		
430		
1,350		

4.21.   Name six methods to measure molecular weight.

4.22. Briefly describe ebulliometry. Cryoscopy. Vapor pressure osmometry. Membrane osmometry.

4.23. Briefly describe SEC (GPC).

4.24. Explain the chemistry example of a condensation polymerization reaction given in the text.

4.25. Explain the chemistry example for an addition polymerization reaction given in the text.

4.26. What is a dimer? A trimer?, An Oligomer?

4.27. What is the meaning of monofunctional? Bifunctional? Trifunctional? Which leads to thermoplastic polymers? Which leads to a thermoset? Can a monofunctional molecule be polymerized?

4.28. What does the term 6,6 in Nylon 6,6 mean?

4.29. What is interfacial condensation? Give an example.

4.30. What are the three features of most addition type polymerizations?

4.31. What is a polyolefin? Name two.

4.32. What are some of the difficulties associated with making polymers on a large scale?

4.33. What do the terms batch and continuous processes mean?

4.34. Would you recommend a step-growth polymerization for automotive assembly lines?

4.35. What are single and multiphase processes?

4.36. What is suspension polymerization? Emulsion polymerization?

4.12. Problems

4.1. Calculate the number average and the weight average molecular weights for the data shown below,

i	M_i	N_i
Interval No.	g/mole of chains in interval	No. of chains in interval
1	2,000	2
2	5,000	4
3	15,000	5
4	30,000	3
5	50,000	2
6	60,000	1

5. Differential Constitutive Equations

A review of the basic definitions of stress and strain was given in Chapter 2. It was noted that a linear elastic solid in uniaxial tension or pure shear obeys Hooke's laws given by,

$$\sigma = E\varepsilon \tag{5.1}$$

$$\tau = G\gamma \tag{5.2}$$

where σ (or τ) is the applied stress, ε (or γ) is the resulting strain, and E (or G) is the elastic modulus and is applicable for many materials under certain circumstances of environment for small stresses and small strains.

For polymers, the torsion test is often the test of choice because, as discussed in Chapter 2, the time dependent (viscoelastic) behavior of polymers is principally due to the deviatoric (shear or shape change) stress components rather than the dilatoric (volume change) stress components. Typically, constant strain rate tests are often used for either tension, compression or torsion as discussed in Chapter 3. If the material is linear elastic, the stress rate is proportional to the strain rate as the modulus is time independent. That is,

$$\frac{d\sigma}{dt} = E\frac{d\varepsilon}{dt} \tag{5.3}$$

On the other hand, if the modulus is time dependent a term must be added for the time derivative of the modulus. In fact, in Chapter 3 it was found that the differential equation for the elementary Maxwell model (where μ is viscosity) was given by

$$\sigma + \frac{\mu}{E}\dot{\sigma} = \mu\dot{\varepsilon} \tag{5.4}$$

in which both stress rate and strain rate appear.

Elementary creep and relaxation tests as a means to experimentally characterize polymers were discussed in Chapter 3. Further, elementary mechanical models and the related differential equations were discussed as a means to phenomenologically understand creep, relaxation and constant strain rate tests. Virtually no material exactly obeys these simple models. As a result, more general approaches are needed to adequately model the time dependent behavior of polymers. This chapter develops the methodology by which the governing differential equations for general mechanical models can be developed. The differential equations are used to obtain modulus and compliance functions under quasi-static and dynamic response conditions. The following chapter develops an integral equation approach to constitutive modeling.

5.1. Methods for the Development of Differential Equations for Mechanical Models

The Maxwell and Kelvin elements introduced in Chapter 3, while typically not able to represent real polymer behavior alone, can however be used as the building blocks of more general models. Any number of mechanical models can be created by assembling Maxwell and Kelvin elements together with free springs and dampers in series and/or parallel. One motivation to proceed in this manner is provided by relaxation/retardation times. Recall that the relaxation/retardation time of a Maxwell or Kelvin element is defined by the ratio of μ/E and that therefore there is a single relaxation/retardation time associated with a Maxwell or Kelvin element. From the discussion of polymers in Chapter 4, however, it is clear that entangled networks of polymer chains will exhibit more complicated time behavior to mechanical load and that in fact different segments of chain lengths and different side groups will offer a wider spectrum of relaxation times. The concept of relaxation spectra is discussed in more detail in Chapters 6 and 7. Assembling mechanical models from multiple Maxwell and/or Kelvin units will therefore enable the models to better mimic polymer behavior by providing multiple relaxation/retardation times.

As with the simple models from Chapter 3, each different mechanical model can be described by a differential equation. The differential equation governing the response for any mechanical model may be obtained by considering the constitutive equations for each element as well as the overall equilibrium and kinematic constraints of the network. Once the differential equation is obtained, the response of the model to any desired loading can be examined by solving the differential equation for that particular loading. The solution for simple creep or relaxation loading will provide the creep compliance or the relaxation modulus for the given model. In this section, we provide by way of example a general method to obtain the governing differential equation for any mechanical model.

As a first example, consider the three-parameter model (sometimes known as the Voigt-Kelvin model) shown in **Fig. 5.1**. This model is best approached as a combination of a spring and a Kelvin model acting in series. The three sets of equations then become, where the subscripts 0 or s indicate the value of quantities in the free spring, the subscripts 1 or k indicate the value of quantities in the Kelvin element, and unsubscripted σ and ε are the remote values of stress and strain (the total stress and strain carried by the three-parameter solid).

	Spring	Kelvin Model	Three-Parameter Model
Equilibrium Equations	$\sigma_0 = \sigma_S$	$\sigma_1 = \sigma_{sl} + \sigma_{dl} = \sigma_k$	$\sigma = \sigma_s = \sigma_k$
Kinematic Equations	$\varepsilon_0 = \varepsilon_s$	$\varepsilon_1 = \varepsilon_{sl} = \varepsilon_{dl} = \varepsilon_k$	$\varepsilon = \varepsilon_s + \varepsilon_k$
Constitutive Equations	$\sigma_s = E_0 \varepsilon_s$	$\sigma_k = E_1 \varepsilon_k + \mu_1 \dot{\varepsilon}_k$	To be determined

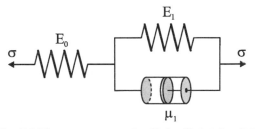

Fig. 5.1 Three-parameter (or Voigt-Kelvin) solid.

The objective is to find the constitutive equation (governing differential equation) for the three-parameter model. The kinematic equation for the three-parameter solid is,

$$\varepsilon = \varepsilon_s + \varepsilon_k \tag{5.5}$$

From equilibrium, the stress in the free spring, σ_s, and the stress in the Kelvin element, σ_k, are the same as the remote stress, σ. To find the differential equation it is convenient to write the Kelvin constitutive equation as,

$$\sigma_k = E_1\varepsilon_k + \mu_1\frac{d\varepsilon_k}{dt} = E_1\varepsilon_k + \mu_1 D\varepsilon_k \tag{5.6}$$

where $D = \dfrac{d}{dt}$ is a differential operator. Note that D^2, D^3, ... indicate the second, third, ..., derivatives with respect to time. Since differential operators obey the fundamental rules of algebra, they may be manipulated as algebraic terms in polynomial expressions by factorization, multiplication, etc. **Eq. 5.6** can now be solved for the Kelvin strain,

$$\varepsilon_k = \frac{\sigma_k}{E_1 + \mu_1 D} \tag{5.7}$$

Recognizing that $\sigma_k = \sigma_s = \sigma$, and substituting **5.7** and the constitutive law for the spring into **Eq. 5.5**, after simplification one obtains,

$$\sigma + p_1\dot{\sigma} = q_0\varepsilon + q_1\dot{\varepsilon} \tag{5.8}$$

where,

$$p_0 = 1 \quad p_1 = \frac{\mu_1}{E_0 + E_1} \quad q_0 = \frac{E_0 E_1}{E_0 + E_1} \quad q_1 = \frac{\mu_1 E_0}{E_0 + E_1}$$

Differential equations for viscoelastic polymers are most often given in the standard form as shown in **Eq. 5.8**. The first stress term is not differentiated and the coefficient is taken as one.

The differential equation governing the relationship between stress and strain for a given mechanical model is quite valuable, but needs to be solved in order to determine the model response to specific loading conditions. Fundamental viscoelastic properties such as the creep compliance or relaxation modulus can be found by solution of the differential equation to the appropriate loading. For example, the creep compliance can be determined using the conditions for a creep test of constant stress, as shown in **Fig. 5.2**.

The determination of the initial conditions is best accomplished by inspection of the physical model. Since the input stress is constant for the creep test, the stress rate is zero, $\dot{\sigma} = 0$ and the differential equation for the three-parameter solid, **Eq. 5.8**, becomes,

$$\dot{\varepsilon} + \frac{q_0}{q_1}\varepsilon = \frac{1}{q_1}\sigma_0 H(t) \tag{5.9}$$

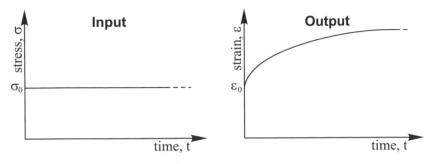

Fig. 5.2 Creep test.

The quantity $\mathbf{H(t)}$ is the Heavyside or unit step function (See appendix A) and is defined to be,

$$H(t) = \begin{Bmatrix} 1, & t \ge 0 \\ 0, & t < 0 \end{Bmatrix} \tag{5.10}$$

Eq. 5.9 is a nonhomogeneous equation whose solution is the sum of the homogeneous and particular solutions given by,

$$\varepsilon(t) = \sigma_0 \left[\frac{1}{E_0} + \frac{1}{E_1}(1 - e^{-t/\tau}) \right] \tag{5.11a}$$

where $\tau = \dfrac{\mu_1}{E_1}$ is the retardation time of the Kelvin element. The creep compliance of the Three-Parameter solid is therefore

$$D(t) = \frac{1}{E_0} + \frac{1}{E_1}(1 - e^{-t/\tau}) \tag{5.11b}$$

Referring to the solution under creep for a Kelvin material given in Chapter 3, quite obviously the solution of the three-parameter model for the case of creep is simply the superposition of the solution for creep of a spring and creep of a Kelvin solid.

Solutions of the differential equation for the conditions of relaxation, constant strain or stress rate and other conditions can be obtained in a similar manner to that followed above.

Using the procedure presented for the Three-Parameter Solid, the differential equation for a four-parameter fluid model (**Fig. 5.3**) can be shown to be,

$$\sigma + p_1\dot{\sigma} + p_2\ddot{\sigma} = q_1\dot{\varepsilon} + q_2\ddot{\varepsilon} \tag{5.12}$$

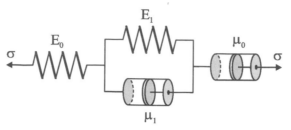

Fig. 5.3 Four-parameter fluid.

Note that this Four-Parameter Fluid model is composed of a Kelvin element (subscripts 1) and a Maxwell element (subscripts 0). Thus, the constitutive laws (differential equations) for the Kelvin and Maxwell elements need to be used in conjunction with the kinematic and equilibrium constraints of the system to provide the governing differential equation. Again, treating the time derivatives as differential operators will allow the simplest derivation of **Eq. 5.12**. The derivation is left as an exercise for the reader as well as the determination of the relations between the p_i and q_i coefficients and the spring moduli and damper viscosities (see problem 5.1).

The solution of **Eq. 5.12** for the Four-Parameter Fluid for the case of creep can be shown to be,

$$\varepsilon(t) = \sigma_0 \left[\frac{1}{E_0} + \frac{1}{E_1}\left(1 - e^{-t/\tau}\right) + \frac{t}{\mu_0} \right]$$

$$\qquad\qquad \uparrow \qquad\qquad \uparrow \qquad\qquad \uparrow \qquad\qquad \text{(5.13)}$$

Inst. Elastic   Delayed Elastic   Flow

Term   Term   Term

Again, the solution is left as an exercise for the reader (see problem 5.4). However, it should be noted that the solution of the differential equation for a four-parameter fluid in the case of creep is the superposition of creep of a Maxwell fluid and creep of a Kelvin solid (refer to Chapter 3).

The creep and creep recovery behavior of a four-parameter fluid is shown in **Fig. 5.4** and is recognized as the response of a thermoplastic type polymer as given earlier in **Fig. 3.13**. The three stages of instantaneous elasticity, delayed elasticity and flow represents the most general type behavior possible for a linear viscoelastic material. **Note: Some texts do not include the flow term as a viscoelastic component, preferring instead** to define viscoelastic behavior only for models with no free damper or flow term.

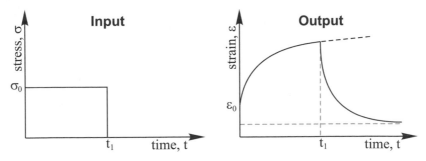

Fig. 5.4 Creep and creep recovery of a four-parameter fluid.

By eliminating various elements in the four-parameter model, the response of a Maxwell fluid, Kelvin solid, three-parameter solid (a Kelvin and a spring in series) can be obtained and the model can be used to represent thermoplastic and/or thermoset response as illustrated in **Fig. 3.13**. For example, the creep response of a three-parameter solid is obtained by eliminating the free damper in **Eq. (5.13)** and gives the creep and creep recovery response shown in **Fig. 3.13** for a crosslinked polymer. The four-parameter fluid can also be evaluated in relaxation or other loading conditions again by solving the differential equation for each case.

5.2. A Note on Realistic Creep and Relaxation Testing

The testing of polymers requires unique understanding of the viscoelastic nature of polymers. For example in a creep test it is required to suddenly apply a constant tensile, compression, or torsion stress to a bar of material. The most common description of a uniaxial tensile creep test is shown in **Fig. 5.5(a)**. Several questions may arise one of which is: How is the load to be applied suddenly without causing dynamic effects. One answer is for the load to be applied as ramp input as shown in **Fig. 5.5(b)**. Obviously, the latter case is not a correct creep test. How big an error is involved? A solution of the differential equation representative of the material for the ramp input of **Fig 5.5(b)** can be obtained and it can be shown that the error in the strain output is negligible if the loading time, t_0, is small compared to the retardation time of the material, τ.

Similarly, the same difficulty occurs in a relaxation test. That is an ideal relaxation test is one where a sudden input of strain is required as shown in

Fig. 5.5c. Again, however, to avoid dynamic effects it is usual to use a ramp input of strain as shown in **Fig. 5.5d** and it can be shown that the error is negligible if the ramp time, t_0, is small compared to the relaxation time of the material, τ (see homework problem 5.5).

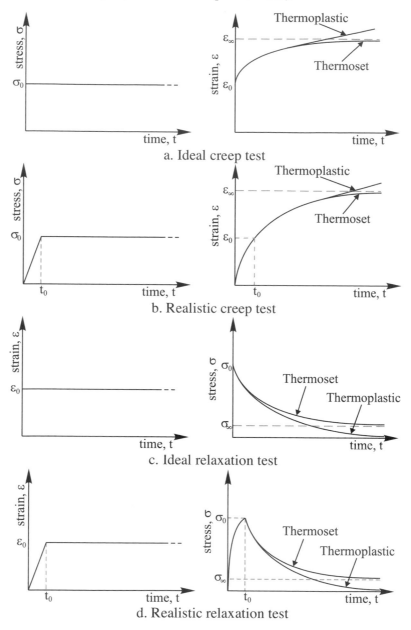

Fig. 5.5 Comparison of ideal and realistic creep and relaxation tests.

A further concern for creep and relaxation occurs due to the stiffness of the polymer tested. If a very soft material is tested in creep, the cross-sectional dimensions or area may change as the material creeps and, therefore, the test may not be a true creep test. For this case, the load must be changed with time such that the amount of load divided by the changing area remains a constant. Before the advent of modern testing machines a number of ingenious methods were developed by which the load would vary in proportion to the area such that the input stress would remain constant. Using a closed loop servo-hydraulic testing system similar to the one shown in **Fig. 5.6**, it is easy to monitor the change in area and use the new area in the computer load control so that the stress remains a constant.

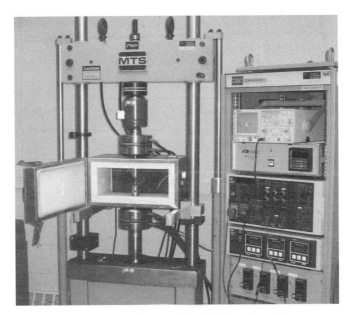

Fig. 5.6 Closed-loop servo-controlled hydraulic testing system.

In relaxation testing, the stiffness of the specimen must be small compared to the stiffness of the load cell and testing machine. Of necessity the specimen is in series with both the load cell and testing machine and, therefore, the deformation in the specimen, the load cell and the testing machine are additive. As the load in the specimen decreases or relaxes, even in a fixed grip circumstance, the load also decreases in the load cell and/or the test machine. The deformation will then actually increase in the specimen to allow a decrease in the deformation (and load) in the load cell. For a very stiff specimen (such as a fiber reinforced composite), the change in load (or stress) recorded may reflect a redistribution of deformation from the load cell and testing machine to the specimen resulting in a non-constant deformation or strain in the specimen. Again, a computer-controlled machine such as the one shown in **Fig. 5.6** can be programmed to sense the change in strain in the specimen and to have the "stroke" or displacement of the test machine altered to keep the strain in the specimen constant. Another example where care must be taken in the interpretation of the relaxation stress response to a constant deformation input to the specimen is in adhesive testing such as often obtained using a lap joint specimen. See Sancaktar (1990) to observe data indicating both stress relaxation and creep occurring simultaneously in the adhesive when a typical lap specimen is tested with constant deformation input. These examples suggest that the relaxation response of any multiphase system must be analyzed with caution.

5.3. Generalized Maxwell and Kelvin Models

As indicated earlier, single Maxwell or Kelvin elements are of limited utility in representing the actual stress-strain response of polymers. A more realistic mathematical model can be developed, however, by considering a series of Maxwell elements in parallel. Consider, first just two Maxwell elements in parallel as in **Fig. 5.7**.

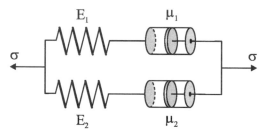

Fig. 5.7 Two Maxwell elements in parallel.

The equilibrium and kinematic equations are,

$$\sigma = \sigma_1 + \sigma_2$$

$$\varepsilon = \varepsilon_1 = \varepsilon_2$$

(5.14)

The constitutive equations for each Maxwell element are,

$$\sigma_1 + \frac{\mu_1}{E_1}D\sigma_1 = \mu_1 D\varepsilon_1$$

$$\sigma_2 + \frac{\mu_2}{E_2}D\sigma_2 = \mu_2 D\varepsilon_2$$

(5.15)

where $\mathbf{D} = \mathbf{d/dt}$ is again the differential operator. Solving each equation in **5.15** for the stress, substituting into **Eq. 5.14**, recognizing that the strain in each element is the same as for the system, and rearranging gives the following differential relation between the applied stress and strain.

$$\sigma + \left(\tau_1 + \tau_2\right)\dot{\sigma} + \tau_1\tau_2\ddot{\sigma} = \left(\mu_1 + \mu_2\right)\dot{\varepsilon} + \left(\mu_1\tau_2 + \tau_1\mu_2\right)\ddot{\varepsilon}$$ (5.16a)

The standard form of (5.16a) is

$$\sigma + p_1\dot{\sigma} + p_2\ddot{\sigma} = q_1\dot{\varepsilon} + q_2\ddot{\varepsilon}$$ (5.16b)

Since the Maxwell elements are connected in parallel, if strain $\varepsilon(t)$ is given, one can either solve the pair of linear first order **Eqs. (5.15)** or the single second order equation **(5.16b)** to find the solution for $\sigma(t)$. As an example, consider the case of stress relaxation in which a constant strain history is applied, $\varepsilon(t) = \varepsilon_0 H(t)$. Due to the kinematic constraint, each Maxwell element sees the same global strain history and the solution for $\sigma_1(t)$ and $\sigma_2(t)$ from **Eqs. 5.15** are as given earlier in **Eq. 3.17**.

$$\sigma_1(t) = \varepsilon_0 E_1 e^{-t/\tau_1}$$

$$\sigma_2(t) = \varepsilon_0 E_2 e^{-t/\tau_2}$$

(5.17a)

From the equilibrium constraint, the solution for the overall stress in the system is a simple superposition of the stresses in each element

$$\sigma(t) = \varepsilon_0\left(E_1 e^{-t/\tau_1} + E_2 e^{-t/\tau_2}\right)$$ (5.17b)

The second order differential equation **(5.16b)** can also be solved to obtain the same solution **Eq. 5.17b**.

Three Maxwell elements in parallel would give a differential relation between stress and strain that contains first, second and third derivatives (see homework problem 5.7) as given below,

$$\sigma + p_1\dot{\sigma} + p_2\ddot{\sigma} + p_3\dddot{\sigma} = q_1\dot{\varepsilon} + q_2\ddot{\varepsilon} + q_3\dddot{\varepsilon}$$ (5.18)

Obviously, as the number of elements increase, the order of the highest derivative increases. After obtaining the differential equation for three Maxwell elements, it is possible to develop a recursion relation to obtain the appropriate coefficients (in terms of E_i's and μ_i's, as in **Eq. 5.6**) for any number of elements so desired.

It is usually not possible to represent the behavior of a polymer under the condition of relaxation with only one or two Maxwell elements in parallel. Rather, as many as 5 to 15 or more elements may be necessary. A model with many elements is called a *Generalized Maxwell Model* and is shown in **Fig 5.8**. The differential equation for a generalized Maxwell model may be expressed as,

$$\sigma + p_1\dot{\sigma} + p_2\ddot{\sigma} + \cdots p_n \overset{n\bullet}{\sigma} = q_1\dot{\varepsilon} + q_2\ddot{\varepsilon} + \cdots q_n \overset{n\bullet}{\varepsilon} \qquad (5.19)$$

where $\overset{n\bullet}{\sigma} \equiv \mathbf{D}^n\sigma$, p_0 is taken to be unity and n is the number of parallel Maxwell elements in the particular model. Mechanical models constructed from springs, dampers and Maxwell and Kelvin elements can in general be represented by a differential equation of the *standard form*

$$\sum_{k=0}^{n} p_k \frac{d^k\sigma}{dt^k} = \sum_{k=0}^{m} q_k \frac{d^k\varepsilon}{dt^k} \qquad (5.20)$$

where n=m and $q_0=0$ for the generalized Maxwell model. As will be mentioned subsequently, the number of derivatives of stress and strain is not the same for a series of Kelvin elements which provides the rationale for the different indices n and m on the summation in **Eq. 5.20**. Some might be tempted to avoid using E_i's and μ_i's and instead develop a generalized model by choosing p's and q's. However, as discussed in the next section, the p's and q's for a differential equation of a particular order may not be chosen arbitrarily and still represent physically meaningful behavior.

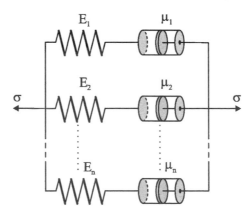

Fig. 5.8 Generalized Maxwell fluid.

As with the two-element example, the solution of a Generalized Maxwell Model for a given strain input, $\varepsilon(t)$, can be found either by superposition of n first order differential equation solutions or by solution of the single n^{th} order differential equation. The n first order equations are all of the form of **Eqs. 5.15**,

$$\sigma_i + \tau_i D\sigma_i = \mu_i D\varepsilon_i(t) \tag{5.21a}$$

where i ranges from 1 to n. The kinematic constraint again provides that the strain in each element is the same as the global strain, $\varepsilon_i(t) = \varepsilon(t)$. And the equilibrium constraint provides that the solution for the global stress is simply a sum of the individual stresses, $\sigma(t) = \sigma_1(t) + \sigma_2(t) + \cdots + \sigma_n(t)$. For the condition of stress relaxation, $\varepsilon(t) = \varepsilon_0 H(t)$, the solution of these linear differential equations can again easily be found by superposition to be,

$$\sigma(t) = \varepsilon_0 \sum_{i=1}^{n} E_i e^{-t/\tau_i} \tag{5.21b}$$

Therefore the relaxation modulus of a Generalized Maxwell Model is given by

$$E(t) = \sum_{i=1}^{n} E_i e^{-t/\tau_i} \tag{5.21c}$$

This type of representation is sometimes called a Prony series and such an exponential expansion is often used to describe the relaxation modulus of a viscoelastic material even without reference to a mechanical model.

The generalized model given above can only be used to represent a thermoplastic if all of the μ_i values are nonzero. In order to represent a thermoset a free spring is sometimes included with the result known as the Wiechert model shown in **Fig 5.9**.

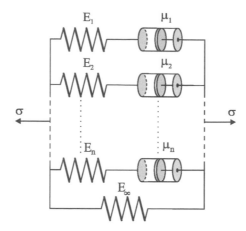

Fig. 5.9 Wiechert Model.

The solution for stress relaxation and the relaxation modulus then become,

$$\sigma(t) = \varepsilon_0 \left(\sum_{i=1}^{n} E_i e^{-t/\tau_i} + E_\infty \right)$$

(5.22)

$$E(t) = \sum_{i=1}^{n} E_i e^{-t/\tau_i} + E_\infty$$

where E_∞ is the equilibrium modulus.

For a Generalized Maxwell Model, whether a solution for a given problem is found by solving the nth order differential equation or the system of n first order equations depends on the particular loading history applied. For the case of stress relaxation, the superposition of solutions of the first order equations is certainly the simpler route. For more complicated strain histories, the method of choice may also depend on whether the solution is to be obtained numerically or analytically. Also, if stress history is applied and strain to be found, use of the single higher order differential equation will likely be more straightforward, since each of the $\sigma_i(t)$ needed in the first order equations are unknown at the outset. Finally, since the relaxation modulus for a Generalized Maxwell Model (**Eq. 5.22**) is known, solutions may also be obtained for given stress or strain histories via an integral constitutive equation approach (instead of solving differential equations), as is shown in the next chapter.

A *Generalized Kelvin Solid* is composed of a number of Kelvin elements in series as shown in **Fig. 5.10a**.

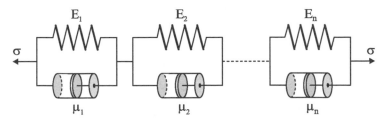

Fig. 5.10a Generalized Kelvin solid.

However, this model still has no instantaneous elasticity and a free spring is normally included in series with the generalized Kelvin solid with the result (sometimes referred to as the Voigt-Kelvin model),

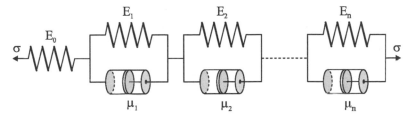

Fig. 5.10b Generalized Voigt-Kelvin solid.

A differential equation for either of the series of Kelvin elements can be found using the same procedure described in developing the differential equation for a series of Maxwell elements. The equilibrium constraint, kinematic constraint and constitutive equations are given by

$$\sigma = \sigma_1 = \sigma_2 = \cdots = \sigma_n$$

$$\varepsilon = \varepsilon_1 + \varepsilon_2 + \cdots + \varepsilon_n \qquad (5.23a)$$

$$\sigma_i = E_i \varepsilon_i + \mu_i D\varepsilon_i, \quad i = 1, 2, \cdots n$$

Proper combination of these equations will result in a governing differential equation in the standard form,

$$\sum_{k=0}^{n} p_k \frac{d^k \sigma}{dt^k} = \sum_{k=0}^{m} q_k \frac{d^k \varepsilon}{dt^k} \qquad (5.23b)$$

where n=m-1 and $p_0=1$. Again, depending on the loading history applied, either the system of n first order **Eqs. (5.23a)** or the single nth order differential **Eq. (5.23b)** can be solved. For the case of simple creep loading, $\sigma(t) = \sigma_0 H(t)$, the solution for the Generalized Kelvin Model can be easily found by superposition of the solutions of the n first order equations to be,

$$\varepsilon(t) = \sigma_0 \left[\frac{1}{E_0} + \sum_{i=1}^{n} \frac{1}{E_i} \left(1 - e^{-t/\tau_i} \right) \right] \tag{5.24a}$$

where the creep compliance is therefore defined to be

$$D(t) = \frac{1}{E_0} + \sum_{i=1}^{n} \frac{1}{E_i} \left(1 - e^{-t/\tau_i} \right) \tag{5.24b}$$

These equations can be used to represent a cross-linked material. Although the Generalized Kelvin Model can be solved for the case of relaxation, due to the forms of the differential equations and ease of solution, Maxwell elements in parallel are typically used for relaxation while Kelvin elements in series are used for creep.

A free damper as well as a free spring can be placed in series with a number of Kelvin elements as given in **Fig. 5.11**,

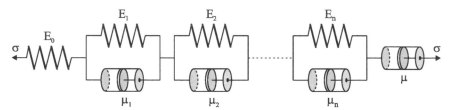

Fig. 5.11 Generalized Voigt-Kelvin solid with a free damper.

The creep compliance will then become,

$$D(t) = \left[\frac{1}{E_0} + \sum_{i=1}^{n} \frac{1}{E_i} \left(1 - e^{-t/\tau_i} \right) + \frac{t}{\mu_0} \right] \tag{5.24c}$$

and can be used to represent a thermoplastic material. As with the Generalized Maxwell Model, the creep compliance found for the Generalized Kelvin Model can be used to characterize a viscoelastic material and then can be used in integral constitutive laws (Chapter 6) to determine the response of the material to any type of stress or strain loading history without solving the differential equations for that given loading history.

An example of creep deflection in a tensile bar for an epoxy at different temperatures is shown in **Fig 5.12**. It will be noticed that the creep response for a temperature of 155° C still has a positive slope after seven hours. Without knowing the type of material, one might expect the response to be that of a viscoelastic fluid. The creep response for 165° C and 170° C clearly have reached a limit and has the character of a thermoset. Because of the nature of the response, the epoxy could be best characterized by a viscoelastic fluid model such as the four-parameter fluid for both the 155° C and 160° C data. On the other hand, the epoxy could best be characterized by a viscoelastic solid model such as the three-parameter solid for temperatures above 160° C. To characterize the material over all time and temperature ranges would require a generalized model with a large number of elements. Methods to accomplish this will be discussed in subsequent sections.

The glass transition temperature for this material is unknown but is likely above 155° C. Assuming such is the case, the material at 155° C is in the glassy region while the material above 170° C is in the rubbery region. In fact, if the load could be applied instantaneously (without inertia effects), the initial elastic strain would be nearly the same for each. The major difference would be the time to reach the limit strain. At 155° C, the time required to reach a strain equivalent to the limiting rubbery value would be very long, perhaps days, weeks or even months. But at 170° C the limiting rubbery strain is reached in a few minutes or less.

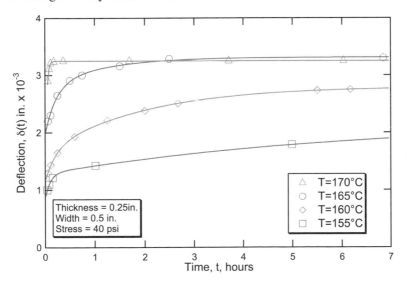

Fig. 5.12 Creep of an araldite epoxy.

Also, it should be noted that the deflection (or strain) reaches a higher limiting value at 165° C than at 170° C. This might be considered an artifact of the experiment at first. However, in reality this is confirmation of the Joule effect mentioned in Chapter 1. More evidence of this phenomenon will be given later.

5.3.1. A Caution on the Use of Generalized Differential Equations

Sometimes in numerical studies it is tempting to attempt to understand how a particular boundary value problem might be affected by the order of the differential equation representing the relationship between stress and strain. For example, the general equation,

$$\sum_{k=0}^{n} p_k \frac{d^k \sigma}{dt^k} = \sum_{k=0}^{m} q_k \frac{d^k \varepsilon}{dt^k} \qquad (5.25a)$$

or

$$\sigma + p_1 \dot{\sigma} + p_2 \ddot{\sigma} + \cdots p_n \overset{n\bullet}{\sigma} = q_1 \dot{\varepsilon} + q_2 \ddot{\varepsilon} + \cdots q_n \overset{m\bullet}{\varepsilon} \qquad (5.25b)$$

might be truncated after the first, second derivative or higher derivative to obtain a workable equation. Care must be taken when generating arbitrary differential constitutive equations in this manner. For example, consider truncation after the first derivative to obtain,

$$\sigma + p_1 \dot{\sigma} = q_0 \varepsilon + q_1 \dot{\varepsilon} \qquad (5.25c)$$

This equation, is in fact the same as the equation for the three-parameter solid and may be written as,

$$\sigma + \frac{\mu_1}{E_0 + E_1} \dot{\sigma} = \frac{E_0 E_1}{E_0 + E_1} \varepsilon + \frac{\mu_1 E_0}{E_0 + E_1} \dot{\varepsilon} \qquad (5.25d)$$

Now consider the relationship between the coefficients p_1, q_0, and q_1 in the form,

$$\frac{q_1}{p_1} - q_0 = \frac{\dfrac{\mu_1 E_0}{E_0 + E_1}}{\dfrac{\mu_1 E_1}{E_0 + E_1}} - \frac{E_0 E_1}{E_0 + E_1} = \frac{E_0^2}{E_0 + E_1} > 0 = \text{positive quantity} \qquad (5.25e)$$

Obviously, in this case the coefficients of the differential equation cannot be selected arbitrarily and must satisfy the above inequality in order to be physically meaningful.

Refer to Flugge (1974) for additional discussion on this subject and other inequalities.

5.3.2. Description of Parameters for Various Elementary Mechanical Models

The methods previously discussed in this chapter can be used to determine the differential equations, solutions and parameters for a number of mechanical models using a variety of combinations of springs and damper elements. **Table 5.1** is a tabulation of the differential equation, parameter inequalities, creep compliances and relaxation moduli for frequently discussed basic models. Note that the equations are given in terms of the p_i and q_i coefficients of the appropriate differential equation in standard format. The reader is encouraged to verify the validity of the equations given and is also referred to Flugge (1974) for a more complete tabulation.

Table 5.1 (Part 1) Differential equations, solutions and parameters

Model	Name	Differential Equation / Inequalities	Creep compliance D(t)
	Elastic Solid	$\sigma = q_0 \varepsilon$	$1/q_0$
	Viscous Fluid	$\sigma = q_1 \dot{\varepsilon}$	$1/q_1$
	Maxwell Fluid	$\sigma + p_1 \dot{\sigma} = q_1 \dot{\varepsilon}$	$(p_1 + t)/q_1$
	Kelvin Solid	$\sigma = q_0 \varepsilon + q_1 \dot{\varepsilon}$	$\dfrac{1}{q_0}(1 - e^{-\lambda t}), \ \lambda = \dfrac{q_0}{q_1}$
	3-parameter Solid	$\sigma + p_1 \dot{\sigma} = q_0 \varepsilon + q_1 \dot{\varepsilon}$ $q_1 > p_1 q_0$	$\dfrac{p_1}{q_1} e^{-\lambda t} + \dfrac{1}{q_0}(1 + e^{-\lambda t}),$ $\lambda = \dfrac{q_0}{q_1}$
	3-parameter Fluid	$\sigma + p_1 \dot{\sigma} = q_1 \dot{\varepsilon} + q_2 \ddot{\varepsilon}$ $p_1 q_1 > q_2$	$\dfrac{1}{q_1} + \dfrac{p_1 q_1 - q_2}{q_1^2}(1 - e^{-\lambda t}),$ $\lambda = \dfrac{q_1}{q_2}$
	4-parameter Fluid	$\sigma + p_1 \dot{\sigma} + p_2 \ddot{\sigma} = q_1 \dot{\varepsilon} + q_2 \ddot{\varepsilon}$ $p_1^2 > 4 p_2$ $p_1 q_1 q_2 > p_2 q_1^2 + q_2^2$	$\dfrac{t}{q_1} + \dfrac{p_1 q_1 - q_2}{q_1^2}(1 - e^{-\lambda t})$ $+ \dfrac{p_2}{q_2} e^{-\lambda t}, \ \lambda = \dfrac{q_1}{q_2}$
	4-parameter Solid	$\sigma + p_1 \dot{\sigma} = q_0 \varepsilon + q_1 \dot{\varepsilon} + q_2 \ddot{\varepsilon}$ $q_1^2 > 4 q_0 q_2$ $q_1 q_1 > q_0 p_1^2 + q_2$	$\dfrac{1 - p_1 \lambda_1}{q_2 \lambda_1 (\lambda_2 - \lambda_1)}(1 - e^{-\lambda_1 t})$ $+ \dfrac{1 - p_1 \lambda_2}{q_2 \lambda_2 (\lambda_1 - \lambda_2)}(1 - e^{-\lambda_2 t})$ *where* λ_1, λ_2 *are roots of* $q_2 \lambda^2 - q_1 \lambda + q_0 = 0$

Table 5.1 (Part 2) Differential equations, solutions and parameters.

Relaxation Modulus E(t)	Complex Compliance	
	Real Part $E_1(\omega)$	**Imaginary Part $E_1(\omega)$**
q_0	$1/q_0$	0
$q_1^{\delta(t)}$	0	$-\dfrac{1}{q_1\omega}$
$\dfrac{q_1}{p_1}e^{-t/p_1}\,\dfrac{q_1}{p_1}e^{-t/p_1}$	$\dfrac{q_1}{p_1}$	$-\dfrac{1}{q_1\omega}$
$q_0 + q_1^{\delta(t)}$	$\dfrac{q_0}{q_0^2 + q_1^2\omega^2}$	$-\dfrac{q_1\omega}{q_0^2 + q_1^2\omega^2}$
$\dfrac{q_1}{p_1}e^{-t/p_1} + q_0(1 - e^{-t/p_1})$	$\dfrac{q_0 + p_1 q_1\omega^2}{q_0^2 + q_1^2\omega^2}$	$-\dfrac{(q_1 + q_0 p_1)\omega}{q_0^2 + q_1^2\omega^2}$
$\dfrac{q_2}{p_1}\delta(t) + \dfrac{1}{p_1}\left(q_1 - \dfrac{q_2}{p_1}\right)e^{-t/p_1}$	$\dfrac{p_1 q_1 - q_2}{q_1^2 + q_2^2\omega^2}$	$-\dfrac{q_1 + p_1 q_2\omega^2}{(q_1^2 + q_2^2\omega^2)\omega}$
$\dfrac{1}{\sqrt{p_1^2 - 4p_2}}\left|(q_1 - \alpha q_2)e^{-\alpha t} - (q_1 - \delta q_2)e^{-\delta t}\right|,$ $\left.\begin{matrix}\alpha\\ \beta\end{matrix}\right| = \dfrac{1}{2p_2}\left(p_1 \pm \sqrt{p_1^2 - 4p_2}\right)$	$\dfrac{(p_1 q_1 - q_2) + p_2 q_2\omega^2}{q_1^2 + q_2^2\omega^2}$	$-\dfrac{q_1 + (q_2 p_1 - p_2 q_1)\omega^2}{(q_1^2 + q_2^2\omega^2)\omega}$
$\dfrac{q_2}{p_1}\delta(t) + \dfrac{q_1 p_1 - q_2}{p_1^2} -$ $- \dfrac{1}{p_1^2}(q_1 p_1 - q_0 p_1^2 - q_2)(1 - e^{-t/p_1})$	$\dfrac{q_0 + (p_1 q_1 - q_2)\omega^2}{q_0^2 + (q_1^2 - 2q_0 q_2)\omega^2 + q_2^2\omega^4}$	$-\dfrac{(q_1 - p_1 q_0)\omega + q_2 p_1\omega^3}{q_0^2 + (q_1^2 - 2q_0 q_2)\omega^2 + q_2^2\omega^4}$

5.4. Alfrey's Correspondence Principle

It is possible using transform methods to convert viscoelastic problems into elastic problems in the transformed domain, allowing the wealth of elasticity solutions to be utilized to solve viscoelastic boundary value problems. Although there are restrictions on the applicability of this technique for certain types of boundary conditions (discussed further in Chapter 9), the method is quite powerful and can be introduced here by building on the framework provided by mechanical models. Recall the differential equation for a generalized Maxwell or Kelvin model,

$$\sum_{k=0}^{n} q_k \frac{d^k \sigma}{dt^k} = \sum_{k=0}^{m} q_k \frac{d^k \varepsilon}{dt^k} \tag{5.26}$$

which can also be written compactly in terms of differential operators, **P** and **Q** as

$$P\sigma = Q\varepsilon \tag{5.27}$$

The Laplace transform represented by,

$$\pounds\{f(t)\} = \bar{f}(s) = \int_{0}^{\infty} e^{-st}{}_{f(t)dt} \tag{5.28}$$

can be used to convert differential equations into algebraic equations. Taking the Laplace transform of **Eq. 5.26(a)** changes the differential equation to an algebraic expression in the transform parameter **s** and, due to the simple form of **5.26(a)**, may be expressed as,

$$\sum_{k=0}^{n} p_k s^k \bar{\sigma}(s) = \sum_{k=0}^{m} q_k s^k \bar{\varepsilon}(s) \,^* \tag{5.29}$$

or

$$\bar{P}(s)\bar{\sigma}(s) = \bar{Q}(s)\bar{\varepsilon}(s) \tag{5.30}$$

See the Appendix B for fundamentals on the Laplace transform. Since the transformed stress and transformed strain are no longer part of the summations, the expression may be further rewritten as

---

* The reader is cautioned that Eq. 5.29 must be used with care in order to include all initial conditions properly. Significant differences arise depending upon whether the time begins at 0+ or 0-. In most circumstances used herein, f(t) = 0 for t < 0 but in creep or relaxation the jump discontinuity at t = 0 must be included.

$$\overline{\sigma}(s) = \left(\frac{\displaystyle\sum_{k=0}^{m} q_k s^k}{\displaystyle\sum_{k=0}^{n} p_k s^k}\right)\overline{\varepsilon}(s) = \frac{\overline{Q}(s)}{\overline{P}(s)}\overline{\varepsilon}(s) \tag{5.31}$$

The quotient of operators can be thought of as an elastic modulus in transform space and the above equation can be written as,

$$\overline{\sigma}(s) = \overline{E}^*(s)\overline{\varepsilon}(s) \tag{5.32}$$

This result of the same form as Hooke's law for a linear elastic material under uniaxial load and is sometimes called Alfrey's Correspondence Principle[1]. The quantity, $\overline{E}^*(s)$, in transform space is analogous to the usual Young's modulus for a linear elastic materials. Here, the linear differential relation between stress and strain for a viscoelastic polymer has been transformed into a linear elastic relation between stress and strain in the transform space. It will be shown in the next chapter that the same result can be obtained from integral expressions of viscoelasticity without recourse to mechanical models, so that the result is general and not limited to use of a particular mechanical model. Therefore, the simple transform operation allows the solution of many viscoelastic boundary value problems using results from elementary solid mechanics and from more advanced elasticity approaches to solids such as two and three dimensional problems as well as plates, shells, etc. See Chapters 8 and 9 for more details on solving problems in the transform domain.

5.5. Dynamic Properties - Steady State Oscillation Testing

Viscoelastic properties are often determined with steady state oscillation or vibratory tests using small tensile (compressive) bars, thin cylinders or flat strips in torsion, beams in bending, etc. The approach is usually referred to as dynamic mechanical analysis (DMA) testing or sometimes dynamic mechanical thermal analysis (DMTA). The latter term is more appropriate as properties are often determined and expressed in terms of temperature

---

[1] What is now known as the correspondence principle for converting viscoelastic problems in the time domain into elastic problems in the transform domain was first discussed by Turner Alfrey in 1944. As a result, the principle is sometimes referred to as Alfrey's correspondence principle. Later in 1950 and in 1955 the principle was generalized and discussed by W.T. Read and E. H. Lee respectively. (See bibliography for references.)

as well as frequency. Here, sinusoidal tensile testing of a uniaxial bar will be used as an example. However, the results will apply, in general, to all types of dynamic testing. As with the Laplace transform approach for the correspondence principle above, the differential equation obtained from general mechanical models will be used to motivate and describe the dynamic properties here, but we will also see in the next chapter that again the results are general (not dependent upon use of a mechanical model) and can be obtained from integral equation methods.

Assume a small uniaxial sample is loaded with a strain input,

$$\varepsilon(t) = \varepsilon_0 e^{i\omega t} \tag{5.33}$$

In practice only the real (or imaginary) part, a cosine (or sine) wave, will be input but the algebra associated with the exponential function is easier to manipulate and will be used for a general derivation. Note also that the discussion here only considers the steady-state dynamic response. Transient terms associated with starting up an oscillatory loading have decayed and are neglected as are inertial terms. Given the form of the differential **Eq. 5.26** for a general mechanical model of a viscoelastic material, an exponential input as in **5.33** will result in a stress output also of exponential form

$$\sigma(t) = \sigma^* e^{i\omega t} \tag{5.34}$$

where ω is the frequency and σ^* is a complex quantity. The σ^* can be further defined

$$\sigma^* = \varepsilon_0 E^*(i\omega) \tag{5.35}$$

such that the stress can be written as

$$\sigma(t) = \varepsilon_0 E^*(i\omega) e^{i\omega t} \tag{5.36}$$

Here $E^*(i\omega)$ is defined as the complex modulus and can be decomposed into real and imaginary parts as

$$E^*(i\omega) = E'(\omega) + iE''(\omega) \tag{5.37}$$

The real part is defined as the **storage modulus**, $E'(\omega)$, and the imaginary part is defined as the **loss modulus**:, $E'(\omega)$. It will be shown later that these respective quantities can be related to the energy stored and dissipated in a loading cycle.

Note that by combining **Eqs. 5.33** and **5.36**, the complex modulus directly relates the time dependent stress to time dependent strain for the case of oscillatory loading

$$\sigma(t) = E^*(i\omega)\epsilon(t) \qquad (5.38)$$

If the input strain and output stress from **Eqs. 5.33** and **5.36** are inserted into the differential equation for a general mechanical model, **Eq. 5.26(a)**, after simplification an expression very similar to **Eq. 5.31** results and the complex modulus is found to be

$$E^*(i\omega) = \frac{\displaystyle\sum_{k=0}^{m} q_k (i\omega)^k}{\displaystyle\sum_{k=0}^{n} p_k (i\omega)^k} \qquad (5.39)$$

Similarly, considering the case of an oscillatory stress as input with a corresponding complex output of strain, the complex compliance can be derived as

$$D^*(i\omega) = \frac{\displaystyle\sum_{k=0}^{n} p_k (i\omega)^k}{\displaystyle\sum_{k=0}^{m} q_k (i\omega)^k} \qquad (5.40)$$

which can be decomposed as

$$D^*(i\omega) = D'(\omega) + iD''(\omega) \qquad (5.41)$$

where the real part is the storage compliance, **D'(ω)**, and the imaginary part is the loss compliance, **D'(ω)**. As before, the relationship between stress and strain is given by the complex compliance as

$$\epsilon(t) = D^*(i\omega)\sigma(t) \qquad (5.42)$$

from which one sees that the complex compliance is simply the inverse of the complex modulus. To further understand the response of a viscoelastic polymer to oscillatory loading, consider a simple Kelvin element with the associated differential equation,

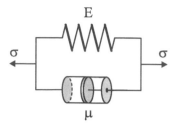

Fig. 5.13 Kelvin solid.

$$\sigma = q_0\varepsilon + q_1\dot{\varepsilon} \qquad (5.43)$$

Application of **Eq. 5.39** can be used to find the complex modulus,

$$E^*(i\omega) = \frac{q_0 + iq_1\omega}{p_0} = E + i\mu\omega = E(1 + i\tau\omega) \qquad (5.44)$$

with the storage and loss moduli,

$$E'(\omega) = E \quad E''(\omega) = \mu\omega \qquad (5.45)$$

Using complex conjugates to invert **Eq. 5.44**, the complex compliance can be found,

$$D^*(i\omega) = \frac{1}{E^*(i\omega)} = \frac{p_0}{q_0 + iq_1\omega} = \frac{p_0(q_0 - iq_1\omega)}{q_0^2 + q_1^2\omega^2} = \frac{E - i\mu\omega}{E^2 + \mu^2\omega^2} \qquad (5.46)$$

with the storage and loss compliances given by,

$$D'(\omega) = \frac{E}{E^2 + \mu^2\omega^2} \quad D''(\omega) = \frac{-\mu\omega}{E^2 + \mu^2\omega^2} \qquad (5.47)$$

These results could also be obtained by solving the differential equation for the Kelvin model using an input condition of, $\varepsilon_0[\cos(\omega t)]$. However, for higher order differential equations, use of **Eq. 5.39** and **5.40** would obviously be advantageous. Note that the above storage and loss compliances are also given in **Table 5.1**, using p_i and q_i coefficients, along with creep and relaxation properties. The reader is urged to use the methods given above to verify the accuracy of the quantities given.

To obtain a physical understanding of polymer response to oscillatory loading and the complex, storage and loss moduli, reconsider input and output stresses and only use the real part of each quantity,

$$\varepsilon(t) = \Re\left(\varepsilon_0 e^{i\omega t}\right) = \varepsilon_0\cos(\omega t) \qquad (5.48)$$

$$\sigma(t) = \Re\left[\varepsilon_0 E^*(i\omega)e^{i\omega t}\right]$$

$$= \Re\left\{\varepsilon_0\left[E'(\omega) + iE''(\omega)\right]\left[\cos(\omega t) + i\,\sin(\omega t)\right]\right\} \qquad (5.49)$$

$$= \varepsilon_0\left[E'(\omega)\cos(\omega t) - E''(\omega)\sin(\omega t)\right]$$

These conditions then represent subjecting a polymer to an oscillatory (cosine) strain input. The stress output is also oscillatory, but is out of phase with the strain input. To visualize, see the input and output results shown in **Fig. 5.14** at a single frequency. The total input and total output are plotted, as well as the in phase and out of phase portions of the stress output.

If stress data is obtained for a real polymer subjected to a cosine strain input, analysis of the resulting plots similar to **Fig. 5.14** will allow the determination of complex modulus, storage modulus and loss modulus. Comparing the amplitudes of the in phase and out of phase outputs to the amplitude of the strain input gives the storage, $E'(\omega)$, and loss, $E''(\omega)$, moduli respectively and the complex moduli, $E^*(i\omega)$, can then be obtained using **Eq. 5.37**. Note however that what is obtained from analysis of **Fig. 5.14** is the value of the moduli at a single frequency; in order to obtain the moduli as a function of frequency a series of such plots must be analyzed. In practice, the DMA testing machines therefore perform a frequency sweep to obtain moduli within a bounded frequency range (limited by the equipment), usually several decades. Chapter 7 will discuss use of tests at different temperatures to extend the range of moduli functions so that a more complete picture of the behavior of the polymer from glassy (high frequency) to rubbery (low frequency) response can be obtained.

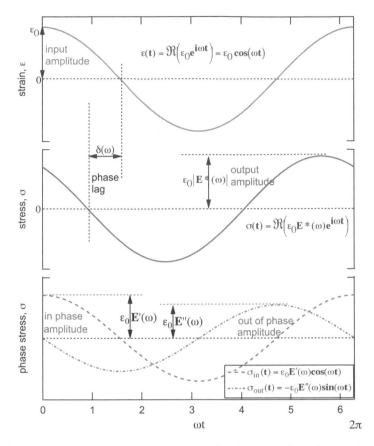

Fig. 5.14 Input and output for a steady state vibration test of a polymer simulated with Kelvin model.

The origin of the phase lag seen in **Fig. 5.14** can also be understood by expressing the complex moduli with a magnitude and phase angle in the complex plane as shown in **Fig. 5.15**.

$$\mathbf{E}^*(i\omega) = E'(\omega) + i\,E''(\omega) = \left|\mathbf{E}^*(i\omega)\right|e^{i\delta(\omega)} \tag{5.50a}$$

where

$$\left|\mathbf{E}^*(i\omega)\right| = \sqrt{\left(E'(\omega)\right)^2 + \left(E''(\omega)\right)^2} \tag{5.50b}$$

and

$$\tan\delta(\omega) = \frac{E''(\omega)}{E'(\omega)} \tag{5.50c}$$

The output stress for a strain input $\varepsilon(t) = \varepsilon_0 \cos\omega t$ can therefore also be written in the form,

$$\sigma(t) = \Re\left\{\varepsilon_0 \left|E^*(i\omega)\right| e^{i(\omega t + \delta(\omega))}\right\}$$

$$= \varepsilon_0 \left|E^*(i\omega)\right| \cos(\omega t + \delta(\omega)) \tag{5.51}$$

where the stress clearly lags the strain input by the material parameter $\delta(\omega)$, which is referred to by one of several common names in the literature as the "loss angle", "loss coefficient", "tan delta" or "damping ratio".

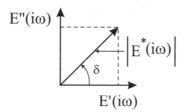

Fig. 5.15 Storage and loss moduli as components of the complex modulus.

As mentioned at the beginning of this section, special dynamic mechanical analysis (DMA) testing systems are commercially available for the rapid evaluation of complex, storage and loss modulus as well as phase angle or damping ratio. Given in **Fig. 5.16(a-c)** are photographs of portions of a typical DMA system showing a polymer specimen, the linear actuator loading mechanism and specimen grips as well as the housing for the electromagnetic coils. Also shown is the monitor of the computer used to control the testing and on which typical damping and storage modulus data are displayed. It is interesting to note that early DMA designs used eccentric cam mechanical loading devices instead of magnetic coils and the harmonic input and output data was often displayed on a dual pen strip-chart recorder as illustrated by the schematic in **Fig. 5.16(d)**. The phase shift was easily visualized by noting the amount the input and output curves were shifted. Data found with the current electro-magnetic digital systems is much more accurate than with earlier mechanical systems but there is not the same easy visualization of the nature of the phase lag as demonstrated analytically in **Fig. 5.14** as the live harmonic data is not typically displayed in the accompanying software.

Another method to visualize the phase lag in older test methods was to feed both input and output into an oscilloscope to obtain a hysteresis loop also shown schematically in **Fig. 5.16(d)**. The amount of energy loss per cycle is the area within the stress-strain loop and is called the dissipation. How the hysteresis loop is obtained is best visualized by plotting the stress versus the strain at corresponding times on the input and output curves as shown **Fig. 5.17**. As the peak input strain, ε, at A begins to decrease, the lagging output stress is still less than the peak output stress. In other words it is the time lag between input strain and output stress that gives rise to the hysteresis loop.

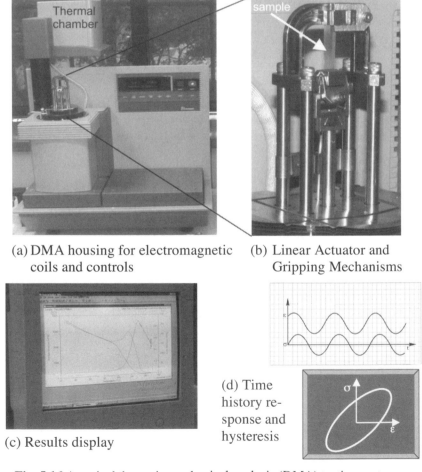

(a) DMA housing for electromagnetic coils and controls

(b) Linear Actuator and Gripping Mechanisms

(c) Results display

(d) Time history response and hysteresis

Fig. 5.16 A typical dynamic mechanical analysis (DMA) testing system.

Determining hysteresis plots manually by plotting strain vs. time input and the stress vs. time output on mutually perpendicular axis and combining respective points in time as shown in **Fig. 5.17** is tedious and is rarely done. However, the concept gives a good physical understanding of how the phase lag in a steady-state vibration test leads to the hysteresis loops routinely obtained with the aid of an oscilloscope. Some use rotating vectors to explain the relation between phase lag and energy loss (e.g., see Flugge, (1974) or Aklonis and McKnight, (1983)).

A plot of the data for a Kelvin model in **Fig. 5.14** at common times will yield a hysteresis loop for the chosen frequency just as illustrated in **Fig. 5.17**. Note that if stress and strain are completely in phase with one another (as is the case for an elastic material), a straight line is obtained as indicated by the dashed diagonal line in **Fig. 5.17**. For a given viscoelastic material, the degree of phase lag and the breadth of the hysteresis loop will depend greatly on the frequency (and temperature) at which the test is performed. For example, at a frequency/temperature where the material behaves in a glassy, elastic manner, phase lag, hysteresis or loss of energy will be small to nonexistent.

Using the DMA (**Fig. 5.16**), steady state viscoelastic response over a wide range of temperatures and frequencies can be found by "sweeping" a range of frequencies at a single temperature or "sweeping" a range of temperatures at a single frequency to generate master curves using the time-temperature superposition principle (TTSP) that will be discussed at length in Chapter 7.

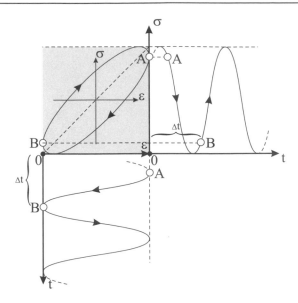

Fig. 5.17 Formation of the hysteresis loop for a polymer as visualized by graphical combination of the stress and strain values parametrically. The dashed line inside the hysteresis loop represents purely elastic response.

As mentioned, the area inside the hysteresis loop represents of the energy lost or dissipated during cyclic deformation. The dissipation can be shown to be proportional to the loss modulus using the basic relationships between work and energy. Recall that the work per unit volume of a stressed material is given by

$$W = \int \sigma d\varepsilon = \int_0^t \sigma \dot{\varepsilon} dt \qquad (5.52)$$

If a material behaves in a perfectly elastic manner, the deformation energy supplied to the material during loading is stored in the stretching of the molecular/atomistic configuration changes and subsequently recovered completely upon unloading: there is no energy dissipated. Therefore for a single complete cycle of oscillatory loading of any material (elastic or not), the net energy stored is zero, as the material is loaded and unloaded symmetrically. The amount of energy dissipated in a single cycle of oscillatory loading can thus be calculated by integrating **Eq. 5.52** over a complete cycle:

$$D = \oint \sigma d\varepsilon = \int_0^{2\pi/\omega} \sigma \dot{\varepsilon} dt \qquad (5.53)$$

For a perfectly elastic material, Hooke's law is obeyed, $\sigma = E\,\varepsilon$. This implies that the width of the hysteresis loop is zero (the dashed line in **Fig. 5.17**) and evaluation of the integral in **Eq. 5.53** results in identically zero. For a viscoelastic material, we can write the stress as a function of the strain via the complex modulus (**Eq. 5.38**) and then rearrange in terms of the storage and loss moduli

$$\sigma(t) = E^*(\omega)\varepsilon(t)$$

$$= E'(\omega)\varepsilon(t) + \frac{E''(\omega)}{\omega}\,i\omega t\varepsilon(t) \qquad (5.54)$$

$$= E'(\omega)\varepsilon(t) + \frac{E''(\omega)}{\omega}\,\dot{\varepsilon}(t)$$

To calculate the energy dissipated over a cycle, **Eq. 5.54** can be substituted in **Eq. 5.53**. Using a sinusoidal strain ($\varepsilon(t) = \varepsilon_0 \sin \omega t$), it can be shown:

$$D = \int_0^{2\pi/\omega} \sigma\dot{\varepsilon}\,dt$$

$$= \int_0^{2\pi/\omega} \left(E'(\omega)\varepsilon(t) + \frac{E''(\omega)}{\omega}\,\dot{\varepsilon}(t) \right)\dot{\varepsilon}\,dt \qquad (5.55)$$

$$= \varepsilon_0^2 \omega\pi E''$$

So we see that the dissipated energy is indeed proportional to the loss modulus. In the glassy or rubbery regions where the loss modulus is infinitesimal, the dissipation is therefore minimal.

5.5.1. Examples of Storage and Loss Moduli and Damping Ratios

If the storage: and loss moduli and damping ratios are found for a Maxwell model, the result will be as shown in **Fig. 5.18(a)** and **5.18(b)**. This result can be found algebraically and then plotted using a spread-sheet or graphics program. The behavior of real polymers is sometimes similar to the results for a Maxwell fluid as is the case for polycarbonate as given **Fig. 5.20**. Notice the characteristic "S" shape of the storage modulus and characteristic "bell" shape of the loss in the experimental data. Note that the Maxwell Fluid shows a loss tangent (damping ratio) unrealistic for a solid polymer as there is no peak (the loss tangent grows without bound at low frequencies). Including a free spring ensures a bell-shaped loss tangent similar to the experimental data. Results for a simple 2 element Wiechert model (a solid containing two Maxwell elements connected in parallel with a spring) is shown in **Fig. 5.21** where the loss tangent peak can be clearly seen. With respect to the transition region, note that the decay in the storage modulus is relatively rapid for a single Maxwell element, limited to about a decade in frequency around the inverse of the single relaxation time, τ. By moving to a model with two Maxwell elements (**Fig. 5.21**), the transition region of the storage and loss moduli are expanded, becoming more like a real polymer, extending around the inverse of the two relaxation times. Not all polymers have such simple shaped moduli functions and the reader is referred to excellent texts such as (Ferry, (1980) and Tobolsky, (1962)) for further examples.

In **Fig. 5.18(b)** results are given versus the inverse of frequency, as this would correspond roughly to the time scale. The modulus, **E(t)**, is also plotted in this figure for comparison, where the values for time along the bottom axis for this curve are identical to the inverse frequency values given. The glassy modulus, $\mathbf{E_0}$, is located at short times for **E(t)** and at long frequencies for $\mathbf{E_1(\omega)}$. Note that the time dependent modulus is quite similar in form to the storage modulus plotted versus inverse frequency. Dynamic results found in the literature are sometimes plotted versus frequency and sometimes versus inverse frequency.

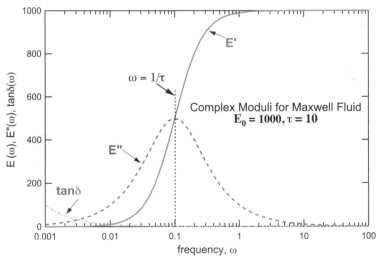

Fig. 5.18 Variation of storage and loss moduli for a Maxwell fluid with frequency.

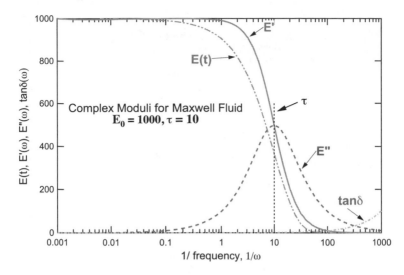

Fig. 5.19 Variation of storage and loss moduli for a Maxwell fluid with inverse frequency; Variation of modulus, E(t), where the time scale on the horizontal axis is numerically identical to the inverse frequency scale shown.

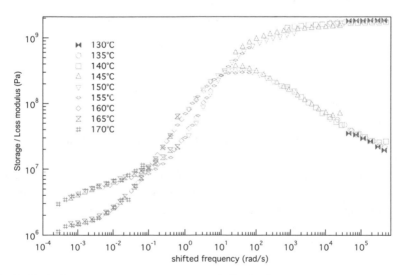

Fig. 5.20 Variation of storage and loss moduli with frequency for polycarbonate.

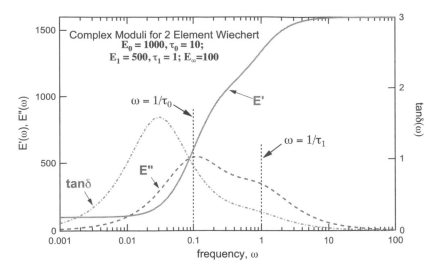

Fig. 5.21 Variation of storage and loss moduli for a 2 element Wiechert model (a solid) with frequency.

In addition to frequency dependent mechanical properties, as mentioned earlier, a DMTA can also be used at constant frequency to determine temperature dependence of properties. In this manner, one can probe the glass-transition temperature (T_g), assess changes in molecular structure due to additional curing upon heating, the effect of crystallinity on properties etc.. The variation of storage modulus and **tan δ** with temperature for a typical polymer is shown in **Fig. 5.22**. The glass-transition temperature is indicated as the temperature where the peak in **tan δ** occurs. Notice the similarity of property changes in temperature to changes with frequency in **Fig. 5.20**. Shown in **Fig. 5.23** is a depiction of the variation of **tan δ** of a polymer over a wide range in temperature with not only the α transition (T_g) indicated but also the β, γ and δ transitions. (The δ transition should not be confused with the larger α transition or **tan δ**. Perhaps to avoid confusion a different terminology should be adopted but the tradition for the names of the various transitions are well established and a change would likely lead to even more confusion.)

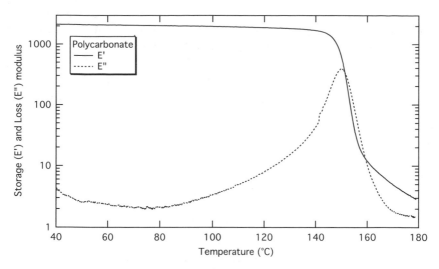

Fig. 5.22 Variation of storage modulus and Tan δ with temperature as determined with a DMTA for polycarbonate. The **T$_g$** is located at the peak of the **Tan δ** curve.

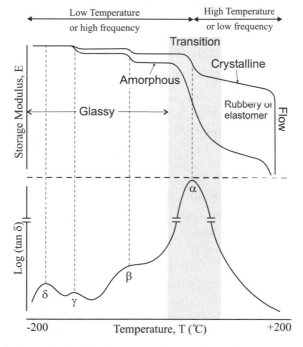

Fig. 5.23 Variation of **Tan δ** with temperature for polystyrene as determined with a DMTA. Here α, β, γ and δ transitions are shown. (Data from Arridge, (1985) by permission of Oxford University Press.)

While all polymers have characteristics similar to the above examples, there is considerable variation among different classes of polymers. To observe this diversity, the reader is referred to the extensive study of the steady state response of many polymer types given by Ferry (1980). In particular, he gives an excellent description of the results for eight categories including dilute polymer solutions, low and high molecular weight amorphous polymers and lightly and highly cross-linked systems as well as highly crystalline polymers.

5.5.2. Molecular Mechanisms Associated with Dynamic Properties

The behavior given in the above examples for polymer response variation with time and temperature under steady-state dynamic loading is directly related to the deformation mechanisms associated with the long chain nature of polymer molecules. As illustrated in **Fig. 5.23**, low frequency response is similar to high temperature (rubbery) response, and high frequency response is similar to low temperature (glassy) response. The basic mechanical responses therefore relate across the time and temperature scales, as do the underlying molecular mechanisms. A brief description of these mechanisms follows. (For more detailed information the reader is referred to (Lazan, (1968)) and (Menard, (1999)).

As described in Chapter 4 the long molecular chains form a tangled mass that might be analogous to a similarly tangled mass of long earthworms. This illustration is especially appropriate due to the constant motion of individual atoms and segments of chains even at a very low temperature. It is especially important to note that the entanglement points between individual chains in thermoplastic polymers act very much like the covalent cross-linked sites in thermosets at low temperatures. Therefore, the behavior of thermoplastics and thermosets are often very similar for temperatures well below the glass-transition temperature. As a result, for low temperatures near the delta and gamma transitions (see **Figs. 5.23**), local motions or bending and stretching of primary valence bonds are the primary mechanisms that contribute to macroscopic deformations in all classes of polymers. At somewhat higher temperatures near the beta transition, but still below the glass transition temperature T_g, side group motions occur and coupled with the bending and stretching of primary bonds leads to larger deformations but the polymer is still glassy and quite brittle. Damping in this regime is small and hysteresis negligible. The delta, gamma and beta transitions are often identified as secondary transitions and depend on the character of the monomeric structure of the polymer. Near the glass transition temperature damping and stiffness properties are governed by the chain segments between entanglement sites in thermoplastics and chain segments between cross-links in a thermoset. These chain segments are much smaller than the macromolecule but are large compared to the chain length of the monomer group. The coiling and uncoiling of these segments are quite slow just below the glass transition temperature and are quite rapid as the rubbery range is approached. In this transition range, damping is quite pronounced and hysteresis in stress-strain is prominent. Crystallinity tends to reduce the intensity of the glass-transition as compared to an amorphous polymer as illustrated in **Fig. 5.23**. In the rubbery region ($T_g < T < T_m$), polymer-damping properties are insensitive to temperature and damping is again negligible. In fact for macromolecules having a three-dimensional cross-linked structure the stiffness may actually increase slightly in the rubbery range. Near the melt temperature of thermoplastic polymers entire chains begin to slip past one another and the polymer properties are similar to those of other highly viscous liquids. On the other hand, thermosets are prevented from such gross chain motions by the cross-links between chains. At very high temperature well above the glass-transition temperature, thermosets tend to char and properties will substantially decrease due to molecular degradation. Cross-linked sites may be broken and then reformed to give the appearance of flow. (See Tolboslky (1962) for a discussion of these mechanisms). Examples of mechanical properties of thermosets that demonstrate these characteristics will be given in Chapter 7.

The recognition of the roles of the various micromechanisms discussed above are important for the development of damping properties that are needed for a specific engineering application. Such design typically begins with the selection of an appropriate monomeric species to control the glass transition temperature T_g, crystalline melt temperature T_m, and secondary transition properties. By controlling the polymerization process the same polymer can be produced in different forms and with different properties. Side group configuration and their influence on crystallinity, the degree of chain branching, crosslinking, etc. are well understood and together with blending, plasticization and the addition of fillers allow a high degree of flexibility in producing a polymer "tailored" for specific engineering requirements.

5.5.3. Other Instruments to Determine Dynamic Properties

There are many types of tests from which steady state (or dynamic) properties can be obtained including the vibrating reed, steady state torsion among others. A relatively simple and easy to build free or unforced vibration test of a flat strip in torsion (torsional pendulum) shown in **Fig. 5.24** is sometimes used to obtain storage and loss moduli and damping rations. The damping factor (or phase angle) can be found from the logarithmic decrement which is related the decrease in amplitude oscillations and the shear modulus can be determined from the period of oscillation. For a more extensive discussion of this test method see (Nielsen, (1965)) or (Nielsen and Landel, (1995)).

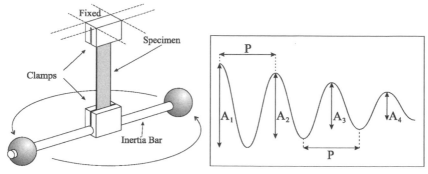

Fig. 5.24 Torsional pendulum for free vibration test (A = amplitude, P = period, logarithmic decrement = ln $A_1/A_2 = A_2/A_3 = A_3/A_4$, etc.)

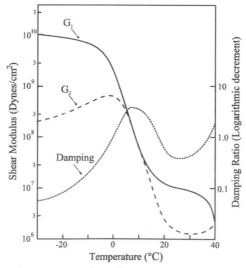

Fig. 5.25 Shear data for a styrene/butadiene copolymer developed using a torsional pendulum. (Data from Nielsen, (1965)).

5.6. Review Questions

5.1   Describe realistic creep and relaxation tests. Illustrate your answer with sketches of the input and output curves.

5.2   Assuming the input stress (strain) in a creep (relaxation) test is a ramp followed by a constant stress (strain), describe under what conditions the test will approximate an ideal creep (relaxation) test.

5.3   Discuss a proper testing procedure to insure that a constant stress and not a constant load is applied in a creep test.

5.4   Discuss a proper testing procedure to insure that a constant strain is applied in a relaxation test.

5.5.   Give sketches for generalized Maxwell and Kelvin models. Label all elements.

5.6.   Give an equation that would represent the relaxation response for a generalized Maxwell Fluid.

5.7.   Give an equation that would represent the creep response for a generalized Kelvin Solid.

5.8.   Describe how one would find the storage modulus, loss modulus and tan δ from experimental data.

5.9.   Explain, describe and/or derive the rationale behind Alfrey's correspondence principle.

5.10.   Describe the molecular mechanisms associated with the regions of response in a steady state oscillation test.

5.11   Name four transition temperatures that can be found using a DMTA. In which region is aging likely to occur?

5.7. Problems

5.1.   Develop the differential equation for a four-parameter fluid.

5.2.   Obtain the solution for creep of a three-parameter solid by solving the differential equation.

5.3.   Obtain the solution for relaxation of a three-parameter solid by solving the differential equation.

5.4.   Obtain the solution for creep of a four-parameter fluid by solving the differential equation.

5.5.   Show that the effect of the initial ramp loading in a realistic relaxation test as given in **Fig. 5.5** d of a material that can be represented by a Maxwell fluid is negligible if the time of the ramp load t_0 is small compared to the relaxation time, τ.

5.6.   Develop the differential equation for two Maxwell elements in parallel.

5.7.   Develop the differential equation for three Maxwell elements in parallel.

5.8.   Develop the differential equation for two Kelvin elements in series.

5.9.   Develop the differential equation for three Kelvin elements in series.

5.10. Obtain the solution for relaxation of two Maxwell elements in parallel by solving the differential equation.

5.11. Find all the parameters necessary to fit the behavior of the 160° C curve given in **Fig 5.12** with a three-parameter solid. Give results on a graph comparing the analytical curve fit to the given data. Discuss the quality of fit using this simple model.

5.12. Find all the parameters necessary to fit the behavior of the 155° C curve given in **Fig 5.12** with a four-parameter fluid. Give results on a graph comparing the analytical curve fit to the given data. Discuss the quality of fit using this simple model.

5.13. Under steady state vibration test conditions:

a. Prove that the phase shift is zero for a Hookean elastic material.

b. Prove that the phase shift for a Newtonian fluid is $\pi/2$.

5.14. Develop expressions for $E^*(i\omega)$, $E'(\omega)$, $E''(\omega)$ for a Maxwell model and plot results as a function of $1/\omega$.

6. Hereditary Integral Representations of Stress and Strain

As discussed previously, the relation between stress and strain for linear viscoelastic materials involves time and higher derivatives of both stress and strain. While the differential equation method can be quite general, a hereditary integral method has proved to be appealing in many situations. This hereditary integral equation approach is attributed to Boltzman and was only one of his many accomplishments. In the late nineteenth century, when the method was first introduced, considerable controversy arose over the procedure. Now, it is the method of choice for the mathematical expression of viscoelastic constitutive (stress-strain) equations. For an excellent discussion of these efforts of Boltzman, see Markovitz (1977).

6.1. Boltzman Superposition Principle

In previous chapters, relaxation and creep testing was introduced and the relaxation modulus and creep compliance were defined as the stress output for a constant strain input (relaxation) and the strain output for a constant stress input (creep). A question naturally arises as how the output could be found if a variable input of either strain or stress were to occur. One could, of course, attempt to solve a general differential equation if the variation is specified but such an approach could, in some cases, be quite tedious.

The Boltzman superposition principle (or integral) is applicable to stress analysis problems in two and three-dimensions where the stress or strain input varies with time, but first the approach will be introduced in this section only for one-dimensional or a uniaxial representation of the stress-strain (constitutive) relation. The superposition integral is also sometimes referred to as Duhamel's integral (see W.T. Thompson, *Laplace Transforms*, Prentice Hall, 1960).

Consider a variable stress input as shown in **Fig. 6.1** with the thought of seeking a method to find the strain output. First assume that the variable input can be represented by a series of step inputs each of which begins at different time as shown. Thus

$$\sigma(t) = \sigma_0 H(t) + (\sigma_1 - \sigma_0)H(t - t_1) + \cdots + (\sigma_n - \sigma_{n-1})H(t - t_n) \qquad (6.1)$$

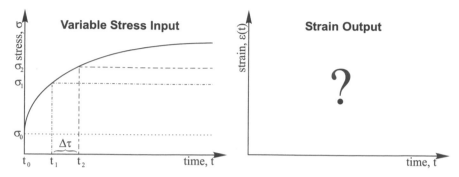

Fig. 6.1 Variable stress input.

Obviously, if sufficiently small steps are selected over corresponding small time intervals, the curve can be fitted to any degree of accuracy desired.

Recall from Chapter 3 that the creep response can be represented by a creep compliance due to a step input at time zero as,

$$\varepsilon(t) = \sigma_0 D(t) \quad \text{for} \quad \sigma(t) = \sigma_0 H(t) \qquad (6.2)$$

Similarly, creep response for any single step input shifted from the origin can be written as,

$$\varepsilon(t) = \sigma_1 D(t - t_1) \quad \text{for} \quad \sigma(t) = \sigma_1 H(t - t_1) \qquad (6.3)$$

Because it is assumed that the material is linear viscoelastic, the strain output for a general varying stress input can be represented as a sum of the output for each individual step in the following manner (see Appendix A for a discussion of the unit step function),

$$\varepsilon(t) = \sigma_0 D(t)H(t) + (\sigma_1 - \sigma_0)D(t - t_1)H(t - t_1) +$$
$$(\sigma_2 - \sigma_1)D(t - t_2)H(t - t_2) + \qquad (6.4)$$
$$(\sigma_3 - \sigma_2)D(t - t_3)H(t - t_3) +$$
$$\cdots + (\sigma_n - \sigma_{n-1})D(t - t_n)H(t - t_n)$$

or in series form,

$$\varepsilon(t) = \sigma_0 D(t)H(t) + \sum (\sigma_n - \sigma_{n-1})D(t - t_n)H(t - t_n) \qquad (6.5)$$

Upon multiplying and dividing by the time increment between each step, $\Delta\tau$, and taking the limit as $\mathbf{n}$ approaches infinity and $\Delta\tau$ approaches zero obtain,

$$\varepsilon(t) = \sigma_0 D(t)H(t) + \lim_{\substack{n\to\infty \\ \Delta\tau\to 0}} \sum \frac{(\sigma_n - \sigma_{n-1})}{\Delta\tau} D(t - t_n)H(t - t_n)\Delta\tau \qquad (6.6)$$

or

$$\varepsilon(t) = \sigma_0 D(t)H(t) + \int_{0^+}^{t} D(t - \tau)\frac{d\sigma(\tau)}{d\tau}d\tau \qquad (6.7)$$

The integral equation is most often written as

$$\varepsilon(t) = \int_{0}^{t} D(t - \tau)\frac{d\sigma(\tau)}{d\tau}d\tau \qquad (6.8)$$

where it is understood that the lower limit is from $t = 0^-$ or includes the jump discontinuity in stress at the origin and the stress is understood to be expressed as $\sigma(t) = \sigma(t)H(t)$. That is,

$$\varepsilon(t) = \int_{0^-}^{t} D(t - \tau)\frac{d[\sigma(\tau)H(\tau)]}{d\tau}d\tau \qquad (6.9)$$

Differentiation of the product of the stress and the Heavyside function gives,

$$\begin{aligned} \frac{d[\sigma(\tau)H(\tau)]}{d\tau} &= \frac{d[\sigma(\tau)]}{d\tau}H(\tau) + \sigma(\tau)\frac{d[H(\tau)]}{d\tau} \\ &= \frac{d[\sigma(\tau)]}{d\tau}H(\tau) + \sigma(\tau)\delta(\tau) \end{aligned} \qquad (6.10)$$

or

$$\varepsilon(t) = \int_{0^+}^{t} D(t - \tau)\left\{\frac{d[\sigma(\tau)]}{d\tau}H(\tau) + \sigma(\tau)\delta(\tau)\right\}d\tau \qquad (6.11)$$

Due to the sifting property of the Dirac Delta function, $\delta(t)$, (see Appendix), **Eq. 6.11** reduces to **Eq. 6.7**.

Using this approach, the output for a more complicated variable stress input as given in **Fig. 6.2** (with $\sigma(t)$ specified) can be found by integration. Note that one must take care in the expression of $\sigma(t)$ and its differentiation in **Eq. 6.7** so that the jump discontinuities at $t = t_1$ and $t = t_2$ are explicitly included as is the jump discontinuity at $t = 0$. Examples with simple $f_i(t)$ functions are provided in homework problems 6.4-6.6.

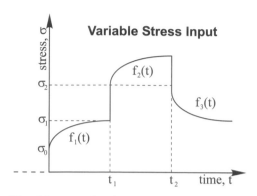

Fig. 6.2 Example of a variable stress input.

An analogous derivation of the stress output for a variable strain input yields the equation,

$$\sigma(t) = \int_0^t E(t - \tau) \frac{d\varepsilon(\tau)}{d\tau} d\tau \qquad (6.12)$$

Because all events over the history of a viscoelastic material contribute to the current state of stress and strain, the lower limit of the hereditary integral is most often taken to be $-\infty$ and **Eqs. 6.8** and **6.12** therefore become,

$$\varepsilon(t) = \int_{-\infty}^t D(t - \tau) \frac{d\sigma(\tau)}{d\tau} d\tau \qquad \sigma(t) = \int_{-\infty}^t E(t - \tau) \frac{d\varepsilon(\tau)}{d\tau} d\tau \qquad (6.13)$$

Some might suggest that no need exists for a lower limit of negative infinity as the instant of first loading is known for most structures. However, in the case of polymer structures, it is especially necessary to carefully consider all previous events including polymerization and production processes. Further, the previous history may include temperature or other environmental changes which could lead to residual stresses that would create changes to the molecular structure and hence need to be included in any realistic stress analysis. Indeed, most structural polymers used in industry are quenched which not only gives rise to residual stresses, but also creates excess free volume at the molecular level that significantly influences the viscoelastic properties of a material. Two such important effects that occur as a result of excess free volume are physical and chemical aging. Such concepts will be discussed at greater length in a later chapter.

Several examples are in order to demonstrate the utility of the Boltzman superposition principle.

Example 1: Assume it is desirable to find the strain output in a creep and creep recovery test shown in **Fig. 6.3(a)**. First note that the stress can be easily represented by two step inputs as illustrated schematically in **Fig. 6.3(b)** and given by

$$\sigma(t) = \sigma_0 H(t) - \sigma_0 H(t - t_1) \tag{6.14}$$

As a result, the response can be written as,

$$\varepsilon(t) = \sigma_0 D(t) H(t) - \sigma_0 D(t - t_1) H(t - t_1) \tag{6.15}$$

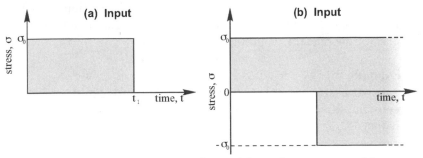

Fig. 6.3 Creep-recovery stress input **(a)** can be represented by superposition of two step inputs **(b)**.

Applying this example to a material which is well represented by a Kelvin solid, where,

$$D(t) = \frac{1}{E}(1 - e^{-t/\tau}), \tag{6.16}$$

and substitution in **Eq. 6.15** gives,

$$\varepsilon(t) = \frac{\sigma_0}{E}(1 - e^{-t/\tau})H(t) - \frac{\sigma_0}{E}(1 - e^{-(t-t_1)/\tau})H(t - t_1) \tag{6.17}$$

For $t < t_1$:
$$\varepsilon(t) = \frac{\sigma_0}{E}(1 - e^{-t/\tau}) \tag{6.18}$$

and for $t > t_1$:
$$\varepsilon(t) = \frac{\sigma_0}{E}e^{-t/\tau}\left(e^{t_1/\tau} - 1\right) \tag{6.19}$$

The resulting output is represented graphically as,

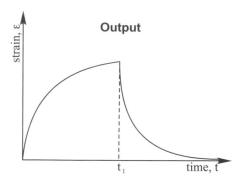

Fig. 6.4 Output for the two step stress input given in **Fig. 6.3**.

which agrees with physical intuition for a Kelvin solid. To accomplish the same result by solving the differential equation for a Kelvin solid would be somewhat more cumbersome. This is left as an exercise for the reader (see problem 6.1).

Example 2:
Another useful example is to consider a Maxwell Fluid subjected to a constant strain rate input as in **Fig. 6.5(a)**, and determine the stress output.

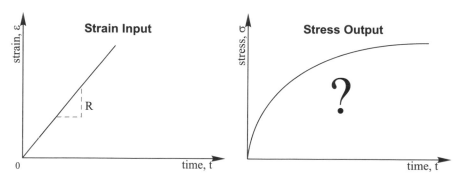

Fig. 6.5(a) Constant strain-rate input for a Maxwell fluid.

Because the strain is given as, $\varepsilon(t) = Rt$, the strain rate is constant, $\dfrac{d\varepsilon}{dt} = R$, and **Eq. 6.12** gives,

$$\sigma(t) = R \int_0^t E(t - \tau)d\tau \qquad (6.20)$$

Differentiating and rearranging will give,

$$E(t) = \frac{1}{R} \frac{d\sigma}{dt} \qquad (6.21)$$

From this result it is apparent that the relaxation modulus can be found from a constant strain-rate test by dividing the slope of the stress output by the strain-rate. Similarly, the creep compliance can be found from an constant stress-rate test by dividing the strain output by the stress-rate,

$$D(t) = \frac{1}{R} \frac{d\varepsilon}{dt} \qquad (6.22)$$

To obtain the output for a Maxwell fluid in a constant strain rate test, the relaxation modulus, $E(t) = Ee^{-t/\tau}$ must be inserted as $E(t-\psi) = Ee^{-(t-\psi)/\tau}E$ and **Eq. 6.20** becomes after changing the dummy variable to ψ,

$$\sigma(t) = RE \int_0^t e^{-(t-\psi)/\tau} d\psi \qquad (6.23)$$

Upon evaluation, **Eq. 6.23** reduces to,

$$\sigma(t) = \tau RE(1 - e^{-t/\tau}) \qquad (6.24)$$

and is the same result obtained by solving the differential equation for a Maxwell fluid and given in Chapter 3 and is plotted in **Fig. 6.5(b)**. Thus, the same conclusions are reached concerning linearity for a constant strain rate test as discussed in the previous chapter.

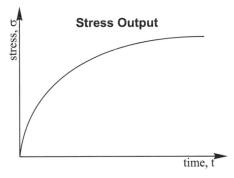

Fig. 6.5(b) Constant strain-rate output for a Maxwell fluid.

6.2. Linearity

It is important to note that the condition of linear viscoelasticity requires both superposition and proportionality. It is necessary for the responses to stresses applied at any time to be superposable (as described in **Fig. 6.1** (and **Eq. 6.4**)) and for responses to different stress levels to be proportional as was illustrated using isochronous stress-strain curves from creep or relaxation tests discussed in Chapter 3 (e.g. **Fig. 3.19**) for an arbitrary constant time $t = t_1$. These are often referred to as separate conditions of linearity with *superposition* referring to the former and *proportionality* referring to the latter. However, the constitutive equations resulting from Boltzman's superposition principle (**Eqs. 6.13**) are quite general and satisfy both conditions for linearity, as can be easily proven. For a more detailed explanation of the mathematical nature of the integral representation of viscoelastic constitutive equations see (Christensen, 1982). Also, the need for time-wise superposition is clearly indicated in Chapter 10 in the development of the Schapery single integral representation for non-linear materials.

6.3. Spectral Representation of Viscoelastic Materials

In the solution of practical boundary value problems it is necessary to have knowledge of the actual creep or relaxation properties of the material. Sometimes experimental data in discrete form can be used in numerical solutions but most often measured values of $E(t)$ or $D(t)$ need to be represented mathematically. The most frequent mathematical approach to represent data is with exponential (Prony) series. The use of exponential series was well understood by early polymer scientist and polymer physicists who considered the need to mathematically represent data. However, as their focus was to develop understanding between macroscopic properties and molecular structure, they sought other general approaches that could be applied in a relatively simple fashion. While the resulting spectral approach may not appear simple, it has been widely used in polymer literature.

To introduce the spectral approach, consider the relaxation modulus for a generalized Maxwell model,

$$E(t) = \sum_{i=1}^{n} E_i e^{-\lambda_i t} \qquad (6.25)$$

where λ_i is the reciprocal of the relaxation time, $\dfrac{1}{\tau_i}$. In **Eq. (6.25)**, we can visualize discrete values of relaxation times, τ_i, being superposed on the time scale. If the number of elements in the generalized Maxwell model become infinite in the limit, i.e.,

$$E(t) = \lim_{n \to \infty} \sum_{i=1}^{n} E_i e^{-\lambda_i t} \tag{6.26}$$

then each point on the time scale is also represented by a relaxation time. Therefore, multiplying and dividing the argument of **Eq. 6.26** by an increment of τ_i or $\Delta\tau_i$ will give,

$$E(t) = \lim_{\substack{n \to \infty \\ \Delta\tau_i \to 0}} \sum_{i=1}^{n} \frac{E_i}{\Delta\tau_i} e^{-\lambda_i t} \Delta\tau_i \tag{6.27}$$

The quantity $E_i/\Delta\tau_i$ is similar to a Dirac delta function or singularity function (see appendix for a discussion) and is defined as $H(\tau_i)$. Taking the limit, **Eq. 6.27** becomes,

$$E(t) = \int_{0}^{\infty} H(\tau)e^{-\lambda t} d\tau \tag{6.28}$$

The quantity $H(\tau)$ is a continuous function defined as the *relaxation spectrum* and is often used in polymer literature. (Note this term should not be confused with the Heavyside step function used earlier to represent a step input.)

If for example, the relaxation spectrum is assumed to be,

$$H(\tau) = E_0 \delta(\tau - \tau_0) \tag{6.29}$$

the relaxation modulus will become upon evaluating **Eq. 6.28**,

$$E(t) = E_0 e^{-t/\tau_0} \tag{6.30}$$

Eq. 6.30 is, of course, the equation for a single Maxwell model and would only provide a very simple approximation of material behavior. The relaxation spectrum for a generalized Maxwell model for which

$$E(t) = \sum_{i} E_i e^{-t/\tau_i} \tag{6.31}$$

is simply

$$H(\tau) = \sum_i E_i \delta(\tau - \tau_i) \tag{6.32}$$

and for a sufficient number of elements can represent a real polymer. If $H(\tau)$ is thought of as many delta functions continuously distributed along the time scale there are essentially an infinity of relaxation times and hence the integration over relaxation time in **Eq. 6.28**.

Comparison with a Fourier or Laplace transform (see appendix) suggest that $H(\tau)$ can be found using the inversion integral,

$$H(\tau) = \frac{1}{2\pi i} \int_{c-i\infty}^{c+i\infty} E(t)e^{\lambda t} dt \tag{6.33}$$

where **t** is complex. The relaxation spectrum $H(\tau)$ is also referred to as a function for the distribution of relaxation times. Note that the units of $H(\tau)$ are psi/sec and can be thought of as a density function of the relaxation modulus over time.

A distribution of retardation times based on a generalized Kelvin model leads to a retardation spectrum, $L(\tau)$, defined by,

$$D(t) = \int_0^\infty L(\tau)\left(1 - e^{-\lambda t}\right) d\tau \tag{6.34}$$

or

$$L(\tau) = \frac{1}{2\pi i} \int_{c-i\infty}^{c+i\infty} D(t)e^{\lambda t} dt \tag{6.35}$$

The relaxation and creep spectra are widely used in the polymer literature where molecular mechanisms are related to macroscopic properties.

Most often the relaxation or retardation times involved in the viscoelastic spectral representations shown in **Eqs. 6.28** and **6.34** are spread over many decades of time and for this reason the equations are often written in terms of a logarithmic time scale such that,

$$E(\ln t) = \int_{-\infty}^\infty H(\ln \tau)e^{-\lambda t} d(\ln \tau) \tag{6.36}$$

and

$$D(\ln t) = \int_{-\infty}^\infty L(\ln \tau)\left(1 - e^{-\lambda t}\right) d(\ln \tau) \tag{6.37}$$

Experimental data is most often represented using base 10 logarithms instead of natural logarithms.

An example of spectra for a given polymer is shown at the end of this chapter. Calculation of spectra is revisited in Chapter 7 along with experimental data. For a more comprehensive discussion of viscoelastic spectra, see for example Christensen (1982), Tschoegl (1989), Tolbolsky (1960), and Ferry (1980).

6.4. Interrelations Among Various Viscoelastic Properties

Relationship between E(t) and D(t): The Laplace transform of **Eqs. 6.8** and **6.12** are,

$$\bar{\varepsilon}(s) = s\bar{D}(s)\bar{\sigma}(s) \tag{6.38}$$

and

$$\bar{\sigma}(s) = s\bar{E}(s)\bar{\varepsilon}(s) \tag{6.39}$$

Substituting $\bar{\varepsilon}(s)$ from **(6.38)** into **(6.39)** gives,

$$\bar{E}(s)\bar{D}(s) = \frac{1}{s^2} \tag{6.40}$$

which upon using the convolution theorem yields,

$$\int_0^t E(t)D(t-\tau)d\tau = t \tag{6.41}$$

This result clearly shows that (unlike an elastic material) the relaxation modulus and the creep compliance are not reciprocals. That is,

$$D(t) \neq \frac{1}{E(t)} \tag{6.42}$$

In cases where the rate of change of strain or stress is very small, the creep compliance and relaxation modulus may be approximately the inverse of each other. Consideration of simple Maxwell and Kelvin models confirm the condition given by **Eq. 6.42** as in Homework problem 6.7.

Relationship between $\bar{E}^*(s)$ and $\bar{E}(s)$ and between $\bar{D}^*(s)$ and $\bar{D}(s)$:

In Chapter 5, using Alfrey's correspondence principal for a generalized mechanical model it was found that the stress and strain could be related in Laplace transform space as

$$\bar{\sigma}(s) = \bar{E}^*(s)\bar{\varepsilon}(s)$$
$$\bar{\varepsilon}(s) = \bar{D}^*(s)\bar{\sigma}(s)$$
(6.43)

where $\bar{E}^*(s)$ and $\bar{D}^*(s)$ are found from the coefficients of the differential equations describing the system. Comparing **Eq. 6.43** with **Eqs. 6.38-39**, it is clear that the correspondence principle holds generally for a viscoelastic material, not just one represented by a mechanical model. It is also seen that the transformed modulus, $\bar{E}^*(s)$ (compliance, $\bar{D}^*(s)$), is obtained from the Laplace transform of the modulus (compliance), multiplied simply by the transform variable **s**:

$$\bar{E}^*(s) = s\bar{E}(s)$$
$$\bar{D}^*(s) = s\bar{D}(s)$$
(6.44)

Note also that while reciprocity of the modulus/compliance is not valid in the time domain (**Eq. 6.42**), it is valid in the transform domain

$$\bar{D}^*(s) = \frac{1}{\bar{E}^*(s)}$$
(6.45)

Relationship between $E^*(i\omega)$ and E(t) and between $D^*(i\omega)$ and D(t):
Expressions analogous to **Eq. 6.43** were developed in Chapter 5 using a strain (or stress) input of the form $\varepsilon^* = \varepsilon(t) = \varepsilon_0 \varepsilon^{i\omega t}$ (or $\sigma^* = \sigma(t) = \sigma_0 e^{i\omega t}$)

$$\sigma^* = E^*(i\omega)\varepsilon^*$$
$$\varepsilon^* - D^*(i\omega)\sigma^*$$
(6.46)

where then $E^*(i\omega)$ and $D^*(i\omega)$ are termed the complex moduli of the material.

To elucidate the relationship between the time dependent modulus and the complex modulus, substitute $\varepsilon(t) = \varepsilon_0 e^{iwt}$ into the hereditary integral **Eq. 6.13**, to obtain

$$\sigma(t) = i\omega\varepsilon_0 \int_{-\infty}^{t} E(t-\tau)e^{i\omega\tau}d\tau$$
(6.47)

Changing variables by letting, $\mathbf{u = t - \tau}$ or $\mathbf{\tau = t - u}$ and noting that if $\mathbf{\tau = -\infty}$, $\mathbf{u = \infty}$, and if $\mathbf{\tau = t}$, $\mathbf{u = 0}$, we obtain

$$\sigma(t) = \left(\varepsilon_0 e^{i\omega t}\right)\left(i\omega \int_0^\infty E(u)e^{-i\omega u}du\right) \tag{6.48}$$

Examining the terms in brackets on the right hand side, we recognize the applied strain $\boldsymbol{\varepsilon}^*$ and the half-sided Fourier transform of the relaxation modulus. Comparing to **Eq. 6.46** and replacing $\mathbf{u}$ by $\mathbf{t}$ the complex modulus is defined as

$$E^*(i\omega) = i\omega \int_0^\infty E(t)e^{-i\omega t}dt \tag{6.49}$$

Similarly, the complex creep compliance can be found to be

$$D^*(i\omega) = i\omega \int_0^\infty D(t)e^{-i\omega t}dt \tag{6.50}$$

Thus, if the time dependent creep or relaxation properties of a material are known, the complex moduli and compliances can be calculated simply via Fourier transforms (**Eqs. 6.49-6.50**). Comparison back to the Laplace transforms (**Eq. 6.44**), we see that s or $i\omega$ times the Laplace or Fourier transform, respectively, of the time dependent properties provide the transformed properties which can be used in the correspondence principle forms **Eqs. 6.43** and **6.46**.

If the time dependent modulus of the material, $\mathbf{E(t)}$, is expressed in a Prony series (generalized Maxwell model) representation (**Eq. 6.31** or **Eq. 5.22**), then the simple algebraic form of the function leads to explicit expressions for the storage and loss moduli from solution to **Eq. 6.49**.

$$\overline{E}^*(\omega) = E'(\omega) + iE''(\omega)$$

$$= \underbrace{\left[E_\infty + \sum_j \frac{E_j\omega^2}{\dfrac{1}{\tau_j^2} + \omega^2}\right]}_{E'} + i\underbrace{\left[\sum_j \frac{E_j\omega/\tau_j}{\dfrac{1}{\tau_j^2} + \omega^2}\right]}_{E''} \tag{6.51}$$

Naturally, corresponding forms can be found for the complex compliance function for a generalized Kelvin model. Verification of these expressions is left as an exercise for the reader.

Often integral **Eqs. 6.49** and **6.50** are given in a different form. For example, separating the relaxation modulus into two components,

$$E(t) = E_\infty + \hat{E}(t) \tag{6.52}$$

where E_∞ is the equilibrium modulus and $\hat{E}(t)$ is the transient modulus. Using these expressions and separating **Eq. 6.52** into real and imaginary will give the form,

$$E'(\omega) = E_\infty + \omega \int_0^\infty \hat{E}(t) \sin\omega t \, dt \tag{6.53}$$

and

$$E''(\omega) = \omega \int_0^\infty \hat{E}(t) \cos\omega t \, dt \tag{6.54}$$

suggested by Christensen, (1982) (see homework problem 6.8.).

In using **Eqs. 6.53** and **6.54**, it is to be noted that E_∞ for a viscoelastic solid (e.g. a three parameter solid) is a non-zero quantity and the equilibrium modulus for viscoelastic fluid (e.g. a Maxwell fluid) is $E_\infty = 0$. For discussions of the advantages of this form see (Christensen, (1982), Tschoegl, (1989), Tolbolsky, (1960), Ferry, (1980)).

Similarly, the creep compliance can be separated into an instantaneous component and a transient component such that,

$$D(t) = D_0 + \hat{D}(t) \tag{6.55}$$

and equations for the storage and loss compliance analogous to using **Eqs. 6.53** and **6.54** can be developed (see problem 6.9).

Using Fourier transforms (see Appendix B), it can be shown that the relaxation modulus and creep compliance can be found from the complex modulus and the complex compliance respectively, by the equations,

$$E(t) = \frac{1}{2\pi i} \int_{c-i\infty}^{c+i\infty} \left[\frac{E(i\omega)}{i\omega} \right] e^{i\omega t} d(i\omega) \tag{6.56}$$

$$D(t) = \frac{1}{2\pi i} \int_{c-i\infty}^{c+i\infty} \left[\frac{D(i\omega)}{i\omega} \right] e^{i\omega t} d(i\omega) \tag{6.57}$$

Returning to **Eq. 6.49** expressions for $E^*(i\omega)$ can be found by substituting $E(t)$ from **Eq. 6.28** to obtain,

$$E^*(i\omega) = i\omega \int_0^\infty e^{-i\omega t} \int_0^\infty H(\tau)e^{-\lambda t}d\tau dt \qquad (6.58)$$

A similar relation can be found for $D^*(i\omega)$. Development of these relationships is left as an exercise for the reader. A schematic representation of the relationship between various viscoelastic properties is given in **Fig. 6.6**. (Gross, (1953)).

An example of measured relaxation data for polyisobutylene is shown in **Fig. 6.7(a)**. The corresponding calculated relaxation spectrum is shown in **Fig. 6.7(b)**. The relaxation data for polyisobutylene shown in **Fig. 6.7** is spread over about sixteen decades of time in seconds. If a single test were performed to obtain this data, the collection of data would have begun in less than a picosecond and the test would have continued for approximately 12 days. To obtain similar curves for other temperatures would require a large number of tests. Instead, a time-temperature-superposition procedure is used to produce a master curve by performing short-term tests at different temperatures and shifting the measured curves on the time scale to produce a long-time master curve for any one temperature. The master curve can then be shifted to determine the response for any temperature within the given data set. This procedure will be discussed in the next chapter.

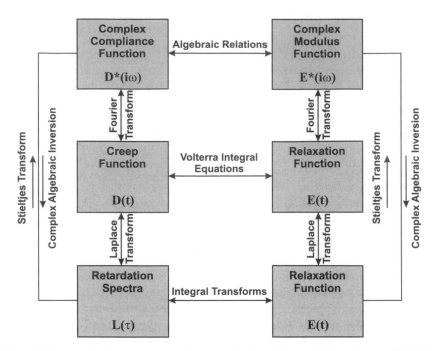

Fig. 6.6 Interrelations among viscoelastic functions. (After Gross, (1953))

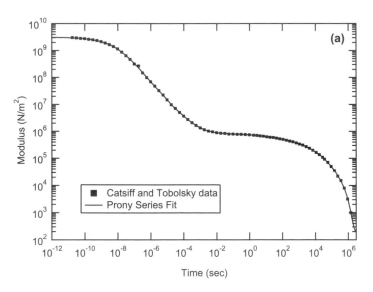

Fig. 6.7(a) Measured relaxation function for polyisobutylene (Original data from Tolbolsky, (1972) and Catsiff and Tobolsky, (1955)) and the series expansion fit.

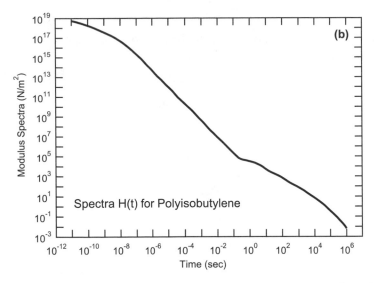

Fig. 6.7(b) Associated relaxation spectra for polyisobutylene. Detail on the original data and the Prony series fit is contained in Chapter 7.

6.5. Review Questions

6.1. Discuss the difference between superposition linearity and proportional linearity and the relation of each to the Boltzman superpositon principle.

6.2. Describe the spectral approach to representing viscoelastic behavior of polymers.

6.3. What is the relationship between the creep compliance and the relaxation modulus.

6.4. Describe the difference between $\overline{D}^*(s)$ and $\overline{D}(s)$.

6.5. Why would one wish to calculate the complex modulus from the relaxation modulus?

6.6. Problems

6.1. Verify the strain output for a two-step stress input given in **Fig. 6.3** by solution of the differential equation for a Kelvin model.

6.2. The relaxation modulus of a Maxwell model is; $E(t) = Ee^{-t/\tau}$. Using the Boltzman superposition principle, find an equation for the stress vs. time in a constant strain rate test.

6.3. The creep compliance of a Kelvin element is $D(t) = \dfrac{1}{E}\left(1 - e^{-t/\tau}\right)$. Using the Boltzman superposition principle, find an equation for the strain vs. time in a constant stress rate test. Sketch your results, i.e., ε vs. t.

6.4. Given the stress input shown below. Give correct expressions of strain for each time interval.

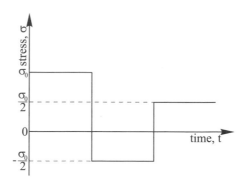

6.5. Using the Boltzman superposition integral, find the strain output for the following stress input for a Maxwell fluid.

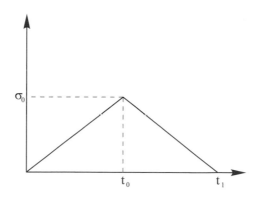

6.6. Using Boltzman's superposition integral, find the strain output for a Kelvin solid for the given stress input.

6.7. Using the relation between compliance and modulus in the Laplace transform space (Eq. 6.40), find the compliance, D(t), for a Maxwell element, given its modulus of $E(t) = Ee^{t/\tau}$. Plot $D(t)$ and the inverse of E(t) to illustrate the validity of **Eq. 6.42**.

6.8. Develop equations for the complex compliance from the creep compliance for a Kelvin model.

6.9. Show that **Eqs. 6.53** and **6.54** can be obtained from using **Eq. 6.49**.

6.10 Show that the complex moduli, E' and E'', can be represented as indicated in **Eq. 6.51** in the case where the time dependent modulus, $E(t)$ is given by a generalized Maxwell model.

6.11 Develop equations relating the storage and loss compliances to the instantaneous and transient compliances analogous to **Eqs. 6.53** and **6.54**.

6.12. Develop equations for the complex modulus from the relaxation modulus for a three parameter model using the integral relationship between the two functions.

6.13. Develop equations for the complex compliance from the creep compliance for a three parameter model using the integral relationship between the two functions.

7. Time and Temperature Behavior of Polymers

One of the most important functions of engineering design is to be able to predict the performance of a structure over its design lifetime. Necessarily the mechanical behavior of materials used in a structure must also be known over the intended life of the structure. For engineering design based upon linear elasticity, it is assumed that no intrinsic change in mechanical properties occurs over time[1]. However, the molecular structure of polymers gives rise to mechanical properties that do change over time.

As engineering structures are often designed to last as long as 20 to 50 years, there is a compelling reason to develop experimental and analytical approaches for polymer based materials that will allow the prediction of long term properties from relatively short term test data. The motivation is even higher when one considers that part of the design process is often that of developing and/or comparing candidate polymeric material systems. Long term testing on the order of years to determine fundamental polymer properties such as the relaxation modulus, $E(t)$, or creep compliance, $D(t)$, are quite impractical. Fortunately, the relationship between property changes of a polymer with time and property changes of a polymer with temperature can be utilized to develop accelerated test methods. The methods discussed in this chapter can assist the design engineer in the difficult task of estimating long-term properties of polymer-based materials from short-term tests. The procedure by which such estimates can be made is known as the time- temperature-superposition principle (TTSP) and is introduced in the following sections.

---

[1] Of course, environmental factors as well as fatigue do influence mechanical properties as a function of time but this degradation of properties due to accumulated damage is quite separate from the inherent time dependence of viscoelasticity considered here.

7.1. Effect of Temperature on Viscoelastic Properties of Amorphous Polymers

In Chapter 3 creep and relaxation testing was discussed as well as a definition of the 10-second modulus. Further, the variation of the 10-second modulus as a function of temperature for various types of polymers was illustrated in **Fig. 3.16** and five regions of viscoelastic behavior were identified as the glassy, transition, rubbery, rubbery flow and liquid flow regions. It was noted that linear polymers exhibit all five regions while the thermoset polymers typically only show the first three regions. However, it was noted that at sufficiently high temperatures thermoset polymers degrade and this can lead to significant changes in properties. These facets will be illustrated later in this chapter.

The general character of the five regions of behavior of thermoset and thermoplastic polymers as a function of temperature given again in **Fig. 7.1** is most often shown in the literature using 10- second modulus data. However, the various regions can also be observed using 30-second data, 5-minute data or even one-hour data depending only on the mechanical characteristics of the polymer being tested and the length that tests are performed. The various regions of behavior can also be observed using creep compliance data such as that shown in **Fig. 7.2**. The data in **Fig. 7.2** is given as the reciprocal of compliance in order for easy comparison to the schematic results for modulus vs. temperature given in **Figs. 3.16** and **Fig. 7.1**. It is to be carefully noted as stated in Chapter 6 that, in general, the relaxation modulus and creep compliance functions are not reciprocals of each other except for regions in which the rate of change of properties is very small. Therefore, data in the glassy and rubbery regions of **Fig. 7.2** can reasonably be interpreted as modulus but those in the transition region may significantly vary from data that would be found in a relaxation test.

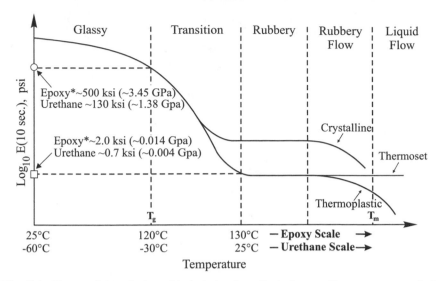

Fig. 7.1 Five regions of viscoelastic behavior of a polymer. Curves are generic in form, but glassy and rubbery data given are for epoxy and urethane (Brinson, 1965, 1968, 1976).[2] For urethane also see (Williams and Arentz, 1964).

As may be seen in **Fig. 7.1** the modulus of a polymer varies considerably with temperature and a polymer of reasonably high molecular weight will be glass-like below the glass transition temperature, T_g. Above the T_g a polymer will have the character of leather in the transition zone, or that of a rubber in the rubbery region, etc. While one polymer may be glass-like at room temperature and another may be rubbery at room temperature their basic behavior relative to the T_g is the same. To illustrate this point glassy (25° C) and rubbery (130° C) moduli values for the epoxy shown in **Fig. 7.2** with a $T_g \sim 120$ ° C (Brinson, (1965)) are included in **Fig. 7.1**. Also included in **Fig. 7.1** are the glassy (~ -80° C) and rubbery (~25° C) moduli values for a polyurethane with a $T_g \sim -30$ ° C (Williams and Arentz, (1964)). Note that the mechanical property response vs. temperature of each are very similar providing the two materials are compared at the same point relative to their respective glass-transition temperatures.

---

[2] The exact point of the beginning of the rubbery region may be somewhat less than the 25° C indicated in **Fig. 7.1**. See Williams and Arentz for details.

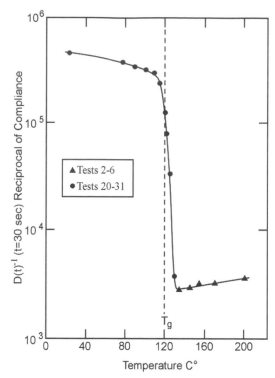

Fig. 7.2 Reciprocal of compliance D(t = 30 sec.)$^{-1}$ vs. temperature for an epoxy, T$_g$=120 °C. (Data from Brinson (1965), (1968).)

The schematic behavior of an epoxy as shown schematically in **Fig. 7.1** is verified by the creep data given in **Fig. 7.2** for an epoxy used in photoelastic investigations, Brinson (1965). It is to be noted that the two polymers identified in **Fig. 7.1** do not have a **T$_m$** but they each will exhibit degradation of properties for temperatures sufficiently above the **T$_g$** as will be demonstrated later in this chapter. Also, the transition region is very sharp in the epoxy as shown in **Fig. 7.2** with a variation of the modulus (compliance) by a factor of 10 for each one degree centigrade change in temperature. Other polymers may display a more moderate-variation in the transition region. Indeed the polyurethane discussed by Williams and Arentz (1964) shows a more gradual variation of modulus with temperature. In this context, it is important to realize that the **T$_g$** is actually a narrow temperature range as opposed to a precise single value. Indeed the various methods to measure **T$_g$** in polymers (see Chapter 3) typically provide similar, but different, numbers from one another. Note also that the modulus (compliance) of the epoxy increases (decreases) slightly with temperature in the rubbery region as shown in **Fig. 7.2**. The latter is evidence of rubberlike behavior or Joule effect and will be discussed further subsequently for both the epoxy and the urethane.

7.2. Development of Time Temperature-Superposition-Principle (TTSP) Master Curves

To illustrate the Time-Temperature-Superposition-Principle consider the short-term creep data for different temperatures shown in **Fig. 7.3** for an epoxy. Data collection for each creep test began at 30 seconds after the initial load was fully applied and the test was terminated after 10 minutes. Each curve for temperatures below 120° C has been shifted to the left so as to form a continuation to shorter times for the 120° C creep compliance curve. Each curve above 120° C has been shifted to the right to form a continuation to longer times for the 120° C compliance curve. The theoretical origin justifying such a shifting procedure will be developed in the next few sections. At present, simply consider that data collected above 120°C must be shifted to the right to represent the longer time needed at 120°C to achieve the same level of creep in the test time frame. Similarly lower temperatures are shifted to the left to represent the shorter time scale for that amount of creep that would be observed at 120°C, providing such measurement could be made. The total curve for 120° C is the "master compliance curve" for that temperature. The master curve data stretches over more than eight decades of time starting at approximately 10^{-4} (0.0001) minutes and ending at close to 10^{+4} (10,000) minutes or nearly a week (6.9 days).

While data was collected from room temperature (~25° C) to 130° C, only the data above 90° C is shown as it was not possible to shift data below this temperature to form a realistic extension to the data shown. Note that the TTSP method is an outgrowth of the kinetic theory of polymers which is only strictly valid above the T_g. While the TTSP is thought to be valid for temperatures below the T_g, the exact lower limit is not well defined. A guiding rule of thumb is that TTSP may be used below the T_g as long as data is shiftable to form a smooth master curve.

The 120° C master curve can now be shifted to determine a new master curve for any temperature between 90° C and 130° C. If shifted to the right to form a master curve for 90° C, the data would span the time from about 0.5 minutes to 10^{+8} minutes (or nearly 200 years). On the other hand, if the data were shifted to the left to form a master curve for 130° C, the data would span the time from about 10^{-8} minutes to 10 minutes. In other words, complete creep (or relaxation) of the epoxy for 90° C would take a very long time while complete creep (or relaxation) would occur very quickly at 130° C. From a practical standpoint obtaining the long time creep data at 90°C would be impractical while at 130°C obtaining the short time creep data would be difficult due to the short timescale. Thus, the power of the TTSP is in the ability to trade off temperature for time and perform mechanical tests of short duration at multiple temperatures to de-termine compliance (modulus) as a function of time spanning many dec-ades in time.

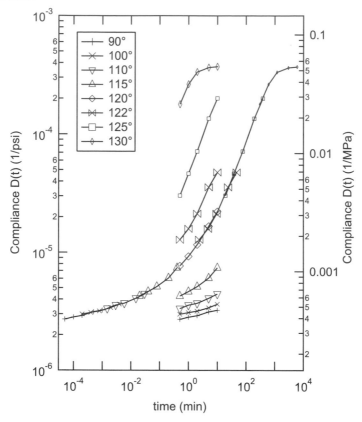

Fig. 7.3 Creep Compliance Master Curve for an Epoxy at 120 ° C. (Data from Brinson (1965).)

Plotting the reciprocal of creep compliance for a time of 0.5 minutes from each of the curves in **Fig. 7.3** with temperature results in the data previ-ously discussed and given in **Fig. 7.2**. This curve verifies the various stages for a polymer as described by **Fig. 3.16** in Chapter 3 and **Fig. 7.1** in this chapter. Here, however, as mentioned earlier no rubbery flow or liquid flow region is observed as the polymer is of the thermosetting type. Further, the "so called" rubbery region in **Fig. 7.2** is not a horizontal rubbery plateau as in **Fig. 7.1** and as often seen in the literature. This is due to the nature of rubber elasticity whose explanation evolved from the kinetic theory of polymers discussed in the next section (Treloar, (1975)).

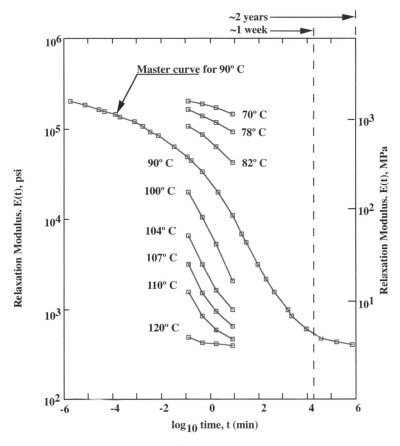

Fig. 7.4 Master curve for a modified epoxy. (Data from Cartner (1978).)

Another example of the application of the TSSP is given in **Fig. 7.4** for a modified (rubber toughened) epoxy adhesive (Renieri, (1976); Cartner and Brinson, (1978); Brinson, (1999)). Here short time relaxation tests of about 10 minutes duration were used for temperatures from 70° C to 120° C to produce a relaxation modulus master curve for 90° C spanning 12 decades of log time from 10^{-6} minutes to two years. The resulting curve, if TTSP is valid, can be shifted to the right by one decade to become the master curve for 87° and the resulting master curve would provide data over 20 years.

An additional decade of shifting to be roughly equivalent to a master curve for 78° would provide data over approximately 200 years. Clearly with this method a prediction of behavior over a design lifetime of 40 or 50 years is possible though no experimental data has ever been collected for such an extended period providing proof that the approach is valid.

A final example of data that is shiftable to form master curves at different temperature is given in **Fig. 7.5**. Here data for polyisobutylene for various temperatures is given with a projection of the master curve for each temperature. While still not a definitive "proof of principle" for the use of the TTSP for reasonable long or short extrapolations, collectively the data provides significant evidence.

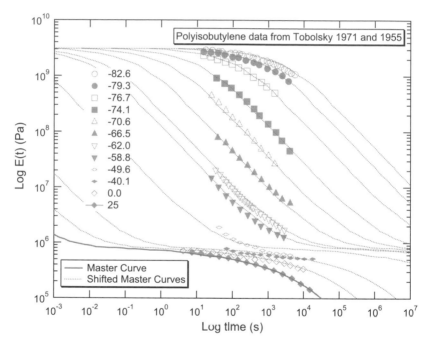

Fig. 7.5   Master curves for polyisobutylene. (after Aklonis and McKnight, (1983); Original data from Tolbolsky, (1972) and Catsiff and Tobolsky, (1955))

7.2.1. Kinetic Theory of Polymers

An examination of **Figs. 7.1-7.4** indicate the similarity between the variation of relaxation modulus (creep compliance) with time and temperature. For this reason, the time and temperature variation of the moduli (or compliances) of a polymer are often said to be related or, in fact pliances) of a polymer are often said to be related or, in fact equivalent. This apparent equivalency led H. Leaderman of National Bureau of Standards (now known as the National Institute for Standards and Technology) to propose the time-temperature-superposition-principle in the early 1940's. Rouse, Zimm, Bueche and others later verified the TTSP using the "kinetic theory of polymers". The method has been extensively studied and extended for many applications by Tobolsky (1960), Ferry (1964), Nielsen (1965), Nielsen and Landel (1994) and many others where extensive references to earlier literature may be found.

The theories of Rouse and Zimm were developed for dilute solutions of polymers above the T_g but Ferry and coworkers essentially extended these to bulk polymers in the rubbery state. In doing so, a number of assumptions were made among which were:

- A bar-mass linkage was used to represent a segment (mer) of a polymer molecule.
- Each polymer segment has a relaxation time and, therefore, the polymer has a very large spectrum of many relaxation times
- It is not possible to calculate an average relaxation time for a polymer based on the molecular structure but it is possible to calculate the relaxation time of the pth segment of a polymer molecule in a dilute solution and then extrapolate to the case for a polymer molecule moving through a viscous medium of its own kind.

The Rouse equation for the relaxation time of an pth (arbitrary) segment is, (see Nielsen (1965), Nielsen and Landel (1994) as well as an article by E. Passaglia and J.R. Knox in Baer (1964)),

$$\tau_p = \frac{6\eta M}{\tau^2 p^2 \rho RT} \qquad (7.1)$$

where η is the viscosity of the of the dilute solution, M is the molecular weight of the segment, p is the number of segments per molecule, ρ is the density of the solution, R is the gas constant and T is the absolute temperature. Rearranging gives,

$$\frac{\tau_p \rho T}{\eta} = \frac{6M}{\pi^2 p^2 R} = \text{constant} \qquad (7.2)$$

and therefore,

$$\frac{\tau_p(T)}{\tau_p(T_0)} = \frac{\eta}{\eta_0}\left(\frac{\rho_0 T_0}{\rho T}\right) \tag{7.3}$$

or the ratio of the relaxation time at one temperature to that at a reference temperature, T_0, is given by Eq. 7.3. If the temperature dependence of the relaxation time is the same for all segments, the ratio of relaxation times may be extrapolated to the case of a molecule moving through a medium of its own kind or that of a bulk polymer and the ratio can be equated to a shift factor a_T,

$$a_T = \frac{\tau(T)}{\tau(T_0)} = \frac{\eta}{\eta_0}\left(\frac{\rho_0 T_0}{\rho T}\right) \tag{7.4}$$

That is, the relaxation times of the bulk polymer at one temperature can be found from that at another temperature by multiplying each relaxation time by the shift factor or,

$$\tau_i(T) = a_T \tau_i(T_0) \tag{7.5}$$

Thus, the shifting of the data demonstrated in **Fig. 7.3** should be represented by **Eq. 7.4**. The term *thermorheologically simple* refers to the key caveat that all relaxation times of the polymer must be affected by temperature in the same way. This assumption has been found to hold for a vast array of homogeneous polymer systems. Typically shift factors are found experimentally or by the WLF equation discussed in the next section.

7.2.2. WLF Equation for the Shift Factor

M.L. Williams, R.F. Landel and J.D. Ferry (1955; see also Ferry, (1980)) applied the TTSP to a large number of polymers and found empirically the following expression for the shift factor,

$$\log_{10} a_T = \log_{10}\frac{\tau(T)}{\tau(T_0)} = \frac{-C_1(T - T_0)}{C_2 + (T - T_0)} \tag{7.6}$$

where the constants C_1 and C_2 had the values of **17.44** and **51.6** respectively if the glass transition temperature T_g is used as the reference temperature T_0.

The development of the above equation has been shown to be of great importance and has become widely used. It is commonly known as the WLF equation and must be one of the most referenced equations ever in the polymer literature. **Eq. 7.6** was thought to be a universal equation for the shift factor for all amorphous glass-forming polymers above the glass-transition temperature. However, further testing proved that different classes of polymers have different constants.

Equation **7.6** can be developed from Doolittle's concept of free volume of a liquid. In Chapter 4 it was noted that the specific volume varies with temperature during quenching as shown in **Fig. 4.27** and can be used to identify the degree of crystallinity as well as the melt temperature, T_m.

The variation of specific volume with temperature is shown again in **Fig. 7.6** where the amount of free volume increases with increasing temperature above the T_g. Free volume can be thought of as the space within a material that is "unoccupied" by atoms and their quantum shells. (Actually, this theory has recently received criticism for not being a good representation of the state of matter. However, the concept of free volume is a useful model to assist in the explanation of the molecular motion of polymers associated with viscoelastic behavior.) The slope of the specific volume curve shown in **Fig. 7.4** is the coefficient of thermal expansion, α_{CTE}, and, in fact, one definition of the T_g is the point at which the coefficient of thermal expansion suffers a discontinuity. The variation in free volume allows for greater mobility of the molecular chains and gives rise to greater time or viscoelastic effects as temperature increases. Sufficiently far above the T_g the polymer can be considered to be a fluid (for thermoplastic polymers). As the polymer is cooled slowly to the T_g it can be considered to be a super-cooled fluid. (Continued slow cooling can suppress the T_g but the times required for molecular equilibrium are quite long. See Ferry, 1980 for a more complete discussion.) For example the T_g is near the upper portion of the transition region for the epoxy shown in **Fig. 7.1** and the free volume is relatively small below the T_g and little viscoelastic response occurs as is illustrated in **Fig. 7.3**. On the other hand, above the T_g free volume is much larger and increases dramatically as temperature increases with resulting increasing viscoelastic effects with temperature again as illustrated in **Fig. 7.3**. When the rubbery region is reached the free volume is so great that time effects occur almost instantaneously in a creep or relaxation test. See also **Fig. 5.12** to visualize the effects of temperature on time effects and hence on the free volume.

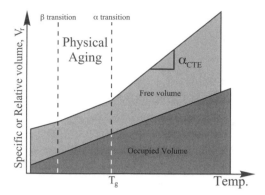

Fig. 7.6 Specific or relative volume vs. temperature for an amorphous polymer.

Doolittle's equation for the viscosity of a liquid is,

$$\ln \eta = \ln A + B\left(\frac{V - V_f}{V_f}\right) \qquad (7.7)$$

where $\mathbf{V}$ is the total volume and $\mathbf{V_f}$ is the free volume. Defining the fractional free volume as,

$$f = \frac{V_f}{V} \qquad (7.8)$$

Doolittle's equation becomes,

$$\ln \eta = \ln A + B\left(\frac{1}{f} - 1\right) \qquad (7.9)$$

Above the $\mathbf{T_g}$ the fractional free volume can be expressed as (see Aklonis and McKnight, (1983)),

$$f = f_g + \alpha(T - T_g) \qquad (7.10)$$

where $\mathbf{f_g}$ is the free volume at $\mathbf{T_g}$. Substituting **Eq. 7.10** into **Eq. 7.9** yields,

$$\ln \eta(T) = \ln A + B\left[\frac{1}{f_g + \alpha(T - T_g)} - 1\right] \qquad (7.11)$$

The ratio of the natural log of viscosity at any temperature to that at the $\mathbf{T_g}$ will give after simplification,

$$\ln \frac{\eta(T)}{\eta(T_g)} = B\left[\frac{1}{f_g + \alpha(T - T_g)} - \frac{1}{f_g}\right] \qquad (7.12)$$

Converting to base 10 logarithms gives,

$$\log_{10} \frac{\eta(T)}{\eta(T_g)} = -\frac{B}{2.303\, f_g} \left[\frac{T - T_g}{\dfrac{f_g}{\alpha_f} + T - T_g} \right] \tag{7.13}$$

or

$$\log_{10} a_T = \frac{-C_1(T - T_g)}{C_2 + (T - T_g)} \tag{7.14}$$

The kinetic theory of polymers and the TTSP are only valid above the glass transition temperature. However, many feel that the procedure, in a modified form, is valid below the glass-transition temperature but exactly how far below is uncertain. The WLF equation, on the other hand is known to be only valid above the T_g because below this temperature the material can no longer be considered a super cooled liquid. In fact, Ferry, (1980) notes that the slope of the shift factor curve should be discontinuous at the T_g for the same reason that the coefficient of thermal expansion suffers a discontinuity at the T_g.

A shift factor below the T_g can be developed using the Arrhenius activation energy equation,

$$\tau(T) = A e^{-E_a / RT} \tag{7.15}$$

where τ is the relaxation time, E_a is the activation energy, R is the gas constant and T is the absolute temperature. Rewriting in logarithmic form,

$$\ln \tau(T) = \ln A - \frac{E_a}{RT} \tag{7.16}$$

and taking the ratio at an arbitrary temperature and the glass transition temperature will give after converting to base 10 logarithms,

$$\log_{10} a_T = \log_{10} \frac{\tau(T)}{\tau(T_g)} = -\frac{E_a}{2.303R} \left(\frac{1}{T} - \frac{1}{T_g} \right) \tag{7.17}$$

Obviously, the shift factor based on activation energy is quite different than the shift factor given by the WLF equation (see HW problem 7.2 and 7.3).

The glass-transition temperature for the epoxy represented in **Figs. 7.2** and **7.3** as determined from relative volume measurements (i.e., by measuring the change in dimensions of a small unstressed specimen at different temperatures) is given in **Fig. 7.7**. As may be observed, the $T_g = 120°$ C. For the above reasons, the master curve of **Fig. 7.3** is given for a temperature of **120° C.**

The shift factors necessary to obtain the master curve of **Fig. 7.3** are given in **Fig. 7.8** and compared to the WLF equation. The measured shift factor data for compliance agrees well with the WLF equation for temperatures above the T_g if the constants are taken to be $C_1 = 17.44$ and $C_2 = 51.6$. Also shown is a best fit of the WLF equation to the shift factor above T_g via a least squares algorithm and the associated constants in **Eq. 7.14**. For the limited data points given, the result is relatively insensitive to modest changes in the constants. Below the T_g the measured data diverges drastically from the WLF equation and the character of the curve changes at the T_g. It is interesting to note, as Ferry (1980) indicated, that the slope of the shift factor vs. temperature curve does suffer a discontinuity of slope at the T_g.

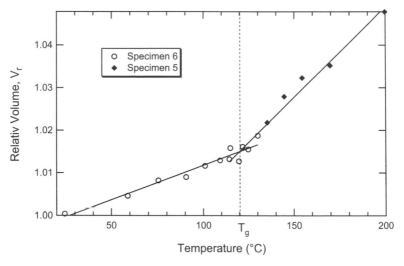

Fig. 7.7 Relative volume vs. temperature for epoxy of **Figs. 7.2** and **7.3**. (Data from Brinson (1965), (1968).)

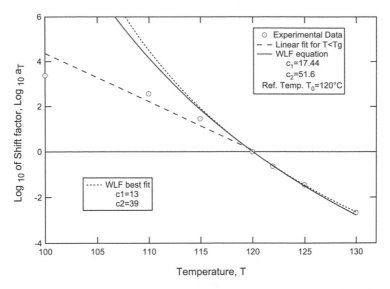

Fig. 7.8 Shift factor from **Fig. 7.3a** compared with the WLF equation. (Data from Brinson (1965), (1968).)

7.2.3. Mathematical Development of the TTSP

An equation to explain the TTSP procedure can be found using an extension to the relaxation spectra definition given by **Eq. 6.28**,

$$E(t) = \int_0^\infty H(\tau)e^{-t/\tau}d\tau \qquad (7.18)$$

Obviously, the relaxation spectra $H(\tau)$ is a function of temperature as is the relaxation modulus $E(t)$ and **Eq. 7.18** should be written as,

$$E(\tau,T) = \int_0^\infty H[\tau(T),T]e^{-t/\tau}d\tau \qquad (7.19)$$

Actually, $H(\tau)$ is more strongly dependent upon the temperature implicitly through the relaxation time rather than explicitly through temperature. In fact, the explicit dependence of $H(\tau)$ on temperature is very weak and can be neglected (Passaglia and J.R. Knox (1960)). The Rouse theory suggests the temperature dependence of the relaxation spectra is such that (see again Passaglia, et al. (1960)),

$$E(t,T) = \rho T \int_0^\infty h(\tau) e^{-t/\tau} d\tau \qquad (7.20)$$

where $h(\tau)$ has no explicit temperature dependence. Multiplying by the factor $(\rho_0 T_0)/(\rho T)$ gives,

$$\left(\frac{\rho_0 T_0}{\rho T} \right) E(t,T) = \rho_0 T_0 \int_0^\infty h(\tau) e^{-t/\tau} d\tau \qquad (7.21)$$

Eq. 7.20 must be valid for any temperature and therefore can be rewritten for the temperature T_0 as,

$$E(t,T_0) = \rho_0 T_0 \int_0^\infty h(\tau_0) e^{-t/\tau_0} d\tau_0 \qquad (7.22)$$

Since the relaxation times are all identically affected by temperature in the same way and the relationship can be expressed as

$$a_T = \frac{\tau}{\tau_0} \qquad \text{or} \qquad \tau_0 = \frac{\tau}{a_T} \qquad (7.23)$$

Eq. 7.22 can be written alternatively using a new time scale as,

$$E(t',T_0) = \rho_0 T_0 \int_0^\infty h(\tau) e^{-t'/\tau} d\tau \qquad (7.24a)$$

where the new time scale, t', is associated with temperature T_0 and

$$t' = t/a_T \qquad (7.24b)$$

Comparing **Eqs. 7.24a** and **7.21** indicates that the right hand side of each are the same providing the time scale in **7.21** is replaced by t'.

In turn the left hand sides are equal under the same condition and

$$E(t',T_0) = \left(\frac{\rho_0 T_0}{\rho T} \right) E(t = a_T t',T) \qquad (7.25a)$$

or equivalently

$$E(t,T) = \left(\frac{\rho T}{\rho_0 T_0} \right) E(t' = t/a_T,T_0) \qquad (7.25b)$$

Equation **7.25** is a formal statement of the TTSP and shows that the modulus at one temperature is identical to that at another temperature after modifying the timescale by a multiplicative factor (and the modulus scale by a small temperature factor). On a logarithmic scale, this multiplicative time factor results in a horizontal shift as demonstrated below. A similar expression can be developed for the creep compliance. **Eq. 7.25** can be used in the process of creating a master curve, in which tests are performed at many temperatures and shifted to the chosen reference T_0 temperature. **Eq. 7.25** can also be used to shift the master curve from its reference temperature to provide the modulus master curve at another temperature.

An alternative approach to develop the TTSP expression (**Eq. 7.25**) is to consider an expression of the viscoelastic modulus as a Prony series as given by **Eq. 5.21b** or **Eq. 6.25** with the temperature dependence now included on the basis of the theories of Rouse and Zimm.

$$E(T_0,t) = \rho_0 T_0 \sum_i E_i e^{-t/\tau_i(T_0)} \tag{7.26}$$

where the relaxation times at the reference temperature T_0 can be related to those at any other temperature via the shift factor

$$\tau_i(T) = a_T \tau_i(T_0) \tag{7.27}$$

The modulus at another temperature **T** can thus be expressed

$$E(T,t) = \rho T \sum_i E_i e^{-t/\tau_i(T)}$$

$$= \rho T \sum_i E_i e^{-t/a_T \tau_i(T_0)}$$

$$= \rho T \sum_i E_i e^{-(t/a_T)/\tau_i(T_0)} \tag{7.28}$$

$$= \rho T \sum_i E_i e^{-(t')/\tau_i(T_0)}$$

$$\frac{\rho_0 T_0}{\rho T} E_T(t) = \rho_0 T_0 \sum_i E_i e^{-(t')/\tau_i(T_0)}$$

and thus comparing **Eq. 7.28** with **Eq. 7.26**, one obtains

$$\frac{\rho_0 T_0}{\rho T} E_T(t) = E(T_0, t' = t/a_T) \tag{7.29}$$

which is equivalent to **Eq. 7.25b**.

Use of **Eq. 7.25** is demonstrated in **Fig. 7.9**. The shift factor is given by **Eq. 7.24b** which upon taking logarithms becomes,

$$\log_{10} t = \log_{10} a_T + \log_{10} t' \tag{7.30}$$

and therefore,

$$\log_{10} t' = \log_{10} t - \log_{10} a_T \tag{7.31}$$

For a reference temperature lower than the test temperature, $T_0 < T$, the WLF equation will give a shift factor on a log scale less than zero, i.e.,

$$\log_{10} a_T = \frac{-C_1(T - T_0)}{C_2 + (T - T_0)} < 0 \tag{7.32}$$

As a result, $E(t', T_0)$ is found from $E(t, T)$ by shifting the data down and to the right as shown in **Fig. 7.9**.

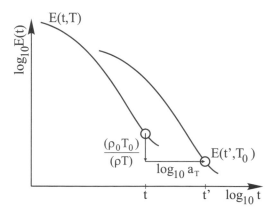

Fig. 7.9 Example for shifting relaxation modulus data for $T_0 < T$. (Vertical shift exaggerated for clarity.)

The change in density is usually very small and much less than the change in temperature. As a result, density changes are often neglected and the TTSP equation is usually written as,

$$E(t', T_0) = \left(\frac{T_0}{T}\right) E(t = a_T t', T) \tag{7.33}$$

In fact in many cases the amount of change due to temperature is also small and the TTSP equation is simply expressed as,

$$E(t', T_0) = E(t = a_T t', T) \tag{7.34}$$

For example the data of **Fig. 7.3** includes only a maximum vertical shift of about 10% due to temperature differences.

The TTSP method is sometimes referred to the "method of reduced variables" because of the necessity of a vertical and horizontal shift due to temperature differences in the collected data.

Eqs. 7.20 or **7.33** can be used to confirm mathematically the Joule effect or the increase of modulus with temperature in the rubbery range (see problem 7.5). The elastic (or 30 second) modulus for the epoxy of **Fig. 7.3** in the rubbery range is shown plotted vs. absolute temperature in **Fig. 7.10**. Obviously, the rubbery modulus does increase linearly with increasing temperature. Even though the extrapolated data does not go through the origin it does serve as confirmation of the Joule effect mentioned in Chapter 1.

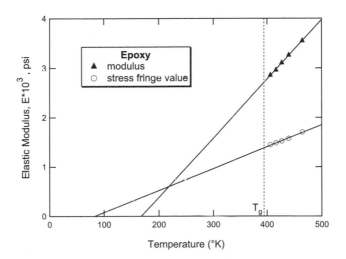

Fig. 7.10 Variation of modulus for the epoxy of **Fig. 7.3** in the rubbery range. (Data from Brinson (1965), (1968).)

Data for a crosslinked polyurethane rubber is shown plotted vs. absolute temperature **Fig. 7.11**. The data shown is for temperatures between **T= 300° K (~25° C)** to **T=425° K(~150° C)** and is well above the **T$_g$** of about -25° C. Up to **T=375° K** the material is quite rubbery. However, beyond this temperature, the transparent (orange color) polymer begins to darken and noticeable creep occurs. In essence this is a rubbery flow region even though cross-linked polymers are not supposed to have such a region. Actually, the temperature is so high that the material begins to physically degrade by compromising some of the primary cross-link bonds which begin to break and reattach leading to a creep mechanism. For example, consider the three chains in **Fig. 7.12**. For a sufficiently high temperature, the crosslink bonds of chain 1 break at site A and reattach at site B of chain 2. This process leads to a permanent deformation which cannot be recovered upon reheating as is normally the case for a thermoset.

Creep of the polyurethane of **Fig. 7.8b** for a temperature of **150° C** is shown in **Fig. 7.13**. The strain varies linearly with time similar to that expected for a Maxwell fluid and reaches nearly 20%. While the strain is increasing, the birefringence remains constant. Since birefringence is caused by the interaction of light with the molecular structure, the latter is an indication that the molecular network is not seeing additional strain. These results then suggest a deformation mechanism as described by **Fig. 7.12.**

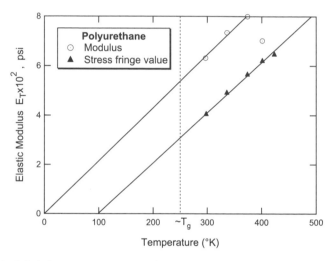

Fig. 7.11 Modulus vs. temperature for a cross-linked polyurethane in the rubbery range. (Data from Brinson (1965), (1976).)

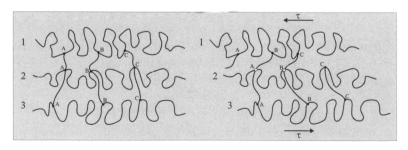

Fig. 7.12 Illustration of creep of a thermoset polymer due to thermally induced degradation.

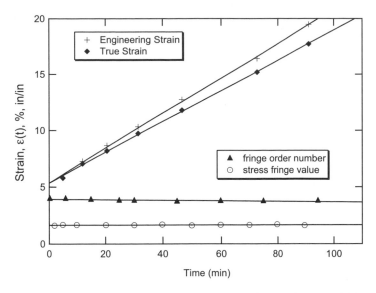

Fig. 7.13 Creep of a polyurethane T =150 °C (Data from Brinson (1965), (1976).)

Since the various methods of representing viscoelastic data including re-laxation modulus, creep compliance, relaxation spectra, creep spectra, complex modulus, complex compliance storage and loss moduli and stor-age and loss compliance are all related as discussed in Chapter 6 and shown schematically in **Fig. 6.6**, the TTSP principle is valid for each. Data of each kind is often generated using the TTSP principle as is illustrated by the storage and loss moduli given in **Fig. 5.19** for polycarbonate.

7.2.4. Potential Error for Lack of Vertical Shift

As noted in the preceding section, often no vertical shift is used when mas-ter curves are formed using the TTSP method. However, the lack of inclu-sion of a vertical shift, even if small, can lead to substantial errors in the prediction of properties over a long time. For example, assume the true master curve for the compliance of a polymer is known and is as given in **Fig. 7.14** for a temperature of T_0. Next assume the original data for a tem-perature $T_1 > T_0$ is as given in **Fig. 7.14** and must be shifted both horizon-tally and vertically to fit the true master curve. Obviously if the data is only shifted horizontally and not vertically, an error in both the time scale and the compliance will result. Not only is the error compounded due to the error in both the time scale and the compliance but also the error is cumulative. As a result, large errors may occur even if the temperature ratio is small. Also, as will be noted in the section on nonlinear behavior, vertical shifts may be necessary for reasons other than the differences in the temperatures for the collected data.

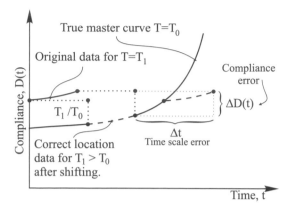

Fig. 7.14 Potential error for lack of a vertical shift (exaggerated for clarity).

7.3. Exponential Series Representation of Master Curves

In the solution of boundary value problems it is often desirable to have a mathematical representation for master curve data for a given polymer over many decades of logarithmic time. A relatively straightforward approach is to use either a generalized Maxwell or Kelvin model with sufficient elements to span the spectrum of relaxation times represented by the transition behavior of the data. Before proceeding, it is instructive to consider again the shape of creep and relaxation representations for a Maxwell model as a simple function of time or logarithmic time. As introduced in Chapter 3, the relaxation modulus for a Maxwell model is given by

$$E(t) = Ee^{-t/\tau} \tag{7.35}$$

while the creep response for a Maxwell model is,

$$D(t) = \frac{1}{E} + \frac{t}{\mu} \tag{7.36a}$$

or

$$D(t) = \frac{1}{E}(1 + \frac{t}{\tau}) \tag{7.36b}$$

where and **E** and **μ** are the spring stiffness and damper viscosity, respectively, and $\tau = \mu/E$.

Variation of the relaxation modulus and creep compliance of a Maxwell model on linear-linear (left) and log-log (right) scales are shown in **Figs. 7.15-7.16**. Notice the rapid decay of the modulus as the time approaches the selected relaxation time and the flow at long times due to the fluid nature of the Maxwell model. The behavior of the modulus and compliance for a simple Maxwell element is similar to that for many polymers in the glassy and transition region.

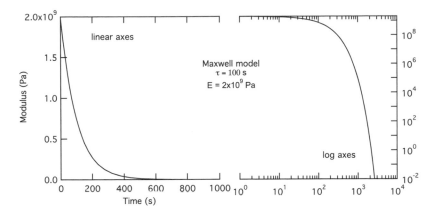

Fig. 7.15   Relaxation modulus of a Maxwell model on linear (left) and loglog (right) scale.

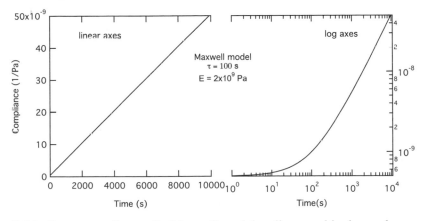

Fig. 7.16   Creep compliance of a Maxwell model on linear and loglog scales.

If several Maxwell models are used in series to represent polymer response, as in the generalized Maxwell model (see Chapter 6), and if the spring moduli and relaxation times are judiciously chosen, the transition region broadens as shown in **Fig. 7.17**. The parameters used for the curves in **Fig. 7.17** are shown in **Table 7.1**. Notice that as the number of elements spanning a time period increases, the transition behavior can be smoothly represented (the two element versus the 5 element case shown here). When a free spring is included, in each case the material model goes from viscoelastic fluid to viscoelastic solid at long times. Clearly, the location of the relaxation times and magnitude of the associated modulus value can be manipulated to produce master curves which represent all five regions of viscoelastic behavior as needed. The two element model shown displays a long rubbery plateau before the flow region, while the five element model suppresses the rubbery plateau for the fluid case.

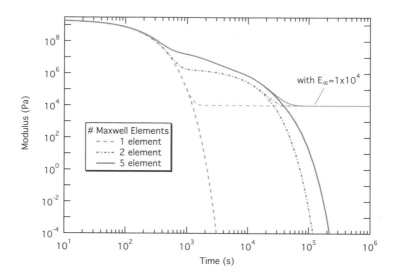

Fig. 7.17 Master curve representation using one, two and five Maxwell elements. Curves without E_∞ decay rapidly at larger times, while curves with the E_∞ term are constant at long time as indicated. Parameters in Table 1.

Table 7.1 Parameters used in generalized Maxwell models for **Fig. 7.17**. Spring constants in Pa, relaxation times in s.

One Element		Two Elements		Five Elements	
τ_i	E_i	τ_i	E_i	τ_i	E_i
100	2e9	100	2e9	100	2e9
		5000	2e6	500	4e8
				1000	3e7
				5000	2e6
				10000	4e5

While the above description suggests that various master curve shapes can be represented by generalized Maxwell or Kelvin models, it does not mean that the determination of the proper values of the spring moduli or relaxation times to obtain a precise fit is trivial.

Examination of the relaxation modulus of a generalized Maxwell model demonstrates the complexity. That is,

$$E(t) = E_1 e^{-t/\tau_1} + E_2 e^{-t/\tau_2} + \cdots + E_n e^{-t/\tau_n} \tag{7.37}$$

and

$$E(t = 0) = E_1 + E_2 + \cdots + E_n \tag{7.38}$$

Obviously, there are **2n** unknowns in **Eq. 7.37**, i.e., **n** moduli and **n** relaxation times. Therefore, at a minimum, **2n** equations are needed to solve for the unknowns. For five Maxwell elements as in **Fig. 7.17**, ten data points would need to be selected in order to write a set of 10 equations. Further, an approach simply fitting to 10 discrete data points from an extensive master curve would not produce a smooth curve that fit the entire data set over time well. In order to obtain a mathematical expression that provides a good fit for an entire master curve from experimental data, a different approach is required. While a number of techniques can be found in the literature, one method will be briefly described in the next section.

7.3.1. Numerical Approach to Prony Series Representation

A general issue in working with viscoelastic materials is representing the measured material properties by an appropriate mathematical function. As indicated earlier, a closed mathematical form facilitates solution of boundary value problems, as well as ease of manipulation of data. While viscoelastic properties can be represented by a number of functional forms, the exponential Prony series

$$E(t) = E_\infty + \sum_{i=1}^{N} E_i e^{-t/\tau_i} \tag{7.39}$$

or

$$D(t) = \frac{1}{E_0} + \sum_{i=1}^{n} \frac{1}{E_i}\left(1 - e^{-t/\tau_i}\right) \tag{7.40}$$

is particularly attractive for a number of reasons. First, the coefficients can be related to simple spring and damper coefficients in a mechanical model (see Chapter 5), facilitating interpretation. Second, a series of simple exponentials is easy to store and manipulate mathematically, as the derivatives and integration of the terms are trivial. Third, in the case where numerical solutions of a boundary value problem are desired, use of the Prony series form for the material modulus enables use of a recursive algorithm for fast and easy solution of the convolution integral constitutive law (Taylor et al., (1970)). This fact is extremely important for calculation of viscoelastic response at long times, as integrating over the long time history requires only retaining terms at the previous time step.

Experimental data for a given material will produce modulus or compliance functions for a polymer as a function of time (for relaxation or creep data) or frequency (for steady state dynamic data from a DMA). As described earlier in the chapter, given the time scale of polymeric response, it is usually necessary to perform separate tests at multiple temperatures in order to obtain the full spectrum of polymer response from glassy to rubbery behavior. The TTSP can then be used to construct a master curve of the data as illustrated in the time domain in **Fig. 7.3** or in the frequency domain in **Fig. 5.19**. Given such a master curve over time or frequency space, the challenge is to find the parameters τ_i, E_i to provide a good fit over all time of the data. This problem has been addressed by a number of methods in the literature including Procedure X (due to Tobolsky and Murakami and discussed by Tschoegl, 1989), the collocation method by Schapery (1962), the multidata method (Cost and Becker, (1970)) and the windowing method (Emri and Tschoegl, (1993)). Here we describe briefly a sign control method developed by Bradshaw and Brinson (1997) which is based on the multidata method.

In the sign control method, as in several approaches, the first step is to select the relaxation times in a reasonable manner based on the time scale of the data. In such a process, the relaxation times are not chosen based on any known polymer structure or derived timescales, but are chosen for mathematical convenience. As real polymers contain a continuous distribution of relaxation times, in this approach a sufficient discrete subset of these relaxations are chosen in order to provide a mathematical function that will fit the material data. For a typical data set, such as that in **Fig. 7.3**, choosing the relaxation times evenly spaced in log time over the data range is reasonable. The number of relaxation times required varies depending on the smoothness of the data, but 10-20 relaxation times over 10 decades of time is a good rule of thumb. To facilitate fitting non-constant values at either end of the data set, one or two relaxation times can be added to lie beyond the time domain data.

Once the relaxation times are selected, the problem reduces to finding the coefficients E_i such that the Prony series function optimally matches the provided time domain data. An obvious procedure to use is a generalized least squares approach, which was done in the multidata method (Cost and Becker, 1970). In this approach, coefficients are found that minimize the χ^2 error between the modulus data (given as P data pairs (E_p, t_p)) and the calculated function, $E(t)$,

$$\chi^2 = \sum_{p=1}^{P} \left(\frac{E(t_p) - E_p}{\sigma_p} \right)^2 \tag{7.41}$$

where $E(t_p)$ is the value of the Prony series function (Eq. 7.39) evaluated at time t_p and σ_p is the standard deviation of the pth data point. However, this method will typically provide values for E_i that are both positive and negative in value. Given the physical relationship between the E_i coefficients and springs in a mechanical model, it is desired for these coefficients to remain positive. In addition, it can be shown that a sufficient condition for the viscoelastic modulus to satisfy all physical and thermodynamic principles for a material is that the Prony coefficients E_i be positive.

Consequently, the sign control method (Bradshaw and Brinson, 1997), modifies the use of the least squares algorithm to ensure that the Prony coefficients be positive. This is accomplished via an iterative Levenberg-Marquadt method based on the first derivatives relative to each unknown coefficient (Press et al, 1992). The method is provided with an initial guess for the coefficients (all positive), uses these to predict a new set of values and then calculate χ^2. If the new set decreases the error, it becomes the current step; otherwise the previous values are used to take a smaller step. The additional constraint that $E_i > 0$ is enforced by setting $E_i = |E_i|$ before calculating the χ^2 error; only those cases that lead a reduction in the χ^2 error are kept. From this procedure, optimal E_i values are found such that the Prony series fits the entire data range.

To illustrate the ability of a generalized Maxwell Model (Prony Series) to fit long term data, consider the master curve data from Fig. 7.5 for polyisobutylene. A complete data set at 25°C was constructed as shown in Fig. 7.18. Thirty relaxation times evenly spaced in log time between 10^{-11} and 10^7 were chosen and the sign control method used to calculate the Prony series representation seen in Fig. 7.19. The modulus E(t) calculated from the Prony terms in **Fig. 7.19** is overlaid on the experimental data in **Fig. 7.18**. It is clear that the Prony series has captured the data well. In addition, with the large number of coefficients taken, the discrete E_i spectrum is approaching a continuous spectra. In the next section, we will compare these coefficients with the spectra found via another method. Note also that the Prony series as a well-behaved mathematical function will provide values for modulus for any time inserted into **Eq. 7.39**. However, as the coefficient values were obtained for data only in a specific range, care should be taken when using these functions to ensure that predictions are made only within the bounds of the known experimental data. As seen in **Fig. 7.18**, the Prony series will predict a value for the modulus at times beyond 10^6 and less than 10^{-11}, however these values are fictitious as they do not correspond to measured experimental data.

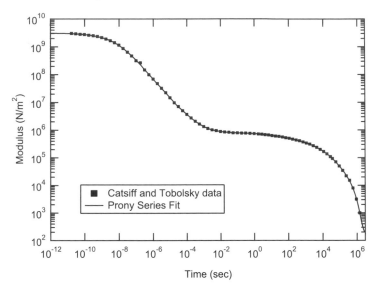

Fig. 7.18 Master curve for tensile modulus of polyisobutylene at 25°C (Original data from Tolbolsky, (1972) and Catsiff and Tobolsky, (1955)). Fit from Prony series shown in **Fig. 7.19**.

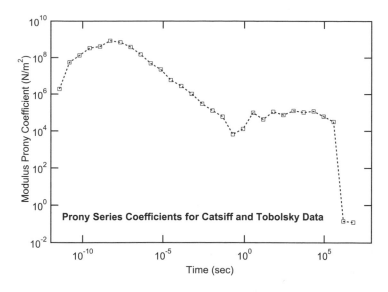

Fig. 7.19 Prony series coefficients used to obtain tensile modulus for Polyisobutylene. Fit to data shown in **Fig. 7.18**.

The sign control method can also be used to fit frequency domain data, which is useful given the prevelance use of DMAs to measure storage and loss moduli for polymers. In this case, the storage and loss moduli as functions of frequency are used as the data to be fit, and the relationship between the Prony coefficients and these functions (**Eq. 6.51**) are used as the functional form. An example of this application is shown in **Fig. 7.20** for Polycarbonate. The storage and loss moduli obtained from DMA data and shifted to form a master curve is shown in **Fig. 7.20** (the temperature data was also shown in **Fig. 5.19**). A 28 element Prony series was used to fit the data and the coefficients are shown in **Fig. 7.21**, while the fit to the experimental data is overlaid on **Fig. 7.20**. Again it is seen that the mathematical representation of a Prony series provides an excellent form to represent measured experimental data for a polymer. Given the Prony series that fits the frequency based data, the time domain modulus can be readily produced using **Eq. 7.36a**. Note that the large oscillation of the Prony coefficients between relaxation times of 1 and 100 is not indicative of an oscillating spectra and these oscillations can be eliminated by refining the chosen relaxation times while still maintaining the quality of the fit to the experimental data.

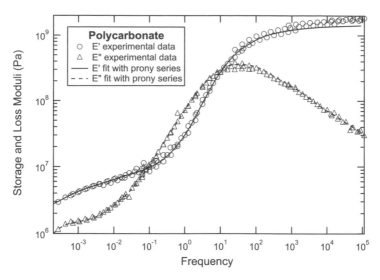

Fig. 7.20 Experimental data from a DMA for polycarbonate shifted to form a master curve. Lines showing the fit of the Prony series from **Fig. 7.21** are overlaid on the plot.

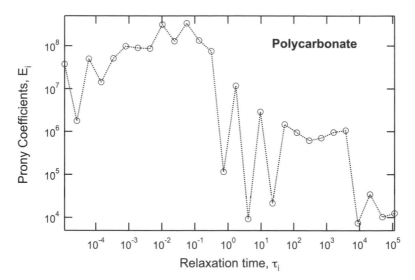

Fig. 7.21 Prony coefficients used to fit the data for Polycarbonate in the frequency domain in **Fig. 7.20**.

Further details on the application of the sign control method can be found in the original paper (Bradshaw and Brinson, (1997)). A software code in C (dynamfit.c) written to perform these calculations is available on the authors' websites at University of Louisville and Northwestern University respectively. This code has been written to perform the calculations described above to find the moduli or compliances of time domain or frequency domain data. In addition, the mathematical fitting procedure described can be extended to perform interconversions between viscoelastic properties as discussed in Chapter 6 (e.g. between modulus and compliance) and the existing code allows for such calculations as well.

7.3.2. Determination of the Relaxation Modulus from a Relaxation Spectrum

Recall from Chapter 6 that the spectrum of relaxation times is defined by the equation,

$$E(t) = \int_0^\infty H(\tau) e^{-t/\tau} d\tau \tag{7.42}$$

One can think of the spring moduli as having been replaced by the spectrum of relaxation times, $H(\tau)$. As a result, **Fig. 7.19** or **7.21** can also be thought of as a discrete representation of $H(\tau)$ as a function of τ.

Sometimes $\overline{H}(\tau) = \tau H(\tau)$ is defined such that,

$$E(t) = \int_0^\infty \frac{\overline{H}(\tau)}{\tau} e^{-t/\tau} d\tau = \int_{\ln \tau = -\infty}^{\ln \tau = +\infty} \overline{H}(\tau) e^{-t/\tau} d\ln \tau \tag{7.43}$$

Tobolosky and his students have used this approach extensively and suggest certain forms for $\overline{H}(\tau)$. Aklonis and McKnight, 1983 (a former student of Tolbolsky) suggests the following data for $\overline{H}(\tau)$ for a viscoelastic fluid with a simple transition region as shown by the curve in **Fig. 7.22**,

Table 7.2 Wedge Approximation of $\overline{H}(\tau)$ for **Fig. 7.22**.

$\overline{H}(\tau) = 0$	$\log_{10} \tau < 0$	or $\tau < 1$
$\overline{H}(\tau) = \dfrac{k}{\tau}$	$0 < \log_{10} \tau < 1$	or $1 < \tau < 10$
$\overline{H}(\tau) = 0$	$\log_{10} \tau > 1$	or $\tau > 10$

Using these values in **Eq. 7.43** results in,

$$E(t) = \frac{k}{t} \left(e^{-t/10} - e^{-t} \right) \qquad (7.44)$$

Fig. 7.22 illustrates the relationship between the wedge distribution, $\overline{H}(\tau)$, given in **Table 7.2** and the exponential series given in **Eq. 7.44**.

Using this approach, it is possible to obtain relatively simple function that can represent a complete master curve. For example, Tolbolsky has suggested that the master curve for polyisobutylene given in **Fig. 7.18** can be found from the "wedge" and "box" distribution shown in **Fig. 7.23**, where the wedge represents the transition and the box the rubbery plateau and flow regions. Note the similarity in form to the Prony coefficients found via the sign control method for the same data as shown in **Fig. 7.19**.

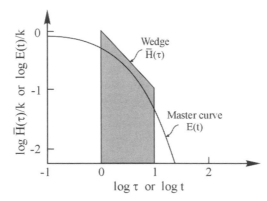

Fig. 7.22 Continuous (wedge) distribution of relaxation times and corresponding relaxation modulus. (After Aklonis and McKnight, (1983), p. 155)

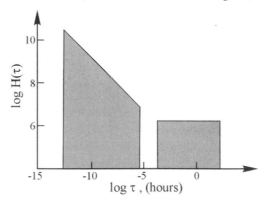

Fig. 7.23 Wedge and box distributions of $\overline{H}(\tau)$ needed to fit the Master Curve for Polyisobutylene; Here $H(\tau)$ in dynes/cm²; to convert to Pa, divide by 10; the magnitude and form match the discrete spectra in **Fig. 7.19**. (After Aklonis and McKnight, (1983), p. 156.)

Alternatively, one can use various approximations (see Ferry, 1980) to determine the relaxation spectra directly from modulus data or a mathematical function fit to the data. A method that works well is known as Alfrey's rule, in which the exponential function in the integral in **Eq. 7.42** is 0 at small τ's and 1 at large τ's and is thus replaced by a step function **H(t-τ)**. With this simplification, **Eq. 7.42** can be differentiated to obtain

$$H(\tau) \cong -\frac{dE(t)}{d\ln t}\bigg|_{t=\tau} \qquad (7.45)$$

And with substitution of the Prony series in for **E(t)**, one obtains the relaxation spectra as

$$H(t) \cong \sum_{j=1}^{N} \frac{t}{\tau_j} E_j e^{-\frac{t}{\tau_j}} \qquad (7.46)$$

Application of Alfrey's rule to the polycarbonate data of **Fig. 7.20** results in a continuous and smooth relaxation spectra as shown in **Fig. 7.24**. The shape and magnitude of the spectra obtained in this fashion corresponds to the discrete Prony elements of **Fig. 7.21** when smoothed so as to eliminate the oscillations obtained in the least squares fitting process. While the spectrum in **Fig. 7.20** still exhibits slight non-smoothness, manipulation of the Prony elements and/or use of a more accurate method to determine **H(τ)** could provide a smoother curve. For most purposes, the spectra shown is adequately smooth.

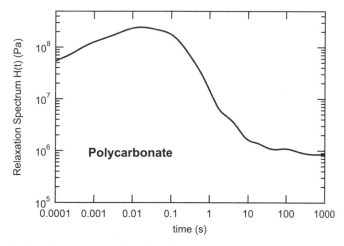

Fig. 7.24 Relaxation spectra of polycarbonate calculated from the Prony series elements in **Fig. 7.21** via Alfrey's rule.

7.4. Constitutive Law with Effective Time

Recall from Chapter 6 that the constitutive law describing the stress and strain relation of a viscoelastic material can be written as

$$\varepsilon(t) = \int_{0^-}^{t} D(t-\tau)\frac{d\sigma(\tau)}{d\tau}d\tau \ \text{ or } \ \sigma(t) = \int_{0^-}^{t} E(t-\tau)\frac{d\varepsilon(\tau)}{d\tau}d\tau \qquad (7.47)$$

If temperature changes during the loading history, clearly the material property inside the integral must change accordingly. In order to be able to account for general thermomechanical loading including spatial and temporal variations in temperature, the relationship between time and temperature developed in this chapter can be utilized. Returning to the TTSP **Eq. 7.25b** and taking the vertical shift to be negligible, we can relate the material modulus at one temperature to the modulus at a reference temperature $\mathbf{T_0}$ by the shift factor, $\mathbf{a_T}$

$$E_T(t) = E_{T_0}(\xi = t/a_T) \qquad (7.48)$$

Thus, one can consider the modulus at temperature **T** and time **t** to be the same as the modulus at a reference temperature $\mathbf{T_0}$ at a reduced time ξ. At a small time increment later, **dt**, the modulus at temperature **T** has changed to a new value correspondingly **dξ** later in reduced time at the reference temperature

$$E_T(t+dt) = E_{T_0}(\xi+d\xi) \qquad (7.49)$$

where

$$d\xi = \frac{dt}{a(T(t))} \qquad (7.50)$$

Physically, this expression represents that all relaxation times in a time increment at temperature **T** are **1/a** times slower/faster than those occurring in the reduced time increment at the reference temperature. Integrating one obtains an expression for reduced time, or *"effective time"* as it is often called, as

$$\xi(t) = \int_{0}^{t} \frac{d\zeta}{a(T(\zeta))} \qquad (7.51)$$

Note that the shift factor is a function of time according to the temperature history: as the temperature changes, so does the shift factor. To account for a temperature history, the constitutive law can thus be written in effective time space as

$$\sigma(\xi) = \int_{0^-}^{\xi} E_{T_0}(\xi - \xi') \frac{d\varepsilon}{d\xi'} d\xi' \tag{7.52}$$

Mapping this constitutive law to the real time domain results in

$$\sigma(t) = \int_{0^-}^{t} E_{T_0}(\xi(t) - \xi(t')) \frac{d\varepsilon}{dt'} dt' \tag{7.53}$$

where

$$\xi(t) = \int_{0}^{t} \frac{d\zeta}{a(T(\zeta))} \quad \text{and} \quad \xi(t') = \int_{0}^{t'} \frac{d\zeta}{a(T(\zeta))} \tag{7.54}$$

Eq. 7.53 is straightforward to apply for a problem with both temperature and strain known as functions of time. The effective time is determined via **Eq. 7.54** for the given temperature history. The modulus function is then evaluated at the effective time and used with the differentiated strain function to determine the stress response history. However, **Eq. 7.53** is no longer a convolution integral and as such can be difficult to solve. Thus, often problems are solved in the effective time domain (**Eq. 7.52**). The constitutive law can be written to find the strain response as a function of stress history analogously as

$$\varepsilon(t) = \int_{0^-}^{t} D_{T_0}(\xi(t) - \xi(t')) \frac{d\sigma}{dt'} dt' \tag{7.55}$$

where the effective time is as given in **Eq. 7.54**.

The effects of a number of environmental factors on viscoelastic material properties can be represented by a time shift and thus a shift factor. In Chapter 10, a time shift associated with stress nonlinearities, or a time-stress-superposition-principle (TSSP), is discussed in detail both from an analytical and an experimental point of view. A time scale shift associated with moisture (or a time-moisture-superposition-principle) is also discussed briefly in Chapter 10. Further, a time scale shift associated with several environmental variables simultaneously leading to a time scale shift surface is briefly mentioned. Other examples of possible time scale shifts associated with physical and chemical aging are discussed in a later section in this chapter. These cases where the shift factor relationships are known enables the constitutive law to be written similar to **Eq. 7.53** with effective times defined as in **Eq. 7.54** but with new shift factor functions. This approach is quite powerful and enables long-term predictions of viscoelastic response in changing environments.

7.5. Molecular Mechanisms Associated with Viscoelastic Response

As discussed in Chapter 4, the forces holding polymer molecules together are primary (covalent) or secondary bonds though even metallic-like and ionic types bonds can be found in certain polymers. However, in general, the bonds between mer units along the backbone chain are covalent as are those within side groups and those connecting side groups to the main chain. In thermoset polymers both covalent and secondary (e.g., dipole, Van der Waal) connect individual chains to each other, while in thermoplastic polymers only secondary bonds connect individual chains to each other. These distinctions between the molecular character of thermoplastic and thermoset polymers dictate the differences seen in behavior throughout the five regions of viscoelastic behavior depicted in **Figs. 3.16,** and **7.1**. However, it should be noted that entanglements in thermoplastic polymers, especially where the chains have extensive side groups, often demonstrate behavior very similar to thermosets especially below the T_g.

Many factors are important in understanding the molecular mechanisms associated with the macroscopic behavior of a polymer. A few of these are the degree of polymerization (see **Table 4.6**), the degree of crystallization (see **Table 4.4**), the relative extent of cross-linking and/or entanglements, complexity of side groups, deformability of bonds and bond angles, the amount of thermal energy with chains or chain segments associated with a particular environment (such as temperature or moisture content). Clearly, the most important single parameter to define the state of a polymer is the temperature, especially the glass transition temperature, T_g, and possibly other transition temperatures such as those for the β, γ and δ transitions as well as the melt temperature, T_m. As stated earlier in the discussion of a particular epoxy and a particular polyurethane (see **Fig. 7.1**), one polymer may be glassy at a temperature where another may be rubbery. Below the T_g, polymers are glass like solids with only a small amount of viscoelasticity (creep or relaxation) within a short (minutes or hours) time frame. Near the T_g increasing amounts of viscoelasticity are encountered as the temperature is increased. These differences can be explained on the basis of thermal agitation or vibrations of individual chains, or perhaps more appropriately chain segments, in the presence of free volume. Obviously, free volume increases with increasing temperature above the T_g (see **Fig. 3.18** and **7.6**). Below the T_g molecular agitation can only occur in very small local regions and any amplitude of vibration must be correspondingly small. As the temperature is raised above the T_g, free volume and chain vibrations increase, resulting in translational and configurational motions of chains with respect to each other. The frequency and amplitude of motion increases with increasing temperature until a state of almost pure Brownian motion occurs.

In the glassy region, deformations associated with instantaneous elasticity are the lengthening and shortening of bond distances and bond angles. If creep occurs, it is often associated with motion of side groups. In the transition region, the molecular motions involved are short range translational and configurational changes due to rotations about bond angles and, to a lesser extent, the same mechanisms encountered in the glassy region. Molecular segments are more flexible as the temperature is increased in the transition region and, in time, are able to slide past one another giving evidence of a fluid-like behavior. Molecular mechanisms in the rubbery regions are much like the ones in the transition region except the time scale is much shorter. That is, creep or relaxation is near instantaneous and viscoelastic behavior can only be observed by dynamic tests such as steady state oscillations and/or impact conditions. In the rubbery and liquid flow regions, pronounced unrecoverable deformation occurs with the mechanisms mostly associated with long range configurational changes. Molecules slide past each other with relative ease and bonds (secondary and, in some cases primary) may be broken and reformed.

It is descriptive here to quote from Aklonis and McKnight, (1983). "It is impossible to describe quantitatively the time ranges that give each type of behavior, since the temperature variable causes all these ranges to be relative. Accordingly ... a plastic (a polymer in the glassy state) would have a modulus of a rubber on a time scale of perhaps a thousand years while a rubber might behave like a plastic on a nanosecond time scale."

7.6. Entropy Effects and Rubber Elasticity

Early molecular theories were unable to describe the high deformation (as much as 1000%) of natural rubber under an applied stress (Treloar, (1975)). An early theory, similar to the theory still used today to describe inter-atomic forces (see Chapter 11), suggesting deformations were associated with the stretching of inter-atomic or intermolecular bonds could only account for a few percent of strain. One method considered to overcome this shortcoming was the two-phase theory of Ostwald that attributed deformations to the molecular network being embedded in a second highly viscous phase. Such a theory was used by photoelasticians to explain the "frozen stress" effect which led to the "stress freezing and slicing method" that was used successfully for many years to experimentally determine the internal stress in three-dimensional bodies (Hetenyi, (1938)). Another approach that could account for larger deformations was the folded chain model that permitted strains up to 300%. However, the foregoing models could not explain the thermo-elastic or Joule effect inspiring further model development that led to the forerunner of what is now called the kinetic theory of polymers that stated that deformations in the rubbery state are directly proportional to the absolute temperature. (See Treloar, (1975) for a complete discussion of the history of the relation between the development of the kinetic theory and the thermo-elastic effect).

As noted in the previous section, deformations below the $\mathbf{T_g}$ are associated with stretching and shortening of bond distances and bond angles while in the rubbery region deformations are associated with rotation about bond angles (see **Fig. 4.4**). The former mechanisms are associated with changes in internal mechanical energy and the latter are associated with changes in internal entropy. Indeed, as stated by Rosen (1993), "… to exhibit significant entropy elasticity, the material must be above its glass transition temperature and cannot have appreciable crystallinity".

A tensile force applied to a linear elastic bar does a certain amount of work as the bar is stretched defined by the relation,

$$dW = f \, dl \qquad (7.56)$$

where f is the force and dl is the amount of axial deformation. The work (or input energy) is transferred to the bar as internal energy. Typically it is normal to assume that the internal energy is only mechanical energy that can be recovered as that in an ideal spring. In fact, part of the input work causes a change in the temperature of the bar and just as with a perfect gas the temperature increases if the bar is compressed and decreases if the bar is stretched. A tensile bar tested adiabatically (no heat flow into or out of the bar) will show a small decrease in temperature that, in turn, will cause the elastic modulus to change slightly. Timoshenko and Goodier, (1970), note that the difference between the adiabatic and isothermal (constant temperature) modulus of iron is only about 0.26% and reference experimental work performed by Kelvin in 1855 to support this small difference.

One method to approximate adiabatic testing is to perform the test rapidly enough that no heat is lost from the sample but not so rapid that dynamic or inertia effects occur. Mueller (1969) gives experimental results obtained by cyclic testing 2.5 cm diameter steel bars in tension, compression and torsion with a loading-unloading cycle of about one minute in duration. In a tensile test he shows a decrease in temperature of approximately one degree centigrade on loading and a similar increase on unloading. In a compression test he shows a similar temperature increase in loading and decrease in unloading. That is, the thermal effect is reversible in either tension or compression as is the mechanical effect. Mueller also shows that in torsion there is no change in temperature on loading and unloading as is expected because torsion or pure shear can be thought of as a combination of equal tension and compression on a differential element as shown in **Fig. 2.19**. On testing a 2.0 cm PVC bar in compression Mueller shows a more significant 6.5°C temperature rise but does not show a reversible thermal effect upon unloading. That is the thermal processes in the PVC were irreversible and can be attributed to viscoelastic and/or flow processes. Further insight to the mathematics of irreversible thermodynamic processes of polymers and other materials can be found in the many papers of Schapery (1964, 1966, 1969), the book by Lubliner (1990) or the book by Fung (1965). A discussion of the thermodynamics of irreversible processes is beyond the scope of this text but the results of Schapery's early irreversible thermodynamic approach for nonlinear viscoelastic materials is presented in Chapter 10.

The study of the amount of heat energy absorbed or released by a polymer as it is heated or cooled is most often accomplished with a calorimeter. For example the differential scanning calorimeter (DSC) is often used to measure the melting point temperature and the heat of melting, the glass transition temperature, curing and crystallization processes. Mueller (1969) describes the development, design and use of a special differential deformation calorimeter that allows the measurement of the amount of heat absorbed or released when a specimen is loaded in simple tension. For the greatest sensitivity he suggests testing only samples with small cross-sectional dimensions such as fibers, wires or films.

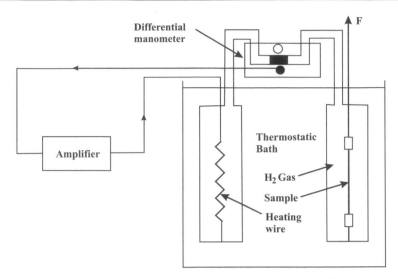

Fig. 7.25 Deformation calorimeter (After Mueller, (1969))

A schematic of the Mueller's deformation calorimeter is given in **Fig. 7.25** and consists of two parallel cylinders immersed in a thermostatic bath. The sample to be tested is centrally located in one cylinder between two polyamide or Teflon clamps with low heat capacities. The top clamp is attached to a thin invar wire through which a load is applied that is resisted by the specimen through a lower clamp attached to the bottom of the cylinder. The second cylinder contains a heating coil and both cylinders are connected to a differential manometer. If the sample gains heat on deformation in tension (as with a rubber) an excess pressure occurs and causes a feedback mechanism attached to the differential manometer to provide heat to the comparison cylinder. Knowing the balancing heat as a function of time allows the determination of the total change in enthalpy. Endothermic effects can be determined by preheating the cylinder that would then be cooled by a sample tested in tension (for samples tested at temperatures below the rubbery range). Also, more recent efforts on the measurement of entropy effects using a deformation calorimeter similar to the one designed by Mueller can be found in papers by Farris (1989) and Kishore and Lesser (2005).

To convert measurements of heat changes in a sample to information about the distribution of input work (energy) into either internal mechanical energy or thermal energy requires the use of basic thermodynamic relationships. The following gives only a small glimpse into the relationships between energy, entropy and temperature and the reader is advised to consult the more elaborate sources found in Mueller (1969), Treloar (1975),

Lubliner (1990), and other texts for a more thorough treatment of the mathematics of the thermodynamic effects related to either reversible of irreversible processes of deformed solids.

The relationship between energy, entropy and temperature for reversible processes is best described using the first and second law of thermodynamics. The first law relates the change in internal mechanical energy, dU, and the change in internal thermal energy (heat), dQ, to the work done on the system by external forces, dW, and is given as,

$$dW = dU - dQ \qquad (7.57)$$

The second law of thermodynamics defines the entropy change, dS, in a reversible process such that,

$$TdS = dQ \qquad (7.58)$$

Combing **Eqs. 7.57** and **7.58** gives,

$$dW = dU - TdS \qquad (7.59)$$

If the Helmholtz free energy* is defined as $A = U - TS$ then changes in the Helmholtz free energy at constant temperature are given by,

$$dA = dU - TdS = dW \qquad (7.60)$$

and states that the Helmholtz free energy is equivalent to the external work done on the system or the difference between the internal mechanical energy and internal heat energy.

For the circumstance where a tensile bar is under a constant hydrostatic pressure (e.g. atmospheric pressure) as well as a tensile load, the total input work would be (Treloar, (1975)),

$$dW^* = dW - pdV = fdl - pdV \qquad (7.61)$$

and **Eq. 7.57** could be written as,

$$dW - pdV = dU - TdS - pdV = dG \qquad (7.62)$$

where dG is the change in Gibbs free energy[†]. Hence, the Gibbs free energy is also equivalent to the external work on the specimen but the exter-

---

[†] Assuming the existence of a strain energy potential, the Hemholtz free energy, (under constant temperature and volume) and the Gibbs free energy, (under constant temperature and pressure) can be written in terms of stress and strain as (see Gittus, (1975)),

$$\sigma_{ij} = \left(\frac{\partial A}{\partial \varepsilon_{ij}} \right) \qquad \sigma_{ij} = \left(\frac{\partial G}{\partial \varepsilon_{ij}} \right)$$

nal work includes a portion associated with the hydrostatic pressure. In terms of enthalpy, $H = U - pV$, and for a constant temperature and pressure **Eq. 7.62** becomes,

$$dG = dW^* = dH - TdS \tag{7.63}$$

For polymers in the rubbery range the volume change dV is small and if the pressure is only the atmospheric pressure the pdV term is so small as to be negligible. The negative sign for the pV term in the enthalpy definition here is due to the fact that the work of atmospheric pressure is in opposition to the positive work of a tensile force on an uniaxial specimen.

Assuming a constant volume **Eq. 7.59** (or **Eq. 7.62**) can be written as,

$$fdl = dU - TdS \tag{7.64}$$

Using **Eq. 7.64** and recognizing that the derivatives are total derivatives Rosen (1993) obtains the following equation for the change in length for a change in temperature for tensile specimen under a constant load and constant volume,

$$\left(\frac{\partial l}{\partial T}\right)_{f,V} = \frac{1}{f}\left(\frac{\partial U}{\partial T}\right)_{f,V} - \frac{T}{f}\left(\frac{\partial S}{\partial T}\right)_{f,V} \tag{7.65}$$

This equation defines the change in length for a change of temperature for the aforementioned conditions and provides an explanation for the classic experiment of Joule for a tensile specimen of rubber heated while hanging under a constant tensile load. The first term on the right represents the change in internal energy and corresponds to the usual effect of positive change in length for an increase in temperature. The second negative term represents the change in entropy for an increase in temperature. For rubber the entropy effect completely dominates the energy effect and, therefore, the length of the rubber specimen contracts when heated under a constant load.

The change in the Hemholtz free energy may be written for nonconstant temperatue as,

$$dA = dU - TdS - SdT \tag{7.66}$$

Using **Eqs. 7.66** and **7.64** Treloar (1975) obtains the following relations,

$$\left(\frac{\partial S}{\partial l}\right)_T = -\left(\frac{\partial f}{\partial T}\right)_l \qquad \left(\frac{\partial U}{\partial l}\right)_T = f - \left(\frac{\partial f}{\partial T}\right)_l \qquad (7.67)$$

The first equation provides a definitive method to determine the entropy change per unit extension and the second equation provides a definitive method to determine the associated energy change per unit extension. Treloar further explains that if a rubber specimen is stretched and held at constant length while the temperature is varied both the entropy and internal energy can be determined. Specifically, if the force temperature diagram found from such a test is linear then both the internal energy and entropy are independent of temperature. If the linear force temperature plot passes through the origin, the internal energy is zero and the elastic modulus of the rubber is only related to the change in entropy. If the force temperature plot intercepts the positive force axis, the departure from the origin represents the contribution of the internal energy to the elastic modulus. A similar analysis to experimentally determine the distribution of external work energy into internal entropy and internal mechanical energy is given by Mueller (1969).

In the test just described the stress can be determined from the force and the strain is a constant. As a result the modulus versus temperature would vary in the same manner as the force versus time. Therefore, the fact that the modulus of the polyurethane shown in **Fig. 7.11** increases linearly with absolute temperature and goes through the origin is indicative that the associated deformation processes were only related to changes in the internal entropy and not to changes in internal energy.

Another approach to verifying that the modulus varies linearly with absolute temperature uses statistics to relate the entropy changes under loading to configurational changes of the molecular chains. This information combined with the second of **Eqs. 7.62** (with the change in energy term taken as zero) yields the following result (Rosen, (1993)),

$$E_{(initial)} = \frac{3\rho RT}{M_c} \qquad (7.68)$$

where ρ is the density, M_c is the number average molecular weight, **R** is the gas constant and **T** is the absolute temperature.

7.7. Physical and Chemical Aging

Even with no applied stress, the mechanical properties of polymers may vary with time due to changes occurring in the molecular structure. Variations due to changes in molecular packing are called physical aging and changes due to modification of the inter/intra-molecular bonding are referred to as chemical aging. Physical aging effects are thermoreversible while chemical aging effects are not.

Physical aging effects can be best explained through an understanding of the molecular packing or free volume changes that take place during cooling after the polymerization processing steps are completed. As discussed in Chapter 3 and shown in **Fig. 3.18**, the specific or relative volume decreases linearly with temperature until the glass transition temperature is reached. At the T_g the rate of change of specific volume decreases as shown again in **Fig. 7.26**. The exact location of the T_g depends upon the rate of cooling. If the rate of cooling is very slow, the T_g will be decreased or if the rate of cooling is very fast, the T_g will be increased as shown in **Fig. 7.26**. Manufacturing processes often involve rapid cooling and thus can lead to significant increased T_g and an excess of free volume below the T_g. Such a state suggests the material can experience increased viscoelastic response to a stress or deformation input, as well as increased physical aging effects.

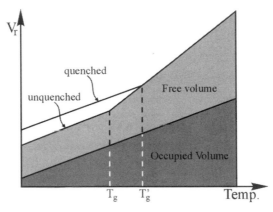

Fig. 7.26 Specific or relative volume vs. temperature for a quenched or unquenched polymer.

In general, as the temperature is decreased, molecular motions decrease and the molecular structure becomes more tightly packed as indicated by the amount of free volume. Above the T_g, molecular reconfigurations to attain the equilibrium volume are accomplished in the experimental time scale of the temperature change. However, below the T_g decreased chain mobility and the small amount of free volume result in a non-equilibrium thermodynamic state. The polymer chains are unable to rearrange to attain their equilibrium volume during the timescale of the temperature change and thus there is continued, slow rearrangement of molecules long after the temperature change as thermodynamic equilibrium is sought. During this approach to equilibrium volume and structure, mechanical and other properties of the polymer change with time in a process called physical aging. In general, as polymers age modulus increases and they become more brittle. The temperature range for physical aging is between the T_g and the first highest secondary transition temperature, T_β (see **Fig. 5.24** and **Fig. 7.6**).

An excellent discourse on the subject of physical aging may be found in Struik, (1969). With very careful measurements, Struik demonstrated that the effect of aging is to continuously decrease the compliance of the material and that the short term aging curves are related to each other by a shift factor along the time axis. Thus, master curves can be constructed that give the effects of physical aging over extended times from measurements of aged compliance over shorter times, very similar to the previously discussed TTSP (time-temperature-superposition-principle). This process is referred to as a time-aging-time-superposition-principle and it is illustrated in **Fig. 7.27** with data for PMMA.

Notice that data at each aging time is related via a simple shift in log space. This shift factor is typically denoted as a_{te}. The aging shift factor takes a particularly simple analytical form

$$a_{tc} = \left(\frac{t_{eref}}{t_e} \right)^\mu \tag{7.69}$$

where t_{eref} is the reference aging time (aging time the curves are shifted to), t_e is the aging time and μ is the shift rate, defined by the slope of the shift factor - aging time curve.

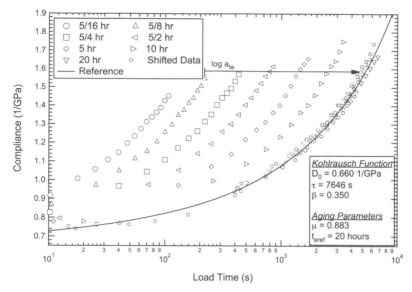

Fig. 7.27 Illustration of the concept of physical aging for PMMA. Material is first rejuvenated above T_g then quenched to 15°C below T_g for isothermal aging. Creep compliance curves are obtained at each aging time and these can be shifted as shown to provide a momentary master curve. A Kohlrausch fitting function is also shown through the shifted data. (Data from Wang (2007).)

Because the material ages continuously, the creep tests performed as in **Fig. 7.27** must be of short duration at each aging time so that the compliance result is representative of viscoelastic properties at that aging time. The individual curves and the shifted master curve (also called the momentary master curve) is typically well fit by a Kohlrausch stretched exponential function

$$D(t) = D_0 e^{(t/\tau)^\beta} \tag{7.70}$$

where β is the stretch parameter which has the effect of creating a spectrum of relaxation times even though only a single parameter τ is used. If a long term creep test is performed, the continued aging during the application of the load leads to a continued stiffening of the material and a rollover in the compliance function as is illustrated in **Fig. 7.28**. It is possible to define an effective time, ξ, based upon the aging shift factor similar to the effective time defined for temperature shift factor in **Eq. 7.51**

$$\xi(t) = \int_0^t a_{te}(\zeta)d\zeta \tag{7.71}$$

where the shift factor evolves with the time of loading as

$$a_{te} = \left(\frac{t_{eref}}{t_e}\right)^\mu = \left(\frac{t_{eref}}{t_{e0}+t}\right)^\mu \qquad (7.72)$$

and t_{e0} is the aging time at t=0. By doing so, long term response accounting for accumulate aging can be predicted using expressions similar to **Eqs. 7.48** and **7.52**.

$$D_{longterm}(t) = D(\xi(t)) \qquad (7.73)$$

where D(t) is the momentary master curve at $t_e=t_{eref}$. Such a prediction is shown compared to the data in **Fig. 7.28**.

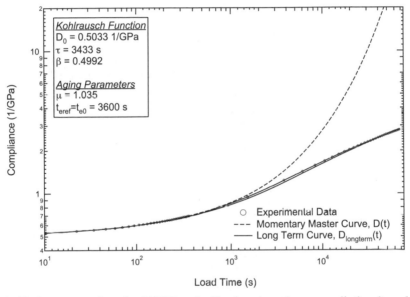

Fig 7.28 Long term data for PEEK and effective time theory prediction based on momentary master curve (based on short term creep experiments). (Data courtesy of R. D. Bradshaw, University of Louisville; long term prediction using shift rate from Guo and Bradshaw (2007).)

The aging superposition process can be combined with the TTSP provide more extensive information of the material response as a function of time, aging time and temperature. In the last two decades many have studied physical aging extensively. Representative references include Wong, et al., (1981), McKenna, (1989, 1994), and Crissman, et al., (1990). A discussion of viscoelasticity and physical aging concepts together with an extensive list of references can be found in Brinson and Gates, (2000).

As shown in **Fig. 7.6**, the rate of change of the specific volume increases slightly for temperatures below the T_g, which indicates that the free volume actually begins to increase slightly with decreasing temperature below the T_g. This same trend is shown in Struik, 1969 and is, perhaps, one reason that the lower limit for physical aging is the beta transition temperature. An interesting study by Adamson, 1983 demonstrates a unique aging process for temperatures below the beta transition. With very careful moisture absorption measures for a Hercules 3501-5 epoxy resin system used as a matrix material for graphite/epoxy composites, he demonstrated that additional water can be "pumped" into the resin and/or the composite by changes in the operating temperature of the material. It was demonstrated that if the epoxy was moved from a 74°C bath after more than 100 days to a 25°C bath that approximately 25% more moisture would be absorbed in the next 50-60 days. Further it was shown that if the sample was returned to the 74°C bath, the sample desorbed moisture to return to the previous moisture content within a few days. This phenomenon is referred to as the reverse thermal spike mechanism and demonstrates that moisture absorption mechanisms can be quite different than expected in both resins and polymer-matrix composites and lead to damage that might be unanticipated.

Struik (1969) defines chemical aging as "thermal degradation, photo-oxidation, etc." It can safely be said that chemical aging is not as mature a subject as physical aging and fewer recent references exist specifically related to chemical aging. Indeed, the authors are not aware of any compendium on chemical aging similar to the outstanding study compiled by Struik (1969) for physical aging. That having been said, it is possible to find a considerable amount about the degradation of polymers beginning with the classic book by Tobolsky (1962) wherein is found a long chapter on "chemical relaxation". This chapter includes a description of vulcanized rubber exhibiting a rather rapid decay to zero stress in a relaxation experiment in the temperature range of 100° C – 150° C. Tobolsky argues that because in a network polymer a relaxation to a non-zero stress is expected the phenomenon can be attributed to a rupture of the rubber network and is due to the presence of molecular oxygen. His text contains numerous experimental results to substantiate this claim. He also indicates that some rubbers in the temperature range of 100° C – 150° C show a softening or modulus reduction while others show a hardening or modulus increase. The former is similar to the behavior of a polyurethane rubber discussed herein in an earlier section of this chapter. The softening process is due to

chain scission or the breaking of a chain resulting in two chains (Rodriguez, (1996)) and the hardening process is due to the occurrence of additional cross-linking taking place at the higher temperatures. Tobolsky,
(1962) describes the process of cleavage (scission) at cross-link cites as
well as along a chain. He further describes the process of chemical permanent set and chemical creep under constant load that would explain the results given in **Fig. 7.13**. More recent results on the response of rubbers at
elevated temperatures, accounting for chain scission and oxygen depletion
can be found in papers by Shaw, Wineman and co-workers (2005).

Epoxy specimens used in the creep experiments resulting in **Figs. 7.2,
7.3** and **7.10** and urethane specimens used in the creep experiments resulting **Fig. 7.11** and **Fig. 7.13** are shown in **Fig. 7.29**. The two specimens on
the left are epoxy and third specimen is polyurethane with the final recti-
linear strip being an untested polyurethane sample. The first epoxy specimen was used in the creep tests with temperatures not exceeding 130° C
(see **Fig. 7.2** and **Fig. 7.3**) while the second epoxy was use in the creep
tests between 130° C and 200° C (see **Fig. 7.2** and **Fig. 7.11**). The change
in color between the two indicates some degree of degradation though any
change in the room temperature modulus was undetectable. The polyurethane specimen before testing was the same color as the untested strip and
had the same dimensions as the epoxy specimens. Clearly, the color
changed significantly during the creep test at 150° C and significant unrecoverable deformation remained. The room temperature modulus was also
significantly lowered for the polyurethane due to chemical aging at the
elevated temperatures. These tests and specimens reinforce the information
given by Tobolsky.

Fig. 7.29   Epoxy (left two) and urethane (right two) specimens showing degradation when tested at high temperatures.

While physical aging occurs between the beta- and glass-transition temperatures, chemical aging effects are most often observed at temperatures significantly above the glass-transition temperature or for other extreme environmental conditions. For example, radiation can lead to additional cross-linking or even to the cross-linking of a linear polymer (Sullivan, (1969). The effect of atomic oxygen (AO) has been studied extensively for spacecraft applications (Pippin, (1995) and setas-www.larc.nasa.gov/LDEF/ATOMIC_OXYGEN/ao_intro.html. Atomic oxygen can lead to the erosion of polymers in space applications and can lead to a breakdown of the chemical structure.

A common occurrence related to resins used in the manufacture of composites is the slow continued crosslinking at high operating temperatures leading to increased brittleness and micro-cracking. An interesting study by Kuhn, et al., (1995), uses a modification of an equation first proposed by Debenedetto (see Kuhn, et al. for reference) to predict changes in the glass-transition temperature due to additional cross-linking for a high-temperature carbon fiber-reinforced polyimide composite. The effects of both physical and chemical aging related to changes in the glass-transition temperature and dimensional changes were documented experimentally.

More information on polymer degradation mechanisms can be found in Rodriguez, (1996); Kumar and Gupta, (1998) and other texts. The various degradation mechanisms discussed include chain scission, depolymerization, side group changes, antioxidants, radiation, moisture. While not all of the included information would be classified as chemical aging, it is a good start to understand many similar mechanisms.

As a final note it is appropriate to point out that not all aging effects can be classified as either physical or chemical. For example, in many polymer-processing operations for consumer items such as art objects, kitchen utensils, souvenir items, or auto parts, plasticizers are used to make parts more pliable during processing and/or to speed up processing time. Often some of the plasticizing agent remains after the process is complete and, over time, the plasticizer desorbs leaving voids or cracks leading to diminished mechanical properties. For transparent objects, the resulting small surface crazes or cracks are clearly visible. Sometimes even very large cracks result and they can be observed even it the object is not transparent. Most reputable manufactures are aware of these problems and adjust their processing operations so as to minimize such concerns.

The subject of physical and chemical aging received a great deal of attention for applications related to the aerospace industry in the late 20th century and deserves more attention by other industries that manufacture structural systems from polymer based materials.

7.8. Review Questions

7.1. Name the five regions of viscoelastic behavior of a polymer and give a sketch of the 10 second modulus vs. temperature for thermoplastic (amorphous and crystalline) and thermoset polymers.

7.2. Describe how the 10 second modulus is determined experimentally using sketches and equations as necessary.

7.3. Define and describe the time-temperature-superposition principle.

7.4. What is a master curve? How is one produced?

7.5. Who was the person that first introduced the use of master curves.

7.6. Describe the kinetic theory of polymers.

7.7. Discuss the meaning of the term "dilute solution" and describe how this is important in the development of the TTSP.

7.8. Under what conditions is the WLF equation valid?

7.9. Define specific volume and how is it measured? What is the fractional free volume?

7.10. What is a relaxation spectrum and how is it related to a relaxation modulus?

7.11. Describe the molecular mechanisms associated with viscoelastic response in the glassy, transition and rubbery regions of behavior.

7.12. Describe the process of physical and chemical aging.

7.13. Describe the thermal spike mechanism associated with moisture absortion.

7.9. Problems

7.1. Given the data below, develop a master curve using TSSP.

Time (min)	D(t), 90C	D(t), 100C	D(t), 110C	D(t), 115C
0.5	2.700E-06	2.990E-06	3.330E-06	4.260E-06
1	2.820E-06	3.080E-06	3.570E-06	4.650E-06
2	2.900E-06	3.170E-06	3.700E-06	5.130E-06
5	3.130E-06	3.390E-06	4.080E-06	6.060E-06
10	3.230E-06	3.640E-06	4.440E-06	7.410E-06

Time (min)	D(t), 120C	D(t), 122C	D(t), 125C	D(t), 130C
0.5	7.690E-06	1.282E-05	3.030E-05	1.786E-04
1	9.260E-06	1.587E-05	4.651E-05	2.632E-04
2	1.163E-05	2.128E-05	7.143E-05	3.333E-04
5	1.667E-05	3.571E-05	1.351E-04	3.636E-04
10	2.222E-05	4.762E-05	2.000E-04	3.704E-04

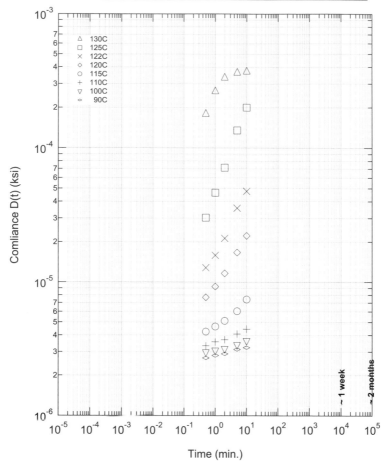

Short term compliance data at eight temperatures for a polymer.

7.2. Determine the shift factors needed to obtain the master curve found in problem 1. Plot your results (Log$_{10}$ a$_T$ vs. temperature) and compare with a plot of the WLF equation.

7.3. Calculate the shift factors for the data of problem 2 using the activation energy approach and compare with the experimentally determined shift factors found below the T$_g$.

7.4. The creep data for an epoxy is given in Chapter 5 on page 135 of these notes. Using the TTSP equation derived in class, determine the maximum (asymptotic) value you would expect the deflection, δ, to be for the temperatures, T= 155° C, 160° C and 165° C.

7.5. Using the TTSP prove that modulus should increase with increasing temperature (the Joule effect) in the rubbery range.

7.6. A master curve for a polymer for a temperature of 100° C is given in the figure below. Assuming the TTSP and the WLF equation is valid, estimate the short and long time response for temperatures of 120° C and 90° C. Sketch the expected master curves for each temperature on the given graph.

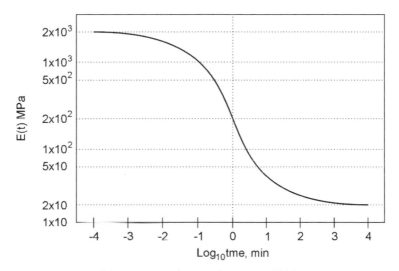

Master curve for a polymer at 100°C.

7.7 A master curve and data are given below for a polymer at $T_g = 100°$ C. Construct an E(0.1 min) vs. temperature curve from the data. (You may neglect a vertical shift for this problem.)

Time (min)	Modulus (GPa)
0.01	2
0.1	1.6
1	0.8
3.5	0.4
10	0.12
35	0.025
100	0.01
1000	0.006
1.0E+04	0.005
1.0E+05	0.005

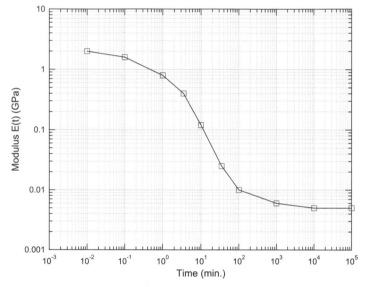

Master curve at 100°C for a polymer associated with tabulated values in problem 7.7.

8. Elementary Viscoelastic Stress Analysis for Bars and Beams

The study of polymer engineering science and viscoelasticity is not complete unless attention is given to the stress (or strain) analysis of important structural problems. These include sets of problems related to viscoelastic materials (e.g., polymers) analogous to those in the first course in solid mechanics (often called strength of materials), courses on structural mechanics (including energy methods, Castigliano's theorems, etc.), the theory of linear elasticity (stress functions, three dimensional problems, etc.), the theory of linear elastic plates and shells, elastic stability and others. While it is not possible to cover all these topics, it is possible to cover selected problems in several areas to demonstrate common methods of approach such that individuals can continue to explore problems unique to their own area of interest. Hopefully, even the brief introduction given here can assist one in solving structural analysis problems for viscoelastic materials provided the necessary background to solve a similar structural analysis problem for an elastic material has been mastered.

8.1. Fundamental Concepts

Generally viscoelastic problems can be solved using relations between internal stresses and external loads subject to the geometry of the structure in a similar manner as for elastic materials in the subject areas mentioned above. For both elastic and viscoelastic materials, the "state of the material" or equations of state must be included. Here elastic and viscoelastic materials are different in that the former does not include memory (or time dependent) effects while the latter does include memory effects. Because of this difference, stress, strain and displacement distributions in polymeric structures are also usually time dependent and may be very different from these quantities in elastic structures under the same conditions.

In Chapter 6, it was shown that the Boltzman superposition principle could be used to derive an integral constitutive law for a linear viscoelastic material as

$$\sigma(t) = \int_0^t E(t-\tau)\frac{d\varepsilon(\tau)}{d\tau}d\tau$$

(8.1)

Taking the Laplace transform of this convolution integral yields

$$\overline{\sigma}(s) = s\overline{E}(s)\overline{\varepsilon}(s)$$

(8.2)

which can also be written as

$$\overline{\sigma}(s) = \overline{E}^*(s)\overline{\varepsilon}(s)$$

(8.3)

where $\overline{E}^*$ **(s)** is s times the Laplace transform of the time dependent modulus of the material, **E(t)**. In Chapter 5 it was also shown that for representation of viscoelastic materials by mechanical models and differential equations, $\overline{E}^*$ **(s)** is the ratio of the transform of the strain and stress differential operators.

Eq. 8.3 is equivalent, in transform space, to Hooke's law for an axially loaded elastic bar or,

$$\sigma = E\varepsilon$$

(8.4)

The stress and strain in an elastic structure may vary with time providing external loads vary with time. Therefore, it is possible to transform time dependent stresses and strains for elastic structures to give,

$$\overline{\sigma}(s) = E\overline{\varepsilon}(s)$$

(8.5)

but since the modulus is time independent, the resulting equation is quite different than **Eq. 8.3**, i.e., **E** in **Eq. 8.5** is a constant but $\overline{E}^*$ **(s)** in **Eq. 8.3** is the Laplace transform of time dependent functions.

The fact that **Eq. 8.3** can be considered as the equivalent of Hooke's law in the transform domain leads to a general method to solve many practical viscoelastic boundary value problems in a simple manner. This procedure is often attributed to Turner Alfrey and is sometimes referred to as **Alfrey's correspondence principle.** Simply stated the procedure is as follows:

- **Find a previously solved linear elasticity boundary value problem with the same geometry, loading, and boundary conditions as the linear viscoelastic boundary problem for which a solution is needed.**

- **Replace all variables in the elastic solution (stresses, strains, displacements, etc.) and the applied loads by their Laplace transform.**

- **Replace all elastic constants by s times the transform of the time dependent moduli (or the ratio of the transform of the analogous differential operators). That is:**

$$E \to \overline{E}^*(s) = s\overline{E}(s)$$

$$D \to \overline{D}^*(s) = s\overline{D}(s)$$

$$G \to \overline{G}^*(s) = s\overline{G}(s) \tag{8.6}$$

$$\nu \to \overline{\nu}^*(s) = s\overline{\nu}(s)$$

- **The resulting expressions are the solution in the transform domain to the viscoelastic boundary value problem. The solution in the time domain can be found upon inversion.**

In this chapter the correspondence principle will be used to solve elementary viscoelastic problems for bars and beams. In the following chapter the principle will be used to solve problems in two-dimensional elasticity. This procedure can only be used for a certain class of problems. In general, the procedure can be used on any problem in which the load functions (including boundary conditions) can be separated into a product function of space and time. These restrictions will be discussed more fully in the following chapter on two and three-dimensional problems. It is also appropriate to note that in addition to the correspondence principle there are, in general, two additional methods that may be used to solve viscoelastic boundary value problems. These are: formulate and solve the problem in the time domain or formulate and solve the problem in the transform domain. The latter two techniques will be discussed and demonstrated in detail in Chapter 9. The reason for mentioning these methods here is that they can best be demonstrated at an elementary level using the derivation of the beam deflection equation for pure bending as discussed later in this chapter.

8.2. Analysis of Axially Loaded Bars

Consider an elastic and a viscoelastic bar in uniaxial tension as shown in **Fig. 8.1** where the axial load may be time dependent.

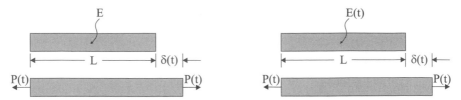

Fig. 8.1 Loads and deformation in elastic (a) and viscoelastic (b) bars.

As noted for an elastic bar in Chapter 2, the average or engineering stress, the average or engineering strain and Hooke's law are given by,

$$\sigma(t) = \frac{P(t)}{A}, \ \varepsilon(t) = \frac{\delta(t)}{L}, \ \sigma(t) = E\varepsilon(t) \tag{8.7}$$

where **P(t)** is the applied load, **L** is the original length, **A** is the original cross-sectional area and **E** is Young's modulus and is a constant. Obviously, the only reason for the variation of stress and strain with time is due to the variation of load with time. The stress-strain equation can be written as,

$$\frac{P(t)}{A} = E\frac{\delta(t)}{L} \tag{8.8}$$

and the axial deformation would be,

$$\delta(t) = \frac{P(t)L}{AE} \tag{8.9}$$

Again, the deformation varies with time only because the load varies with time. Note the cross sectional area used is still the original area and is constant. If true stress were used, the cross section would change but for many polymers under practical loads the variation would be small and can be neglected.

For a viscoelastic bar, a solution for stresses, strains and displacements can be obtained using the correspondence principle by replacing all variables in **Eq. 8.7** by their Laplace transforms and the moduli by s times their Laplace transform,

$$\overline{\sigma}(s) = \frac{\overline{P}(s)}{A}, \quad \overline{\varepsilon}(s) = \frac{\overline{\delta}(s)}{L}, \quad \overline{\varepsilon}(s) = \overline{D}^*\overline{\sigma}(s), \quad \overline{\delta}(s) = \frac{\overline{P}(s)L}{A}\overline{D}^* \quad (8.10)$$

where Hooke's law and the deformation have been written in terms of compliance instead of modulus. The solution for stress in the time domain can be found by finding the inverse Laplace transform and would be,

$$\sigma(t) = \frac{P(t)}{A} \quad (8.11)$$

Comparing **Eq. 8.11** with the first **Eq.** in **8.7** it is seen that the stress is exactly the same as in an elastic beam with a time dependent axial load.

In fact, from this result it is clear that any elastic problem in which no elastic constants appear in the solution will have a counterpart viscoelastic solution that will be identical to the elastic solution.

Solutions for deformations where tractions are prescribed always have material properties included and therefore, displacements in elastic and viscoelastic bodies will be quite different. Noting that,

$$\overline{D}^*(s) = s\overline{D}(s) \quad (8.12)$$

the displacement of the viscoelastic bar given by the last equation in **Eq. 8.10** is rewritten as,

$$\overline{\delta}(s) = \frac{\overline{P}(s)L}{A} \cdot s\overline{D}(s) \quad (8.13)$$

Eq. 8.13 can be inverted using the convolution integral and is,

$$\delta(t) = \frac{L}{A} \int_0^t D(t-\xi)\frac{dP(\xi)}{d\xi}d\xi \quad (8.14)$$

For uniaxial loading, either **Eq. 8.13** or **Eq. 8.14** can be used to solve for the displacement in a viscoelastic bar over time given an explicit loading function, **P(t)** and material compliance, **D(t)**. It is sometimes useful to manipulate the expressions algebraically in the Laplace domain, **Eq. 8.13**, and then simply invert the final expression to the time domain.

In the simple case that the axial load is a constant step input given by,

$$P(t) = P_0 H(t) \qquad \text{and} \qquad \overline{P}(s) = \frac{P_0}{s} \quad (8.15)$$

the stress in the bar (elastic or viscoelastic) will be,

$$\sigma = \frac{P_0}{A} \qquad (8.16)$$

The displacement for a viscoelastic bar can be rewritten as in **Eq. 8.14**,

$$\delta(t) = \frac{L}{A} \int_0^t D(t - \xi) \frac{d\big[P_0 H(\xi)\big]}{d\xi} d\xi \qquad (8.17)$$

or

$$\delta(t) = \frac{P_0 L}{A} \int_0^t D(t - \xi) \delta(\xi) d\xi \qquad (8.18)$$

where, due to unfortunate conventional notation the $\delta(t)$ on the left hand side is the axial deformation, while the $\delta(\xi)$ on the right hand side is the Dirac delta function. **Eq. 8.18** becomes (upon using the result in Appendix A for integrating Dirac delta functions),

$$\delta(t) = \frac{P_0 L}{A} D(t) \qquad (8.19)$$

or as mentioned above, **Eq. 8.13** can be rewritten using **Eq. 8.15** to obtain,

$$\overline{\delta}(s) = \frac{P_0 L}{A} \cdot \overline{D}(s) \qquad (8.20)$$

and inverted to obtain **Eq. 8.19** without recourse to integral equations.

Thus, for the constant load input of **Eq. 8.15**, the resulting displacement for an elastic bar is a constant, **PL/AE**, while the viscoelastic bar exhibits creep and increasing displacements with time.

Note that the result for the case of a step load is quite simple and provides displacements in **Eq. 8.19** that are identical in form to the elastic displacements in **Eq. 8.9** with **1/E** replaced by the elastic compliance **D**. However, for any non-constant load, the integration of **Eq. 8.14** becomes non-trivial, cannot be solved without explicitly stating a form for compliance **D(t)** and yields a very different displacement field in the viscoelastic material over time.

To provide an example of a non-trivial case, consider a bar in uniaxial tension where the load is given by

$$P(t) = p_0 t \qquad \text{and} \qquad \overline{P}(s) = p_0 / s^2 \qquad (8.21)$$

In this case, the stress field for both elastic and viscoelastic bars is

$$\sigma(t) = \frac{p_0}{A}t \tag{8.22}$$

and the displacement field for the elastic bar from **Eq. 8.9** is given by

$$\delta(t) = \frac{p_0 L}{AE}t \tag{8.23}$$

For a viscoelastic bar, **Eq. 8.13** can be used in the Laplace domain, simplified and inverted, or **Eq. 8.14** can be used directly in the time domain as illustrated here. In either case, the time dependent compliance of the material must be chosen to determine the solution. Here, choose a simple Kelvin solid such that

$$D(t) = \frac{1}{E}\left(1 - e^{-t/\tau}\right) \tag{8.24}$$

Using **Eq. 8.20** and **8.23** in **Eq 8.14** yields

$$\begin{aligned}
\delta(t) &= \frac{L}{A}\int_0^t \frac{1}{E}\left(1 - e^{-(t-\xi)/\tau}\right)p_0 d\xi \\
&= \frac{Lp_0}{AE}\left(t - \tau\left(1 - e^{-t/\tau}\right)\right)
\end{aligned} \tag{8.25}$$

which is clearly different in form from the elastic solution **Eq 8.22**. The resulting time dependent displacements can be plotted as shown in **Fig. 8.2**, where it is seen that the displacements in the viscoelastic bar lag the elastic solution due to the delay in the viscous term.

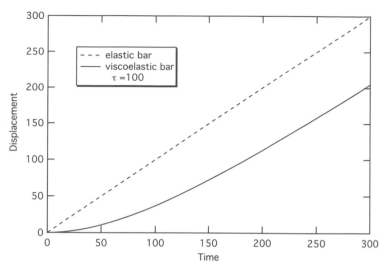

Fig. 8.2 Displacement with time for uniaxial loading with $P(t)=p_0t$. Note difference in elastic and viscoelastic response. Simple Kelvin model used for viscoelastic bar.

8.3. Analysis of Circular Cylinder Bars in Torsion

A viscoelastic bar in torsion can be analyzed in a similar manner as the axial bar in tension or compression. Assume a time dependent end torque is applied to a circular cylindrical bar as shown in **Fig. 8.2**,

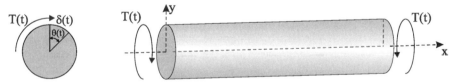

Fig. 8.3 Torsion of elastic and viscoelastic bars.

The stress and angular deformation for an elastic bar is given by,

$$\tau(t) = \frac{T(t)r}{I_p} \qquad \text{and} \qquad \theta(t) = \frac{T(t)L}{I_p G} \qquad (8.26)$$

where I_p is the polar moment of inertia (second moment of area), r is the radius to the location in the cross section where the stress is to be determined, $T(t)$ is the time dependent input torque, L is the length and G is the shear modulus.

For a viscoelastic bar in the transform domain, the solution is found by replacing all variables in elastic solution by their Laplace transform (and moduli by s times their Laplace transform) such that,

$$\bar{\tau}(s) = \frac{\overline{T}(s)r}{I_p} \qquad \text{and} \qquad \bar{\theta}(s) = \frac{\overline{T}(s)L}{I_p} \cdot s\bar{J}(s) \qquad (8.27)$$

where $\bar{J}(s)$ is the transform of the shear creep compliance and $\bar{J}^*(s) = s\bar{J}(s)$.

Inversion of the equation for the transform of shear stress will give the solution for shear stress in the time domain,

$$\tau(t) = \frac{T(t)r}{I_p} \qquad (8.28)$$

which is identical to the elastic solution given in **Eq. 8.26**. Inversion of the transform of the angular displacement provides (using the same procedure as previously for a bar in tension or compression),

$$\theta(t) = \frac{L}{I_p} \int_0^t J(t - \xi) \frac{dT(\xi)}{d\xi} d\xi \qquad (8.29)$$

For a simple step input in torque,

$$T(t) = T_0 H(t) \qquad \text{and} \qquad \overline{T}(s) = \frac{T_0}{s} \qquad (8.30)$$

the solution for stress and angular displacement will become,

$$\tau = \frac{T_0 r}{I_p} \qquad \text{and} \qquad \theta(t) = \frac{T_0 L}{I_p} J(t) \qquad (8.31)$$

As for the uniaxial tension case, while the elastic solution for angular displacement is constant in time for a constant torque input, the viscoelastic bar exhibits increasing displacement from creep over time. Note again that the expression **Eq. 8.31** is quite simple in the step input case and analogous in form to the elastic solution **Eq. 8.26**. For time varying loading, the integration of **Eq. 8.29** is nontrivial and results in a more complex form.

8.4. Analysis of Prismatic Beams in Pure Bending

In general, developing appropriate stress and deformation analysis solutions for the design of complex structures made with viscoelastic polymer-based materials can be very difficult and challenging. However, as discussed in this section, the various analytical approaches mentioned earlier can be illustrated using the elementary analysis associated with beams in pure bending.

8.4.1. Stress Analysis of Beams in Bending

A viscoelastic bar in pure bending can be analyzed in a similar manner as the axial bar in tension or compression and the circular cylindrical bar in torsion. Assume a time dependent bending moment is applied to a bar (with a vertical axis of symmetry) as shown in **Fig. 8.4**,

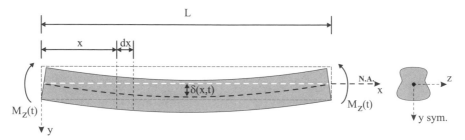

Fig. 8.4 Pure bending of elastic and viscoelastic bars.

The bending stress and deformation for an elastic bar is given by,

$$\sigma_x(y,t) = \frac{M_z(t)y}{I_z}, \quad \frac{d^2\delta(x,t)}{dx^2} = \frac{M_z(t)}{EI_z} \qquad \textbf{(8.32a, 8.32b)}$$

where I_z is the moment of inertia (second moment of area) of the cross-section, y is the vertical distance from the neutral axis to the location in the cross section where the stress is to be determined, $M_z(t)$ is the time dependent input bending moment, L is the length and E is Young's modulus.

For a viscoelastic bar in the transform domain, the solution is found by replacing all variables in the elastic solution by their Laplace transform (and all moduli by s times their Laplace transform) such that,

$$\overline{\sigma}_x(y,s) = \frac{\overline{M}_z(s)y}{I_z} \tag{8.33}$$

Inversion of the **Eq. 8.33** will give the solution for axial stress in the time domain,

$$\sigma_x(y,t) = \frac{M_z(t)y}{I_z} \tag{8.34}$$

which is identical to the elastic solution given in **Eq. 8.32a**. Displacements, on the other hand, will be quite different for elastic and viscoelastic beams in bending as is illustrated in the following sections.

8.4.2. Deformation Analysis of Beams in Bending

Three methods of solving viscoelastic boundary value problems were given early in the Fundamental Concepts section of this chapter. The development of the beam equation serves as a simple method to illustrate these various techniques. Before proceeding with this section, the reader is advised to review the procedure for developing the deflection equation for linear elastic prismatic beams given in elementary texts on solid mechanics.

It may be well to note that while the deflection derivations shown in this section are for pure bending, the equations developed are valid for general loadings (i.e., point, distributed, etc.) as long as shear deformations are negligible as in elastic beams.

(1) Development of beam deflection equation using the correspondence principle: Development of an appropriate equation for the deformation of any viscoelastic beam can be developed using the correspondence principle. That is, the viscoelastic equivalent to the deflection equation given by **Eq. 8.32a** can be developed in the transform domain by replacing the appropriate variables by their Laplace transform,

$$\frac{d^2\overline{\delta}(x,s)}{dx^2} = \frac{\overline{M}_z(s)}{I_z}\overline{D}^*(s) \tag{8.35}$$

or

$$\frac{d^2\overline{\delta}(x,s)}{dx^2} = \frac{\overline{M}_z(s)}{I_z} \cdot s\overline{D}(s) \tag{8.36}$$

Inversion yields,

$$\frac{d^2\delta(x,t)}{dx^2} = \frac{1}{I_z} \int_0^t D(t-\xi) \frac{dM_z(\xi)}{d\xi} d\xi \qquad (8.37)$$

If a step bending moment is input such that,

$$M_z(t) = M_z H(t) = M_0 \qquad \text{and} \qquad \overline{M}_z(s) = \frac{M_0}{s} \qquad (8.38)$$

integration of Eq. 8.31 yields,

$$\frac{d^2\delta(x,t)}{dx^2} = \frac{M_0}{I_z} D(t) \qquad (8.39)$$

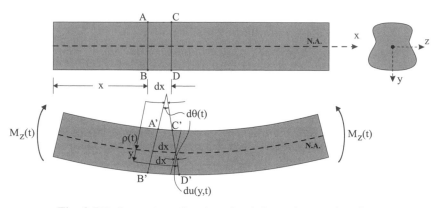

Fig. 8.5 Deformation of a viscoelastic beam in pure bending.

(2) Derivation of viscoelastic beam deflection equation in the time domain: It is instructive to derive the deflection equation for a viscoelastic beam without resorting to Laplace transforms. Consider the undeformed and deformed beam shown in **Fig. 8.5**. Making the assumptions (the same as in elementary solid mechanics) of small deformations, linear behavior, and a non-warping cross-sections (plane sections remain plane) will give the relations,

$$dx = \rho(t)d\theta(t) \qquad du(y,t) = yd\theta(t) \qquad \frac{du(y,t)}{dx} = \frac{y}{\rho(t)} = \varepsilon_x(y,t) \quad (8.40)$$

where $\rho(t)$ is the radius of curvature, $\theta(t)$ is the angular rotation of adjacent cross-sectional plane sections and $u(t)$ is the deformation in the x direction of a point y distance from the neutral axis (NA). The length of the beam at the neutral axis does not change with time but the radius of curvature, $\rho(t)$, decreases with time as the angular deformation increases with time. The relation between axial stress and strain at any point on the cross-section is given by,

$$\varepsilon_x(y,t) = \int_0^t D(t-\xi) \frac{\partial \sigma_x(y,\xi)}{\partial \xi} \partial\xi = \frac{y}{\rho(t)} = y\kappa(t) \tag{8.41}$$

where $\kappa(t)$ is the curvature. Alternatively, writing with stress as the dependent variable gives,

$$\sigma_x(y,t) = y \int_0^t E(t-\xi) \frac{\partial \kappa(\xi)}{\partial \xi} \partial\xi \tag{8.42}$$

Equilibrium of forces in the axial direction on any cross-section gives,

$$\sum F_x = 0 = \int_A y \left[\int_0^t E(t-\xi) \frac{d\kappa(\xi)}{d\xi} d\xi \right] dA$$

$$0 = \left[\int_0^t E(t-\xi) \frac{d\kappa(\xi)}{d\xi} d\xi \right] \int_A y dA \tag{8.43}$$

and indicates that the neutral axis is at the centroid of the cross-section as in elementary beam theory. Equilibrium of moments about the z-axis gives,

$$\sum M_z = 0 = M_z(t) - \int_A y^2 \left[\int_0^t E(t-\xi) \frac{d\kappa(\xi)}{d\xi} d\xi \right] dA \tag{8.44}$$

or

$$M_z(t) = \int_A y^2 \left[\int_0^t E(t-\xi) \frac{d\kappa(\xi)}{d\xi} d\xi \right] dA \tag{8.45}$$

From **Eqs. 8.40**,

$$\varepsilon_x(x,t) = \frac{y}{\rho(t)} = y\kappa(t) \tag{8.46}$$

and noting that,

$$\kappa(t) = \frac{d^2v/dx^2}{\left[1+\left(dv/dx\right)^2\right]^{3/2}} \cong \frac{d^2v}{dx^2} = \frac{d^2\delta}{dx^2} \tag{8.47}$$

where **v** is the displacement in the **y** direction, one obtains,

$$M_z(t) = \int_A y^2 \left[\int_0^t E(t-\xi) \frac{d}{d\xi} \left(\frac{d^2\delta}{dx^2} \right) d\xi \right] dA \qquad (8.48)$$

or

$$\frac{d^2\delta(t)}{dx^2} = \frac{1}{I_z} \left[\int_0^t D(t-\xi) \frac{dM_z(\xi)}{d\xi} d\xi \right] \qquad (8.49)$$

which is the same result obtained by the correspondence principle given by **Eq. 8.37**. The proof that **Eq. 8.49** follows from **Eq. 8.48** is left as an exercise for the reader (See homework problem 8.5).

(3) Derivation of viscoelastic beam deflection equation in the transform domain: The beam loaded as shown in **Fig. 8.5** can be converted to an associated problem in the transform domain by transforming all time dependent parameters and the boundary conditions. Obviously, the deflection equation in the transform plane can be developed following the derivation steps as used in elementary solid mechanics and a result equivalent to **Eqs. 8.35-8.37**. This proof is left as an exercise for the reader (see homework problem 8.6).

8.5. Stresses and Deformation in Beams for Conditions other than Pure Bending

The equations for bending stress and deflection developed in the previous sections may be used for beams with loading conditions other than pure bending. In so doing, the same restrictions apply as in using their elastic counterparts for conditions other than pure bending. A few examples will be given for beams with distributed loads.

Example 1: Consider a simply supported rectangular viscoelastic beam with a step input of a uniformly distributed load as given in **Fig. 8.6(a)**.

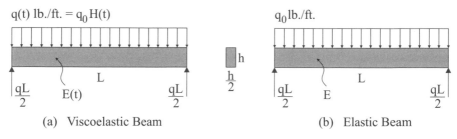

(a) Viscoelastic Beam   (b) Elastic Beam

Fig. 8.6 Uniformly loaded elastic and viscoelastic beams.

The solution for stresses, deformations and strains for an elastic beam of the same geometry and loading is known. For example the maximum bending stress occurs at the outer surface of the beam at mid-span and is given by,

$$\sigma_{max} = \frac{M_{max}y}{I} = \frac{3}{2}\frac{q(t)L^2}{h^3} = \frac{3}{2}\frac{q_0L^2}{h^3} \tag{8.50}$$

Replacing all variables by their Laplace transform gives,

$$\overline{\sigma}_{max}(s) = \frac{3}{2}\frac{\overline{q}(s)L^2}{h^3} = \frac{3}{2}\frac{q_0L^2}{s\,h^3} \tag{8.51}$$

Inversion of this solution will provide the solution for the viscoelastic beam and will be,

$$\sigma_{max}(t) = \frac{3}{2}\frac{q(t)L^2}{h^3} \tag{8.52}$$

or

$$\sigma_{max}(t) = \frac{3}{2}\frac{q_0L^2}{h^3}H(t) \tag{8.53}$$

Again, the viscoelastic solution for stress is exactly the same as the elastic solution stress. As stated earlier, in general, if the linear elastic solution for stresses for a given boundary value problem does not contain elastic constants, the solution for stresses in a viscoelastic body with equivalent geometry and equivalent loads is identical to that for the elastic body. This means that the stress analysis of most problems considered in elementary solid mechanics such as beams in bending, bars in torsion or axial load, pressure vessels, etc. will have the same solution for stress in a linear viscoelastic material as in a linear elastic material. Further, stress analysis of combined axial, bending, torsion and pressure loads can be handled easily using superposition.

Solutions for beam deflections (or stresses induced by deflections such as a sagging supports) will be quite different for elastic and viscoelastic materials. This difference is due to the appearance of elastic constants in the mathematical expressions for deflections and displacements and is demonstrated in the next examples.

Example 2: The maximum elastic deflection for the beam in **Fig. 8.6(b)** will be,

$$\delta_{max} = \frac{5}{384} \frac{q_0 L^4}{I} \frac{1}{E} = \frac{5}{384} \frac{q_0 L^4}{I} D \tag{8.54}$$

where $D = 1/E$ is the elastic compliance. The viscoelastic solution in the transform domain will be,

$$\bar{\delta}_{max}(s) = \frac{5}{384} \frac{\bar{q}(s) L^4}{I} \bar{D}^*(s) \tag{8.55}$$

Recalling that $\bar{D}^*(s) = s\bar{D}(s)$ gives,

$$\bar{\delta}_{max}(s) = \frac{5}{384} \frac{\bar{q}(s) L^4}{I} \cdot s\bar{D}(s) \tag{8.56}$$

Knowing that the transform of $q(t) = q_0 H(t)$ is q_0/s and substituting in **Eq. 8.56** gives,

$$\bar{\delta}_{max}(s) = \frac{5}{384} \frac{q_0 L^4}{I} \bar{D}(s) \tag{8.57}$$

and the viscoelastic solution in the time domain will be,

$$\delta_{max}(t) = \frac{5}{384} \frac{q_0 L^4}{I} D(t) \tag{8.58}$$

Assuming the material to be a thermoset polymer that can be represented by a three parameter solid, the deflection at mid-span would become,

$$\delta_{max}(t) = \frac{5}{384} \frac{q_0 L^4}{I} \left[\frac{1}{E_0} + \frac{1}{E_1}(1 - e^{-t/\tau}) \right] \tag{8.59}$$

and the maximum deflection would vary with time. Again, note that the simple inversion of **Eq. 8.57** and the resulting elastic-like form of **Eq. 8.58** is due to the constant load applied in this example. For time varying loads, the inversion step results in expressions that differ substantially in form from that of the elastic solution. This case will be considered in **Examples 4-5**.

Example 3: Now consider the case of the simply supported viscoelastic beam shown in **Fig. 8.7** which is suddenly given a constant deformation at mid-span. The objective is to find the amount of a center load needed for the beam deformation to remain constant.

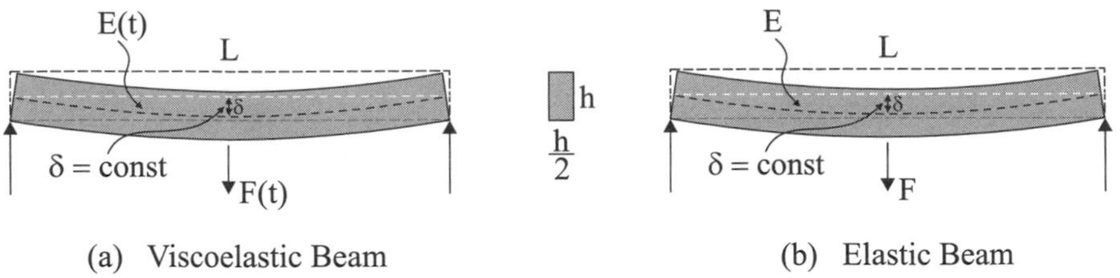

(a) Viscoelastic Beam   (b) Elastic Beam

Fig. 8.7 Elastic and viscoelastic beam.

The deflection produced by a central constant force in the elastic beam will be,

$$\delta = \frac{FL^3}{48EI} \tag{8.60}$$

or

$$F = \frac{48I}{L^3} \delta\, E \tag{8.61}$$

and, of course, this force will remain constant for an elastic beam. For a viscoelastic beam, the force corresponding to a central displacement can be found using the correspondence principle,

$$\overline{F}(s) = \frac{48I}{L^3} \overline{\delta}(s)\, \overline{E}^*(s) \tag{8.62}$$

If the deflection input it is assumed to be stepwise or $\delta(t) = \delta_0\, H(t)$, the Laplace transform will be $\overline{\delta}(s) = \delta_0/s$. Substituting this input condition into **Eq. 8.62** as well as $\overline{E}^* = s\overline{E}(s)$ and inverting gives the viscoelastic solution,

$$F(t) = \frac{48I}{L^3} \delta_0\, E(t) \tag{8.63}$$

The force $F(t)$ will vary with time as the relaxation modulus varies with time. If the beam is made of a thermoplastic polymer and if representation by a Maxwell model is appropriate, the solution will be,

$$F(t) = \frac{48I}{L^3} \delta_0\, Ee^{-t/\tau} \tag{8.64}$$

and the force will vary with time as shown in **Fig. 8.8**.

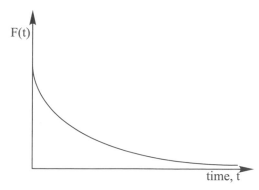

Fig. 8.8 Force need to maintain a constant central deflection a thermoplastic viscoelastic beam.

The force needed to hold the deflection constant will become zero for a
time large compared to the relaxation time. On the other hand if the beam
is made of a thermoset, the force would not decrease to zero regardless of
the time duration.

This example further illustrates that the stresses are not the same in a
viscoelastic beam as in an elastic beam when displacements are prescribed.
That is, the stresses in the beam of this example approach zero as time approaches infinity.

Determining deflections for circumstances when the load is both a function of the spatial coordinates and time requires special attention. The following example demonstrates the appropriate approach.

Example 4: Assume a beam is loaded as shown in **Fig. 8.9** and that the
distributed load is both a function of distance and time. While a general
loading, **p(x,t)**, can be accommodated with the correspondence principle,
only a loading which is a product of two separate functions as shown in
Fig. 8.9., i.e., **p(x,t) = p(x)f(t)** will be discussed here.

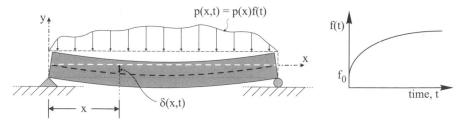

Fig. 8.9 Viscoelastic beam with a spatially varying distributed load.

The solution for an elastic beam with an equivalent load can be written as,

$$\delta(x,t) = \frac{p(x,t)}{EI} g(x) = \frac{p(x,t)}{I} g(x)D = \frac{p(x)f(t)}{I} g(x)D \qquad (8.65)$$

where $\mathbf{D}$ is the constant elastic compliance and $\mathbf{g(x)}$ contains the spatial distribution of the deflection solution beyond $\mathbf{p(x)}$. To clarify the definition of $\mathbf{g(x)}$, consider the deflection equation of an elastic beam with a uniformly varying load, where $\mathbf{p(x,t)=q_0x}$, which can be found in elementary texts to be,

$$\delta(x) = \frac{1}{EI} \cdot \frac{q_0 x}{360L} \left(7L^4 - 10L^2 x^2 + 3x^4\right) \qquad (8.66)$$

Using the notation from **Eq. 8.65**, for this case

$$g(x) = (7L^4 - 10L^2 x^2 + 3x^4) \qquad (8.67)$$

Returning to the general expression **Eq. 8.65**, using the correspondence principle for a polymer beam that is viscoelastic, the solution in the transform domain will be,

$$\overline{\delta}(x,s) = \frac{p(x)}{I} g(x) \cdot \overline{f}(s)\overline{D}^*(s) = \frac{p(x)}{I} g(x)\left[\overline{f}(s) \cdot s\overline{D}(s)\right] \qquad (8.68)$$

Using the definition of the convolution integral, the solution in the time domain will be,

$$\delta(x,t) = \delta'(x) \int_{0^-}^{t} D(t - \tau) \frac{df(\tau)}{d\tau} d\tau \qquad (8.69)$$

where,

$$\delta'(x) = \frac{p(x)}{I} F(x) \qquad (8.70)$$

In **Eq. 8.70** $\delta'(x)$ is the deflection for an elastic beam loaded only with $\mathbf{p(x)}$ but with the Young's modulus removed. That is, Young's modulus is now included in the integrand of **Eq. 8.69** as the compliance in the transform domain. For the loading given in **Fig. 8.9**, it is to be noted that the integral must include the jump discontinuity at the origin by recognizing that the initial value of $\mathbf{f(t)}$ is $f_0 H(t)$. Using the results of Appendix A, the inverse of **Eq. 8.68** can be written as,

$$\delta(x,t) = \delta'(x)\left[f_0 D(t) + \int_{0^+}^{t} D(t-\xi)\frac{df(\xi)}{d\xi}\,d\xi\right]$$ (8.71)

Example 5: To illustrate the use of the results of **Example 4**, consider a beam under a uniformly varying distributed load as shown in **Fig. 8.10** which has a time dependency as given also in **Fig. 8.10**.

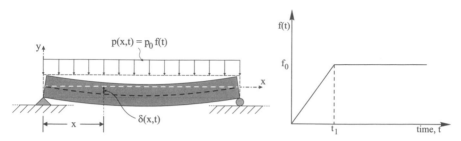

Fig. 8.10 Viscoelastic beam with a uniformly varying distributed load.

The elastic solution for a uniformly loaded beam is,

$$\delta(x,t) = \delta'(x)f(t)D$$ (8.72)

where

$$\delta'(x) = \frac{p_0}{24\,I}\left[x^4 - 2Lx^3 + L^3 x\right]$$ (8.73)

The viscoelastic solution can be found using **Eq. 8.69** noting that the time dependent portion of the load function can be written as

$$f(t) = \frac{f_0}{t_1}tH(t) - \frac{f_0}{t_1}(t-t_1)H(t-t_1)$$ (8.74)

To perform the integration, it will be convenient to separate the integral and thus the function into portions before and after **t_1**. In this form, **f(t)** and it's first derivative may be written:

$$\begin{cases} f(t) = \dfrac{f_0}{t_1}t, & \dfrac{df}{dt} = \dfrac{f_0}{t_1}, & 0 \le t \le t_1 \\[2mm] f(t) = f_0, & \dfrac{df}{dt} = 0, & t > t_1 \end{cases}$$ (8.75)

Using **Eq. 8.71**, the deflection is found to be given by,

$$\delta(x,t) = \begin{cases} \delta'(x)\dfrac{f_1}{t_1}\displaystyle\int_0^{t_1} D(t-\xi)d\xi, & 0 \le t \le t_1 \\[4mm] \delta'(x)\left[\dfrac{f_1}{t_1}\displaystyle\int_0^{t_1} D(t-\xi)d\xi + (0)\int_{t_1}^{t} D(t-\xi)d\xi\right], & t > t_1 \end{cases} \qquad (8.76)$$

Assuming the beam can be represented as a Kelvin solid such that,

$$D(t-\xi) = \frac{1}{E}\left[1 - e^{-(t-\xi)/\tau}\right] \qquad (8.77)$$

the deflection will be,

$$\delta(x,t) = \begin{cases} \dfrac{\delta'(x)}{E}\dfrac{f_0}{t_1}\left[t - \tau\left(1 - e^{-t/\tau}\right)\right], & 0 \le t \le t_1 \\[4mm] \dfrac{\delta'(x)}{E}\dfrac{f_0}{t_1}\left[t_1 - \tau e^{-t/\tau}\left(e^{t_1/\tau} - 1\right)\right], & t > t_1 \end{cases} \qquad (8.78)$$

A non-dimensional plot of the center deflection given by **Eqs. 8.77** and **8.78** are shown plotted vs. time in **Fig. 8.11**. For comparison the response of an elastic material for the time t_1 is also given. If the retardation time, τ, is much larger than the time t_1 as in the upper curve ($t_1=10$), the material behaves essentially elastically during the loading up to t_1 and there is a long transient response before the asymptotic value is reached after the load is held constant. If the retardation time is much less than the time t_1 as in the lower curve ($t_1=500$), there is an initial transient response of the polymer lagging the applied load, but after the test time reaches the relaxation time ($t=100$), the material responds nearly linearly with the rising applied load. After the constant load is reached, the remaining transient response occurs quickly and the asymptotic value is reached after a short time relative to t_1.

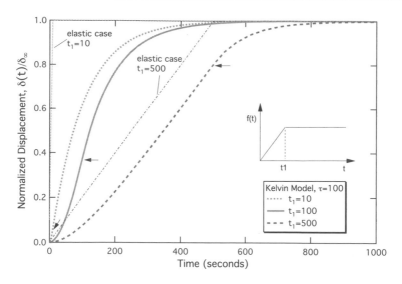

Fig. 8.11 Variation of deflection for the beam given in **Fig. 8.10**. Retardation time of polymer is the same for all cases, but time where load becomes constant, t_1, varies (and is indicated by arrows on each curve). Elastic response also shown for two cases.

8.6. Shear Stresses and Deflections in Beams

The vertical shear stress in an elastic beam is given by,

$$\tau_{xy}(x,t) = \frac{V(x,t)Q}{I_z b} \tag{8.79}$$

where **V(x,t)** is the vertical shear force (and is a function of time if the loading is a function of time) at a section x distant from the left end, **Q** is the first moment of the cross-sectional area about the neutral axis for the area above the point on the cross-section where the stress is desired (for further details, see an elementary solid mechanics text), I_z is the moment of inertia of the cross section about the neutral axis and **b** is the thickness of the beam at the location in the cross-section that the shear stress is desired. Since no properties are contained in the elastic solution, the viscoelastic solution is identical to the elastic solution.

The above solution is only valid for beams whose cross-sectional dimensions are small compared to the length. That is if **b_max/L** is approximately 10 or larger. For beams that do not meet this condition, deflections developed using pure bending theory must include a correction factor to account for the effects of shear. Also, shear stresses need to be corrected as well. The reader is referred to both elementary and advanced solid mechanics texts for the details regarding the shear correction terms.

However, as long as the material is linear, the correspondence principle can be used to obtain viscoelastic solutions from the appropriate elastic solution. It is well to note that such shear corrections are more important for polymeric materials than for metals as moduli are smaller and deformations are correspondingly larger. Therefore, shear corrections are typically more important.

8.7. Review Questions

8.1. Who first introduced the concept of the correspondence principle?

8.2 Describe how to use the correspondence principle to solve stress analysis problems.

8.3 Under what conditions are the solutions for stress identical for elastic and viscoelastic structures?

8.4 Describe three analytical approaches for obtaining solutions viscoelastic boundary value problems.

8.8. Problems

8.1. Determine an expression for the maximum stress in a viscoelastic cantilever beam made of a thermoplastic polymer that can be represented by a Maxwell fluid. Assume the beam of square cross section and is uniformly loaded with a step input.

8.2. Determine the maximum deflection in a cantilever beam made of a thermosetting polymer that can be represented by a three parameter solid. Assume the beam of square cross section and is uniformly loaded with a step input.

8.3. Determine an equation for the deflection of a simply supported beam with a uniform load varying in time similar to the one of **Example 5**. Assume the material to be Maxwellian. Graphically show the resulting deflection in a similar manner as in **Fig. 8.11**.

8.4 Prove that **Eq. 8.49** follows from **Eq. 8.48**. Hint: use Laplace transforms.

8.5 Reformulate the beam in pure bending shown in **Fig. 8.5** to a problem in the transform domain and derive the appropriate deflection solution showing that the result in the time domain is the same as given by the correspondence principle and the derivation given in the time domain.

8.6 Given a beam loaded as shown. Determine an expression for the maximum deflection as a function of time for loading condition in (a) assuming a Kelvin model. Determine an expression for the maximum deflection as a function of time for loading condition in (b) assuming a Kelvin model.

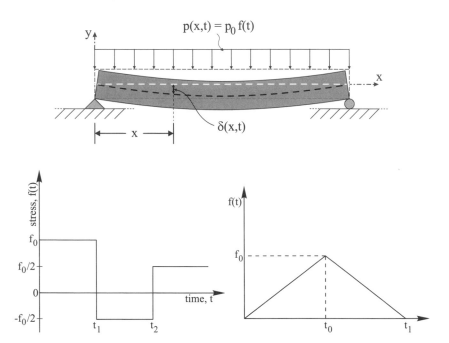

9 Viscoelastic Stress Analysis in Two and Three Dimensions

The various approaches to the solution of viscoelastic boundary value problems discussed in the last chapter for bars and beams carry over to the solution of problems in two and three dimensions. In particular, if the solution to a similar problem for an elastic material already exists, the correspondence principle may be invoked and with the use of Laplace or Fourier transforms a solution can be found. Such solutions can be used with confidence but one must be cognizant of the general equations of elasticity and the methods of solutions for elasticity problems in two and three dimensions as well as any assumptions that might often be applied. To provide all of the necessary information and background for multidimensional elasticity theory is beyond the scope of this text but the procedures needed will be outlined in the following sections.

This chapter will focus on developing the equations, assumptions and procedures one must use to solve two and three dimensional viscoelastic boundary value problems. The problem of an elastic thick walled cylinder will be used as a vehicle to demonstrate how to obtain the solution of a more difficult reinforced viscoelastic thick walled cylinder. In the process, we first demonstrate how the elasticity solution is developed and then apply the correspondence principle to transform the solution to the viscoelastic domain. Several extensions to this problem will be discussed and additional practice is provided in the homework problems at the end of the chapter.

9.1 Elastic Stress-Strain Equations

To this point the relations between stress and strain (constitutive equations) for viscoelastic materials have been limited to one-dimension. To appreciate the procedure for the extension to three-dimensions recall the generalized Hooke's law for homogeneous and isotropic materials given by **Eqs. 2.28**,

$$\varepsilon_{xx} = \frac{1}{E}\left[\sigma_{xx} - \nu\left(\sigma_{yy} + \sigma_{zz}\right)\right], \quad \gamma_{xy} = \frac{\tau_{xy}}{G}$$

$$\varepsilon_{yy} = \frac{1}{E}\left[\sigma_{yy} - \nu\left(\sigma_{xx} + \sigma_{zz}\right)\right], \quad \gamma_{yz} = \frac{\tau_{yz}}{G} \tag{9.1}$$

$$\varepsilon_{zz} = \frac{1}{E}\left[\sigma_{zz} - \nu\left(\sigma_{xx} + \sigma_{yy}\right)\right], \quad \gamma_{xz} = \frac{\tau_{xz}}{G}$$

Most generally stresses, strains and mechanical properties **E**, **G**, and **ν** are time dependent for a viscoelastic material. The relation between shear stress and strain for a viscoelastic material are easy to formulate as they only contain a single property, the shear modulus, **G**. Using the principles developed in preceding chapters, the integral equations and transform methods can be developed. However, equations involving the relationship between normal stresses and strains present a difficulty as two material properties are present and it is unclear how to formulate proper relationships analogous to the one-dimensional differential or integral equations necessary for a viscoelastic material. This difficulty can be overcome by using deviatoric and dilatational components of stress and strain as given in Chapter 2 by **Eqs. 2.58** and **2.62** which can be written for an elastic material as,

$$S_{ij} = \sigma_{ij} - \frac{1}{3}\sigma_{kk}\delta_{ij} \tag{9.2}$$

and

$$e_{ij} = \varepsilon_{ij} - \frac{1}{3}\varepsilon_{kk}\delta_{ij} \tag{9.3}$$

Using these stresses and strains, the **elastic** stress-strain relations given by **Eq. 9.1** can be shown to be,

$$S_{ij} = 2Ge_{ij}$$
$$\sigma_{kk} = 3K\varepsilon_{kk} \tag{9.4}$$

which is the same as **Eqs. 2.63** given in Chapter 2 and where,

$$\sigma_{kk} = 3\sigma_m = 3\sigma = \sigma_1 + \sigma_2 + \sigma_3$$
$$\varepsilon_{kk} = 3\varepsilon_m = 3\varepsilon = \varepsilon_1 + \varepsilon_2 + \varepsilon_3 \tag{9.5}$$

and (from **Table 2.1**),

$$G = \frac{E}{2(1+\nu)} \quad \text{and} \quad K = \frac{E}{3(1-2\nu)} \tag{9.6}$$

By rewriting the constitutive equations as in **Eq. 9.4**, each equation contains only one material property: the deviatoric stress and strain are related by the shear modulus, **G**, while the dilatational stress and strain are related by the bulk modulus, **K**.

9.2 Viscoelastic Stress-Strain Relations

In the elastic constitutive law in **Eqs. 9.4 G** and **K** are constants, with **G** being the elastic shear modulus and **K** being the elastic bulk modulus, again as defined in Chapter 2. The first equation (three equations for the three independent components s_{12}, s_{13}, s_{23}) represents only shape changes while the second equation only contains volume changes. Since only one material property is contained in each equation, the derivations developing viscoelastic constitutive laws using integral or differential equations in Chapters 5 and 6, respectively, can be applied to each equation individually. Thus, Boltzman's superposition principle can be applied to the shear and dilatational strains separately, adding up incremental contributions to the resulting stress components (as in development of **Eqs. 6.13**) resulting in

$$s_{ij}(t) = 2\int_{-\infty}^{t} G(t-\tau) \frac{\partial e_{ij}(\tau)}{\partial \tau} d\tau$$

$$\sigma_{kk}(t) = 3\int_{-\infty}^{t} K(t-\tau) \frac{\partial \varepsilon_{kk}(\tau)}{\partial \tau} d\tau \tag{9.7}$$

where **G(t)** and **K(t)** are the shear and bulk relaxation moduli. Alternately, the viscoelastic stress-strain relations in integral form can be written as,

$$e_{ij}(t) = \frac{1}{2}\int_{-\infty}^{t} J(t-\tau) \frac{\partial s_{ij}(\tau)}{\partial \tau} d\tau$$

$$\varepsilon_{kk}(t) = \frac{1}{3}\int_{-\infty}^{t} B(t-\tau) \frac{\partial \sigma_{kk}(\tau)}{\partial \tau} d\tau \tag{9.8}$$

where **J(t)** and **B(t)** are the shear and bulk creep compliances.

Similarly, the differential operator equations for the shear and dilatational responses of a viscoelastic material may be written analogous to the one-dimensional case in Chapter 5 (**Eq. 5.20** or **Eq. 5.26b**) as,

$$P\left[s_{ij}(t)\right] = 2Q\left[e_{ij}(t)\right]$$
$$\tilde{P}\left[\sigma(t)\right] = 3\tilde{Q}\left[\varepsilon(t)\right]$$

(9.9)

where **P**, **Q**, **P̃** and **Q̃** are differential operators and include the moduli and viscosity of each spring and damper in the mechanical models.

Considering **Eqs. 9.7** and **9.9**, Laplace transforms yield

$$\bar{s}_{ij}(s) = 2\overline{G}^*(s)\bar{e}_{ij}(s)$$
$$\bar{\sigma}_{kk}(s) = 3\overline{K}^*(s)\bar{\varepsilon}_{kk}(s)$$

(9.10)

where $\overline{G}^*(s)$ and $\overline{K}^*(s)$ are related to the transforms of the shear and bulk relaxation moduli via (see also **Eq. 6.44**)

$$\overline{G}^*(s) = s\overline{G}(s)$$
$$\overline{K}^*(s) = s\overline{K}(s)$$

(9.11)

Alternately, the viscoelastic stress-strain relationships in the transform domain may be written as,

$$\bar{e}_{ij}(s) = \frac{1}{2}\overline{J}^*(s)\bar{s}_{ij}(s)$$
$$\bar{\varepsilon}_{kk}(s) = \frac{1}{3}\overline{B}^*(s)\bar{\sigma}_{kk}(s)$$

(9.12)

where $\overline{J}^*(s)$ and $\overline{B}^*(s)$ are similarly related to the transforms of the shear and bulk creep compliances

$$\overline{J}^*(s) = s\overline{J}(s)$$
$$\overline{B}^*(s) = s\overline{B}(s)$$

(9.13)

The two **Eqs. 9.10** may be recombined in the transform domain to obtain an expression relating the total stress and strain tensors, σ_{ij} and ε_{ij}. In doing so, the relationship between Lame's constant, $\overline{\lambda}^*(s)$, and bulk and shear moduli, $\overline{K}^*(s)$ and $\overline{G}^*(s)$, will be recovered, Further manipulations in the transform domain result in the usual relationship between total strain and stress, analogous to **Eq. 2.36**

$$\bar{\varepsilon}_{ij} = \frac{1+\bar{v}^*}{\bar{E}^*}\bar{\sigma}_{ij} - \frac{\bar{v}^*}{\bar{E}^*}\bar{\sigma}_{kk}\delta_{ij} \qquad (9.14)$$

Note that with **Eqs. 9.10, 9.12** and **9.14**, we again have the viscoelastic constitutive law represented in the transform domain in a form equivalent to elasticity. These relationships will then allow us to utilize the correspondence principle as in Chapter 8 to solve 2D and 3D viscoelastic boundary value problems based on elasticity solutions.

9.3 Relationship Between Viscoelastic Moduli (Compliances)

Eqs. 9.10 and **9.12** as well as **9.7** and **9.8** are the viscoelastic equivalent to the generalized Hooke's law for elastic materials.

For isotropic elastic materials, there are only two independent elastic constants and relations exist between various constants as given in **Table 2.1** such as,

$$G = \frac{E}{2(1+v)} \;,\quad K = \frac{E}{3(1-2v)} \;,\quad E = \frac{9GK}{3K+G} \;,\quad v = \frac{3K-2G}{6K+2G} \qquad (9.15)$$

For an isotropic viscoelastic material only two time dependent properties are independent and it is clear from the correspondence principle that similar relationships to **Eqs. 9.15** hold for the Laplace transformed moduli such that,

$$\bar{G}^*(s) = \frac{\bar{E}^*(s)}{2(1+\bar{v}^*(s))} \qquad\qquad \bar{K}^*(s) = \frac{\bar{E}^*(s)}{3(1-2\bar{v}^*(s))}$$

$$(9.16)$$

$$\bar{E}^*(s) = \frac{9\bar{G}^*(s)\bar{K}^*(s)}{3\bar{K}^*(s)+\bar{G}^*(s)} \qquad\qquad \bar{v}^*(s) = \frac{3\bar{K}^*(s)-2\bar{G}^*(s)}{6\bar{K}^*(s)+2\bar{G}^*(s)}$$

Using relations **9.11** will convert the **Eqs. 9.16** to relations between the Laplace transform of the of relaxation moduli, creep compliances. For example,

$$\bar{E}(s) = \frac{9\bar{G}(s)\bar{K}(s)}{3\bar{K}(s)+\bar{G}(s)} \qquad (9.17)$$

or

$$3\overline{E}(s)\overline{K}(s) + \overline{E}(s)\overline{G}(s) = 9\overline{G}(s)\overline{K}(s) \qquad (9.18)$$

By the convolution theorem (Appendix B) **Eq. 9.18** becomes,

$$3\int_0^t E(t-\tau)K(\tau)d\tau + \int_0^t E(t-\tau)G(\tau)d\tau = 9\int_0^t G(t-\tau)K(\tau)d\tau \qquad (9.19)$$

Similar integral equations can be developed for each relationship given in **Eqs. 9.16** or **Table 2.1**.

9.4 Frequently Encountered Assumptions in Viscoelastic Stress Analysis

In solving viscoelastic stress analysis problems, assumptions on the material properties are often essential as gathering accurate time dependent data for viscoelastic properties is difficult and time consuming. Thus, one often only has properties for shear modulus, G(t) or Young's modulus, E(t), but not both. Yet of course for even the simplest assumption of a homogeneous, isotropic viscoelastic material, two independent material properties are required for solution of two or three dimensional stress analysis problems. Consequently, three assumptions relative to material properties are frequently encountered in viscoelastic stress analysis. These are incompressibility, elastic behavior in dilatation and synchronous shear and bulk moduli. Each of the common assumptions defines a particular value for either the bulk modulus or Poisson's ratio as follows.

1. Incompressibility: For small deformation linear elastic problems incompressibility is assured if Poisson's ratio is equal to 0.5, which also means that the bulk modulus is infinite (see **Eq. 9.6**). Under this assumption then, $\nu = 0.5$ and $K_0 = \infty$. Under the same conditions Poisson's ratio for an incompressible viscoelastic material is also a constant 0.5 and,

$$K(t) = \infty \qquad (9.20a)$$

or

$$\overline{K}(s) = \infty \qquad (9.20b)$$

Naturally, this assumption also implies that the dilatational strains are always zero. For computer simulations of viscoelastic problems, this assumption can sometimes cause numerical difficulties. Most standard finite element codes have provisions or options to be used in this case to circumvent the numerical difficulties. The assumption of incompressibility is most reasonable if the polymer under consideration is being considered at temperatures such that it is mostly within the rubbery regime.

2. Elastic in dilatation: In this case, K_0 = constant and,

$$K(t) = K_0 H(t) \qquad\qquad (9.21a)$$

$$\overline{K}(s) = \frac{K_0}{s} \qquad\qquad (9.21b)$$

Since the viscoelastic bulk modulus changes much less with time and temperature than the shear modulus, this assumption is often a good one in cases where one only has characterization data for one viscoelastic property. Note that with this assumption, the Poisson's ratio retains its time dependence.

3. Synchronous shear and bulk moduli: In this case it is assumed that the ratio of the bulk modulus to the shear modulus is a constant such that

$$K(t) = C_1 G(t) \qquad\qquad (9.22)$$

where C_1 is a constant. Thus this case assumes that the time dependence of the two moduli is the same and that the magnitude of their changes through the glass transition are proportional. As was mentioned above, typically the bulk modulus values change significantly less in crossing the glass transition, so the validity of this assumption should be carefully assessed depending on the temperature and time ranges of the problem at hand.

Given assumption (9.22), clearly the transformed moduli are also related by the same constant:

$$\frac{\overline{K}^*(s)}{\overline{G}^*(s)} = \frac{\overline{K}(s)}{\overline{G}(s)} = C_1 \qquad\qquad (9.23)$$

And it can be shown that this assumption then leads to a constant Poisson ratio:

$$\nu(t) = \nu_0 = \text{constant} \qquad\qquad (9.24)$$

The proof of this result is left as an exercise in Problem 9.1c.

9.5 General Viscoelastic Correspondence Principle

In solving simple one dimensional problems as in previous chapters, one could simply convert boundary conditions (applied loadings or displacements) to a known stress (or strain) state and then use the constitutive law to find the corresponding unknown strain (or stress), In two and three dimensional elasticity, it is necessary to be more rigorous in the complete set of governing equations, the application of boundary conditions and their solution. The reader is referred to excellent classical books on elasticity theory such as Timoshenko and Goodier (1970) for the developmental details. Here we summarize the final set of governing equations and list a few solution strategies for elasticity in two and three dimensions. Then we extend this knowledge to solution of viscoelastic problems in two and three dimensions,

9.5.1 Governing Equations and Solutions for Linear Elasticity

The essential governing equations for a *linear elastic body* are given below. In these equations the position variable, x_k, is explicitly shown to emphasize that in multidimensional problems the stress and strain fields vary spatially in the material.

Equations of motion; $\sigma_{ij,j}(x_k) + X_i(x_k) = \rho \dfrac{\partial^2 u_i(x_k)}{\partial t^2}$ $\qquad$ **(9.25)**

where, X_i are body forces and the right hand side is zero for static equilibrium problems.

Strain-displacement equations; $\varepsilon_{ij}(x_k,t) = \dfrac{1}{2}\left[u_{i,j}(x_k) + u_{j,i}(x_k)\right]$ $\quad$ **(9.26)**

Stress-strain equations; $\begin{aligned} s_{ij}(x_k) &= 2Ge_{ij}(x_k) \\ \sigma_{kk}(x_i) &= 3K\varepsilon_{kk}(x_i) \end{aligned}$ $\qquad$ **(9.27)**

Boundary conditions; $\quad \sigma_{ji}(x_k) \cdot n_j = T_i(x_k)$ $\qquad$ **(9.28a)**

(in terms of known tractions, T_j) and/or displacement conditions

$$u_i(x_k) = L_i(x_k) \qquad \textbf{(9.28b)}$$

in terms of known displacements L_i. The tractions, T_i, are applied surface forces, and are related to the stress components at the surface by the Cauchy formula

$$T_i = \sigma_{ij} n_j \qquad \text{(9.28c)}$$

where n_j is the unit normal to the surface. Note the T_i and L_i are applied to the surfaces of the body and for two-dimensional problems there are two conditions per surface, while for three-dimensional problems there are three conditions per surface.

The governing equations **9.25, 9.26, 9.27** comprise 15 coupled partial differential equations in 15 unknowns which are to be solved based upon the boundary conditions (**Eq. 9.28**). As can be seen from **Eqs. 9.28**, there are several types of boundary value problems that can be formed:

1. Traction BVPs: Loading is applied through prescribed surface tractions.
2. Displacement BVPs: Loading is applied through prescribed surface displacements.
3. Mixed BVPs: Loading is applied through a combination of prescribed tractions and prescribed displacements.

Full development of the methods to solve elasticity boundary value problems in either two or three dimensions is beyond the scope of this text. Here we outline the two major approaches.

The first approach is based upon direct solution involving the displacements. In the most basic sense, a strategy can be found to solve the 15 coupled differential equations directly. However, other approaches are more expedient. The most classical approach is to develop the Navier equations by putting the strain-displacement equations (**Eq. 9.26**) into the constitutive equations (**Eq. 9.27**) to obtain the stresses, σ_{ij}, in terms of the displacements, u_i. The result is then inserted into the equilibrium equations (**Eq. 9.25**), yielding three, coupled, second order partial differential equations on the three displacements, u_i. These three equations can then be solved for the displacements. Upon solution the stresses and strains can be found by substitution of the displacements in to the appropriate expressions.

The second approach is based upon solution in terms of the stresses, specifically without use of the displacements directly. While this approach is often more intuitive, allowing calculations of only stresses and strains, caution must be taken to ensure that physically meaningful displacements could be found. Because the strains are calculated by differentiating the displacements, finding displacements necessitates integrating the strain fields. Thus, in stress-based solutions it is essential that the equations of *compatibility* also be satisfied. The equations of compatibility are equations derived from the strain-displacement relations which relate the strain components to one another, imposing conditions upon them. Of the six compatibility equations, only three are independent expressions and these ensure integrability of the resulting strain field to yield continuous displacements. A basic method of solution based on stresses then involves rewriting the compatibility equations in terms of stresses (the Beltrami Michell Equations) and subsequently solving three compatibility equations together with the equations of motion for the six stress components, σ_{ij}. More details on the possible solution methods can be found in many elasticity texts including a nice synopsis of classical methods in A.J. Durelli, E.A. Phillips, C.H. Tsao; "Introduction to the Theoretical and Experimental Analysis of Stress and Strain", McGraw Hill, NY, 1958, p.99.

Another popular and useful approach for many practical engineering problems that can be reduced to two dimensional *plane strain* or *plane stress* approximations involves an auxiliary stress potential. In this approach, a bi-harmonic equation is developed based on the stresses (in terms of the potential) satisfying both the equilibrium equation and the compatibility equations. The result is that stresses derived from potentials satisfying the biharmonic equation automatically satisfy the necessary field equations and only the boundary conditions must be verified for any given problem. A rich set of problems may be solved in this manner and examples can be found in many classical texts on elasticity. In conjunction with the use of the stress potential, the principle of superposition is also often invoked to combine the solutions of several relatively simple problems to solve quite complex problems.

9.5.2 Governing Equations and Solutions for Linear Viscoelasticity

The governing equations for a viscoelastic material are the same as those for an elastic material except all stresses, strains and displacements are time dependent and the stress-strain equations are the integral equations given by **Eqs. 9.7** or **9.8**. The dependent variables x_k and **t** are explicitly included to emphasize that in multidimensional problems the stress and strain fields vary spatially in the material and that for viscoelasticity the fields are also time dependent.

Equations of motion; $\qquad \sigma_{ij,j}(x_k,t) + X_i(x_k,t) = \rho \dfrac{\partial^2 u_i(x_k,t)}{\partial t^2}$ $\qquad$ (9.29)

Strain-displacement equations; $\varepsilon_{ij}(x_k,t) = \dfrac{1}{2}\left[u_{i,j}(x_k,t) + u_{j,i}(x_k,t)\right]$ (9.30)

Stress-strain equations; $s_{ij}(x_k,t) = 2\displaystyle\int_{-\infty}^{t} G(t)\dfrac{\partial e_{ij}(x_k,\tau)}{\partial \tau}d\tau$ (9.31)

$$\sigma_{kk}(x_i,t) = 3\int_{-\infty}^{t} K(t)\dfrac{\partial \varepsilon_{kk}(x_i,\tau)}{\partial \tau}d\tau$$

Boundary conditions; $\sigma_{ji}(x_k,t)\cdot n_j = T_i(x_k,t)$ (9.32a)

(in terms of known tractions, T_j) and/or displacement conditions

$$u_i(x_k,t) = L_i(x_k,t)$$ (9.32b)

in terms of known displacements L_i .

Taking the Laplace transform of the **Eqs. 9.29 – 9.32** gives,

Equations of motion; $\overline{\sigma}_{ij,j}(x_k,s) + \overline{X}_i(x_k,s) = \rho\dfrac{\partial^2 \overline{u}_i(x_k,s)}{\partial t^2}$ (9.33)

Strain-displacement equations; $\overline{\varepsilon}_{ij}(x_k,s) = \dfrac{1}{2}\left[\overline{u}_{i,j}(x_k,s) + u_{j,i}(x_k,s)\right]$ (9.34)

Stress-strain equations; $\begin{aligned} \overline{s}_{ij}(x_k,s) &= 2\overline{G}^*(s)\overline{e}_{ij}(x_k,s)\\ \overline{\sigma}_{kk}(x_i,s) &= 3\overline{K}^*(s)\overline{\varepsilon}_{kk}(x_i,s)\end{aligned}$ (9.35)

Boundary conditions; $\overline{\sigma}_{ji}(x_k,s)\cdot n_j = \overline{T}_i(x_k,s)$ (9.36a)

$$\overline{u}_i(x_k,s) = \overline{L}_i(x_k,s)$$ (9.36b)

Obviously, the above transformed governing equations for a linear viscoelastic material (**Eqs. 9.33- 9.36**) are of the same form as the governing equations for a linear elastic material (**Eqs. 9.25 - 9.28**) except they are in the transform domain. This observation leads to the correspondence principle for three dimensional stress analysis: For a given a viscoelastic boundary value problem, replace all time dependent variables in all the governing equations by their Laplace transform and replace all material properties by **s** times their Laplace transform (recall, e.g., $\overline{G}^*(s) = s\overline{G}(s)$), thereby converting the viscoelastic boundary value problem in the time domain into an associated elastic boundary value problem in the transform domain.

Methods for Solving Viscoelastic Problems: As mentioned in Chapter 8 on bars and beams, three related methods can be used to solve linear viscoelastic boundary value problems. These are:

Method 1: Solve the viscoelastic problem in the time domain using Eqs. 9.29 - 9.32.

Method 2: Solve an associated elasticity problem in the transform domain using Eqs. 9.33 - 9.36. Invert the solution to the time domain.

Method 3: Convert an existing elasticity solution into a viscoelastic solution as follows:

1. **Find an elastic boundary value problem and solution with the same geometry, loading and boundary conditions as the viscoelastic boundary value problem or, if not available, solve an elasticity problem with the same geometry, loading and boundary conditions as the viscoelastic boundary value problem.**

2. **Convert the solution of the elastic problem to the solution of the viscoelastic problem in the transform domain by replacing all variables by their Laplace transform and all elastic constants by their equivalent in the transform domain, i.e.,**

$$\sigma \to \bar{\sigma}(s)$$

$$\varepsilon \to \bar{\varepsilon}(s)$$

$$u \to \bar{u}(s) \, , \, v \to \bar{v}(s) \, , \, w \to \bar{w}(s) \tag{9.37}$$

$$E \to \bar{E}^*(s) \, , \, D \to \bar{D}^*(s) \, , \, G \to \bar{G}^*(s) \, , \text{etc.}$$

$$P \to \bar{P}(s) \, , \text{ all loads, etc.}$$

3. **Invert the solution obtained in 2 to the time domain to obtain the solution to the viscoelastic problem in the time domain.**

Method 3 allows viscoelastic problems to be solved quite easily providing that the analogous elastic solution exists or can be found. **There are im-**

portant restrictions to this procedure such that the method cannot be used directly on mixed boundary value problems (combination of applied displacements and applied tractions) in which the boundary condition regions change with time. Two notable examples where this restriction applies are the stress analysis of bodies containing a crack (the Griffith problem) or contact problems (the Hertz and associated problems). Although beyond the scope of this text, the reader should be aware that an extended correspondence principle was developed by Graham, (1968) that allows such viscoelastic problems to be solved using the elastic solution for the same problem. Note that this is still an active field of research, with recent problems associated with contact problems for thin polymer films on substrates (eg, M. Sakai, (2006)) and in the areas of microelectronics, nanoindentation and MEMS processing such problems are technologically important.

As stated in Chapter 8, the correspondence principle presented here will always be valid when the boundary conditions are a product of separable functions of space and time, e.g. $\mathbf{T_i(x_i,t) = T_i'(x_i)f(t)}$ and $\mathbf{u_i(x_i,t) = u_i'(x_i)g(t)}$.

9.6 Thick Wall Cylinder and Other Problems

While it is beyond the scope of this introductory text to fully develop and solve a wide variety of multidimensional stress analysis problems in viscoelasticity, we provide here a classic example to illustrate the use of the correspondence principle to derive a viscoelastic solution from a practical problem in elasticity. We choose here the problem of a Thick Walled Cylinder, often referred to as the Lame Solution. In the following, we first generate the elasticity solution to the classic Lame problem, then extend this elasticity solution to that for a reinforced thick walled cylinder. Subsequently, we use the latter solution to develop the viscoelastic solution via the correspondence principle.

9.6.1 Elasticity Solution of a Thick Wall Cylinder

The thick wall cylinder shown in **Fig. 9.1** is a good example of a problem which is solved using the stress function approach to the solution of two dimensional plane stress or plane strain problems of engineering importance. In this approach, stress fields are derived from a set of potentials, Φ, which satisfy the biharmonic equation. Stresses found from such potentials automatically satisfy the equilibrium and compatibility equations. Thus only the boundary conditions for the problem of interest must be satisfied.

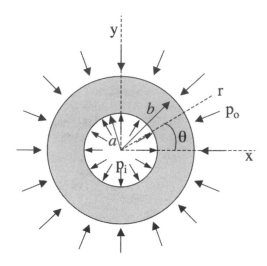

Fig. 9.1 Thick wall cylinder (often known as the Lame problem).

All the elasticity equations given by **Eqs. 9.29 - 9.32** as well as the biharmonic stress function equation can be developed for cylindrical coordinates (see, Timoshenko and Goodier, (1970)). The biharmonic equation is written as,

$$\nabla^4 \phi = 0 \tag{9.38}$$

or in cartesian coordinates,

$$\frac{\partial^4 \phi}{\partial x^4} + 2 \frac{\partial^4 \phi}{\partial x^2 \partial y^2} + \frac{\partial^4 \phi}{\partial y^4} = 0 \tag{9.39}$$

which becomes in polar coordinates,

$$\left(\frac{\partial^2}{\partial r^2} + \frac{1}{r} \frac{\partial}{\partial r} + \frac{1}{r^2} \frac{\partial^2}{\partial \theta^2} \right) \left(\frac{\partial^2 \phi}{\partial r^2} + \frac{1}{r} \frac{\partial \phi}{\partial r} + \frac{1}{r^2} \frac{\partial^2 \phi}{\partial \theta^2} \right) = 0 \tag{9.40}$$

For axisymmetric problems there can be no dependence on the angle θ and **Eq. 9.40** becomes,

$$\left(\frac{d^4 \phi}{dr^4} + \frac{2}{r} \frac{\partial^3 \phi}{\partial r^3} - \frac{1}{r^2} \frac{\partial^2 \phi}{\partial r^2} + \frac{1}{r^3} \frac{\partial \phi}{\partial r} \right) = 0 \tag{9.41}$$

which is known as Cauchy's equation or the equi-dimensional equation. The solution for this equation is well known and is given by,

$$\phi = A\log(r) + Br^2\log(r) + Cr^2 + D \tag{9.42}$$

where **A**, **B**, **C**, and **D** constants that must be determined from the boundary conditions for a particular boundary value problem such as the thick wall cylinder shown in **Fig. 9.1**.

Using the definition of the relationship between the stress function and stresses in polar coordinates, the stresses in any axisymmetric boundary value problem can be found and are,

$$\sigma_{rr} = \frac{1}{r}\frac{\partial\phi}{\partial r} + \frac{1}{r^2}\frac{\partial^2\phi}{\partial r^2} = \frac{A}{r^2} + B[1 + 2\log(r)] + 2C$$

$$\sigma_{\theta\theta} = \frac{\partial^2\phi}{\partial r^2} = -\frac{A}{r^2} + B[3 + 2\log(r)] + 2C \tag{9.43}$$

$$\tau_{r\theta} = \frac{1}{r^2}\frac{\partial\phi}{\partial r} - \frac{1}{r}\frac{\partial^2\phi}{\partial r\partial\theta} = 0$$

Note that the stress function **D** (constant) leads to trivial stresses and is thus subsequently omitted. If there is no hole at the origin, constants **A** and **B** must vanish to avoid singular stresses at the origin. In the solution for the thick wall cylinder, **B** must be zero because although the corresponding stress fields would be admissible in the absence of material at the origin, the resulting displacements are multivalued and not admissible for this geometry. Thus, solution to the Lame problem reduces to finding the constants **A** and **C** from **Eq. 9.43** from the boundary conditions,

$$\sigma_{rr}(r = a) = -p_i \qquad \text{and} \qquad \sigma_{rr}(r = b) = -p_o \tag{9.44}$$

noting that the boundary conditions requiring $\sigma_{r\theta}$ to be be zero on both surfaces are automatically satisfied by **Eq. 9.43**. The general solution to the Lame problem thus is,

$$\sigma_{rr} = \frac{a^2b^2(p_o - p_i)}{b^2 - a^2}\frac{1}{r^2} + \frac{a^2p_i - b^2p_o}{b^2 - a^2}$$

$$\sigma_{\theta\theta} = -\frac{a^2b^2(p_o - p_i)}{b^2 - a^2}\frac{1}{r^2} + \frac{a^2p_i - b^2p_o}{b^2 - a^2} \tag{9.45}$$

where p_0 is the pressure at the outer boundary and p_i is the pressure at the inner boundary. Note that the solution for stresses in an elastic cylinder do not contain elastic constants. Therefore, the solution for stresses is the same in a viscoelastic and an elastic cylinder. The displacements, which can be derived from this solution by using the stress-strain equations (**Eq. 9.27**) and then integration of the strain-displacement equations (**Eq. 9.26**), do however contain the elastic constants. A detailed solution for displacements in a viscoelastic thick wall cylinder with pressure only on the inner surface is given in Flugge (1974).

9.6.2   Elasticity Solution for a Reinforced Thick Wall Cylinder (Solid Propellant Rocket Problem)

Based upon the classic Lame solution above, many other useful problems can be solved in elasticity by variations and extensions. Here we examine a reinforced thick walled cylinder consisting of an inner cylinder of one material and an outer cylinder of another material. The structural analysis of a reinforced thick wall cylinder played an important role in the space program and, indeed, variations of the problem are still important today. A solid propellant rocket can be approximated by such a double cylinder geometry as shown in **Fig. 9.2**. Here the outer shell is an elastic material such as aluminum and the inner cylinder is a polymer composite, typically composed of polyurethane with particulate inclusions to aid in developing maximum thrust during burning. Note that as with many engineering analyses, that considered here is a simplification of a more complex situation. For example, the geometry of an actual rocket propellant typically contains a star shaped inner surface for optimum ablation and thrust. Further, it is a dynamic problem with the rocket accelerating vertically and rotating and the inner boundary moving due to ablation. Nevertheless the ability to obtain an analytically exact solution for a simplified case is extremely valuable in providing checks upon more sophisticated numerical analyses.

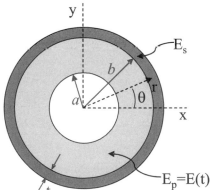

Fig. 9.2 Geometry of the reinforced thick wall cylinder.

Since modulus of the outer metallic cylinder is much larger than that of the inner polymer cylinder, it is reasonable to assume the outer shell is rigid. This assumption provides a further simplification to the problem. In this section, we develop the fully elastic solution based upon the classic Lame solution. Subsequently we will consider the viscoelasticity of the polymer and invoke the correspondence principle to solve the viscoelastic problem,

The elasticity solution to the reinforced thick walled cylinder can be found based upon the stress function given in **Eq. 9.42** except now the constants must be reevaluated. As before, the constant **B** must remain zero, but constants **A** and **C** must be found from the new boundary conditions which for the case of a rigid outer cylinder are

$$\begin{cases} r = a, & \sigma_{rr} = -p \\ r = b, & u_r = 0 \end{cases} \tag{9.46}$$

noting again that the boundary conditions requiring $\sigma_{r\theta}$ to be be zero on both surfaces are automatically satisfied. Based on **Eq. 9.42**, the stresses are again given by **Eq. 9.43**, omitting the terms with coefficients B and D:

$$\sigma_{rr,\theta\theta} = \pm \frac{A}{r^2} + 2C^\dagger \tag{9.47}$$

Since one of the boundary conditions in **Eq. 9.46** is now a displacement boundary condition, we also require the expressions for the displacements in terms of the constants **A** and **C**. These are found first by using the stress-strain-displacement relations and then by integrating the strain components to determine the displacements. The stress-strain-displacement equations for the condition of plane strain in cylindrical coordinates are,

$$\varepsilon_{rr} = \frac{du_r}{dr} = \frac{1}{E}\left[\sigma_{rr} - \nu\left(\sigma_{\theta\theta} + \sigma_{zz}\right)\right]$$

$$\varepsilon_{\theta\theta} = \frac{u_r}{r} + \frac{1}{r}\frac{du_\theta}{d\theta} = \frac{u_r}{r} = \frac{1}{E}\left[\sigma_{\theta\theta} - \nu\left(\sigma_{rr} + \sigma_{zz}\right)\right] \tag{9.48}$$

$$\varepsilon_{zz} = \frac{1}{E}\left[\sigma_{zz} - \nu\left(\sigma_{rr} + \sigma_{\theta\theta}\right)\right] = 0$$

---

† Note that in this and the following equations the comma is used in the subscripts on stress to be able to write both the radial and hoop stresses in one equation. Since the form of these stresses differs only by a minus sign, it is preferred to emphasize their similarity by this nonstandard notation rather than write the two equations separately.

From the last equation, $\sigma_{zz} = v(\sigma_{rr} + \sigma_{\theta\theta})$. Substituting this value for σ_z into **Eq. 9.48** and solving for the radial displacement gives,

$$u_r = \frac{1+v}{E}\left[-\frac{A}{r} + 2Cr(1-2v)\right] \tag{9.49}$$

Note that because of the axisymmetry of the problem, $u_\theta = 0$ and thus the second equation in **Eq. 9.48** provides u_r directly without need for integration.

Using the boundary conditions from **Eqs. 9.46** with **Eqs. 9.47** and **9.49** to obtain values for **A** and **C** results in expressions for stress and radial displacement

$$\sigma_{rr,\theta\theta} = -p\left[\frac{1 \pm \left(\dfrac{b}{r}\right)^2 (1-2v)}{1 + \left(\dfrac{b}{a}\right)^2 (1-2v)}\right] \tag{9.50}$$

$$u_r = \left(\frac{1+v}{E}\right)\left[\frac{pb(1-2v)}{1 + \left(\dfrac{b}{a}\right)^2 (1-2v)}\right]\left[\frac{b}{r} - \frac{r}{b}\right] \tag{9.51}$$

9.6.3 Viscoelasticity Solution for a Reinforced Thick Wall Cylinder (Solid Propellant Rocket Problem)

Since we have the elasticity solution to the reinforced thick walled cylinder problem, we can now find the solution to the viscoelastic problem by applying the correspondence principle (Method 3 from earlier in this chapter). Replacing the variables in **Eqs. 9.50** and **9.51** by the appropriate transforms gives the solution for stresses and displacements of the viscoelastic problem in the transform domain,

$$\overline{\sigma}_{rr,\theta\theta}(r,s) = -\overline{p}(s)\left\{\frac{1 \pm \left(\dfrac{b}{r}\right)^2 \left[1 - 2\overline{v}^*(s)\right]}{1 + \left(\dfrac{b}{a}\right)^2 \left[1 - 2\overline{v}^*(s)\right]}\right\} \tag{9.52}$$

$$\bar{u}_r(r,s) = \frac{1+\bar{v}^*(s)}{\overline{E}^*(s)} \left\{ \frac{\bar{p}(s)b\left[1-2\bar{v}^*(s)\right]}{1+\left(\dfrac{b}{a}\right)^2\left[1-2\bar{v}^*(s)\right]} \right\} \left(\frac{b}{r}-\frac{r}{b}\right) \qquad (9.53)$$

If it is assumed that the polymer is incompressible, $\bar{v}^*(s)=0.5$, the solution is,

$$\sigma_{rr,\theta\theta}(r,t) = -p(t)$$
$$u_r(r,t) = 0 \qquad (9.54)$$

which is identical to the elasticity solution for an incompressible material.

If the material is not incompressible, the solution for stresses in **Eq. 9.52** can be used to obtain the stress field with time for the viscoelastic problem given the material properties. In order to examine a particular loading and material, it is convenient to use **Eqs. 9.16** to obtain the stress solution in terms of the shear and bulk moduli in order to make reasonable assumptions about the material similar to those outlined earlier in **Eqs. 9.20-9.24**. The term including Poisson's ratio can be rewritten as

$$\left(1-2\bar{v}^*(s)\right) = \frac{3\overline{G}^*(s)}{3\overline{K}^*(s)+\overline{G}^*(s)} = \frac{3\overline{G}(s)}{3\overline{K}(s)+\overline{G}(s)} \qquad (9.55)$$

Substitution of **Eq. 9.55** into **Eq. 9.52** leads to

$$\bar{\sigma}_{rr,\theta\theta}(r,s) = -\bar{p}(s)\left[\frac{1\pm\left(\dfrac{b}{r}\right)^2\left(\dfrac{3\overline{G}(s)}{3\overline{K}(s)+\overline{G}(s)}\right)}{1+\beta\left(\dfrac{3\overline{G}(s)}{3\overline{K}(s)+\overline{G}(s)}\right)}\right]$$

$$= -\bar{p}(s)\left[\frac{3\overline{K}(s)+\overline{G}(s)\left(1\pm3\left(\dfrac{b}{r}\right)^2\right)}{3\overline{K}(s)+\overline{G}(s)\left(1+3\beta\right)}\right] \qquad (9.56a)$$

where

$$\beta = \left(\frac{b}{a}\right)^2 \qquad (9.56b)$$

The radial stress expression from **Eq. 9.56a** can also be written as

$$3\overline{K}(s)\overline{\sigma}_{rr}(r,s) + \left(1 + 3\beta\right)\overline{G}(s)\overline{\sigma}_{rr}(r,s)$$
$$= -\overline{p}(s)3\overline{K}(s) - \left(1 + 3\left(\frac{b}{r}\right)^2\right)\overline{p}(s)\overline{G}(s) \qquad (9.57)$$

and the hoop stress expression ($\sigma_{\theta\theta}$) can be written similarly with the appropriate change of sign. These expressions can be simply inverted to convolution integral expressions in the time domain using the properties of Laplace transforms (see Appendix B). For example, the radial stress expression becomes

$$3\int_0^t K(t-\xi)\sigma_{rr}(r,\xi)d\xi + \left(1+3\beta\right)\int_0^t G(t-\xi)\sigma_{rr}(r,\xi)d\xi =$$
$$-3\int_0^t K(t-\xi)p(\xi)d\xi - \left(1+3\left(\frac{b}{r}\right)^2\right)\int_0^t G(t-\xi)p(\xi)d\xi \qquad (9.58)$$

Provided that the loading function, **p(t)**, and the moduli, **G(t)** and **K(t)**, are known, **Eq. 9.56a** can be solved for the stresses in the Laplace domain. These may be inverted to obtain the stresses in the time domain. Alternatively, with suitable numerical and computational skills integral **Eq. 9.58** can be solved numerically directly for the stresses as a function of time.

To illustrate the solution technique for a specific case, we make some simple assumptions for the loading and the material properties. The internal pressure is taken to be a step input in time and the bulk modulus of the polymer is not time dependent (elastic in dilatation behavior) while the shear modulus of the polymer is represented by a single Maxwell model (Maxwellian in shear). These assumptions are summarized as:

$$p(t) = p_0 II(t) \qquad \text{or} \qquad \overline{p}(s) = \frac{p_0}{s}$$

$$\overline{K}^*(s) = K_0 \qquad \text{or} \qquad \overline{K}(s) = \frac{K_0}{s} \qquad (9.59)$$

$$G(t) = G_0 e^{-t/\tau} \qquad \text{or} \qquad \overline{G}(s) = \frac{G_0}{\frac{1}{\tau} + s}$$

Substituting into **Eq. 9.56a** will give after simplification,

$$\overline{\sigma}_{rr,\theta\theta}(r,s) = -p_0 \left\{ \frac{1}{s} + \left(\frac{\dfrac{B(r)-C}{C}}{\dfrac{A}{C}+s} \right) \right\}$$

(9.60a)

where

$$A = 3K_0 \frac{1}{\tau}, \quad B(r) = 3K_0 + G_0\left(1 \pm 3\left(\frac{b}{r}\right)^2\right), \quad C = 3K_0 + G_0(1+3\beta) \quad (9.60b)$$

Inversion of **Eq. 9.60** gives the solution in the time domain as,

$$\sigma_{rr,\theta\theta}(r,t) = -p_0\left[1 + \left(\frac{B-C}{C}\right)e^{-\frac{A}{C}t}\right]$$

(9.61)

A number of features may be pointed out about this solution that are important. First, note that at the inner boundary, **r=a**, **B(r)=C** and thus the radial stress reduces to the applied pressure **p₀**, satisfying the applied boundary condition. Because of the negative sign in **B(r)** for the hoop stress, $\sigma_{\theta\theta}$ is nonzero at the inner boundary.

The limit cases at long and short times are also of interest. At long times, **t → ∞**, the exponential term vanishes, leaving both radial and hoop stresses at all locations in the polymeric material identically equal to the applied pressure **p₀**. At t=0, the exponential term is unity, the relaxation time of the polymer is not involved and the solution is identical to an elastic material with elastic constants **K₀** and **G₀**. These limit cases are reasonable since a Maxwell model is a viscoelastic fluid: at t=0 only the elastic spring can respond, but at long times it is a fluid response and thus yielding the incompressible behavior.

The radial and hoop stresses are plotted versus position in **Fig. 9.3**, where it is easily seen that the boundary condition of $\sigma_{rr}(r=a)=-p_0$ is met on the inner boundary and the limit case of uniform stresses at long time for incompressible behavior is also apparent. The elastic solution is included, which overlays the viscoelastic response at t=0. While the internal pressure applied is compressive leading to compressive radial stresses at all positions and all times, the hoop stress at the inner surface is tensile due to the expansion of the cylinder. While the hoop stress remains tensile for all time for an elastic cylinder, this tensile stress relaxes in the viscoelastic cylinder, ultimately becoming compressive.

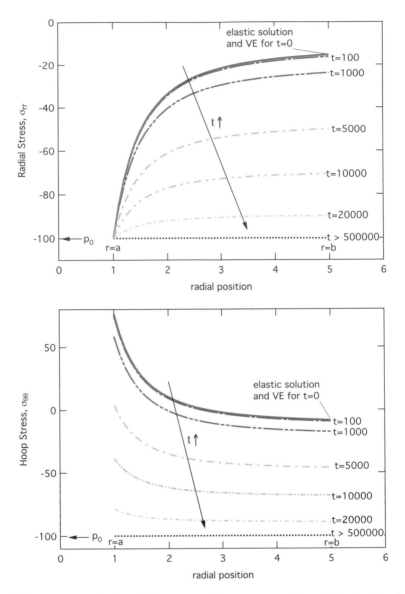

Fig. 9.3   Variation of the radial stress and hoop stress with position in the viscoelastic reinforced cylinder loaded with a step input of internal pressure. Parameters used are $K_0/G_0=3$, $\tau=1000$, where the viscoelastic cylinder has an elastic bulk modulus and is a single Maxwell element in shear modulus. Response parameterized with time from the initial application of load at t=0 to asymptotic response at long times.

The variation of the stresses with time at the inner surface is shown in **Fig. 9.4**, where these effects can be clearly seen. The relaxation and inversion of the hoop stress is particularly interesting because the load applied is a constant stress, which might lead one to expect radial creep, which calculations of the displacements would bear out. However, the interaction of the response in two dimensions combined with the incompressible behavior at long times lead to a pseudo-relaxation response in the hoop direction and ultimately approach to the incompressible stress state.

It is interesting to note that even though the material was assumed to be Maxwellian in shear or a fluid like material such as a thermoplastic, the form of the solution for stresses is similar to what might be expected for a Kelvin solid. The reason, of course, is the interaction of bulk and shear behavior together with the boundary conditions.

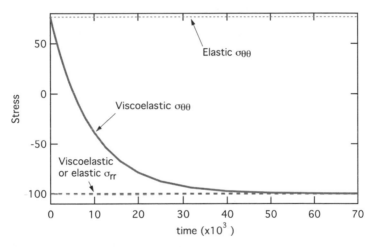

Fig. 9.4   Variation of the stresses at the inner boundary, r=a, for elastic and viscoelastic reinforced cylinder. Parameters and conditions the same as described in **Fig. 9.3**.

Using the same assumptions of the example solved in the Laplace domain (step input in pressure, elastic bulk modulus and Maxwell behavior in shear) with **Eq. 9.58**, the solution of the integral equation (**Eq. 9.58**) will yield the same results. (See homework problem 9.4). Since polymers are such that many Maxwell or Kelvin elements are needed to represent actual behavior, this example shown here is simplistic. However, such simple solutions can show trends in behavior and may give insight to the differences between thermosets and thermoplastics. The next section discusses briefly use of broadband material response functions for more physically realistic polymer representation, as well as the difficulties associated with obtaining properties such as the bulk modulus or Poisson's ration over long time.

The reinforced thick wall cylinder problem with interior pressure has also been solved for the case when the outer shell is assumed to be elastic rather than rigid. (This case obtained by W.B. Woodward and J.R.M. Radok is reported by E. H. Lee in Viscoelasticity: Phenomenological Aspects (J.T. Bergen, Ed.), Wiley, (1960)). The solution for the circumferential stress is,

$$\overline{\sigma}_{\theta\theta}(r,s) = \overline{p}(s) \left[\frac{\dfrac{\alpha(1-\overline{v}^*(s)^2)}{\alpha(1+\overline{v}^*(s)^2)-\overline{E}^*(s)}\left(\dfrac{b^2}{r^2}-1\right)-\left(\dfrac{b^2}{r^2}+1\right)}{\dfrac{\alpha(1-\overline{v}^*(s)^2)}{\alpha(1+\overline{v}^*(s)^2)-\overline{E}^*(s)}\left(\dfrac{b^2}{a^2}-1\right)-\left(\dfrac{b^2}{a^2}+1\right)} \right] \qquad (9.62)$$

where α is a constant prescribing the reinforcement of the outer shell. The solution can be inverted to give the viscoelastic solution in the time domain once the material parameters and loading are determined. E. H. Lee (1963) further discusses how the solution of an elastically reinforced viscoelastic thick walled cylinder can be found by numerically integrating integral equations such as **Eq. 9.58** using measured creep or relaxation functions.

From these examples, it is clear that known solutions in the theory of linear elasticity for two and three dimensional problems including plates and shells can be converted to viscoelastic solutions in the transform domain relatively easily and the solution in the time domain can be found by inversion. Using this method many problems of practical interest can be solved. It is appropriate to note that buckling problems are a special case and the same approach, if not used wisely, can lead to erroneous results.

9.7 Solutions Using Broadband Bulk, Shear and Poisson's Ratio Measured Functions

As discussed in Chapter 7 real material properties extend over many decades of time and for realistic solutions of boundary value problems it is necessary to have methods to incorporate these real measured properties. When material properties can be represented by a Prony series composed of a number of terms, it is possible to obtain solutions for more practical representation of polymers. Examples of the use of Laplace transforms for such an approach may be found in Christensen, (1982). Additionally, methods to numerically integrate convolution integrals such as that given by **Eq. 9.58** using Prony series expansions for the properties were discussed in Chapter 7.

However, it is also necessary to discuss how broadband bulk, shear and Poisson's ratio are measured. The measurement of the broadband shear modulus is easily accomplished using the time-temperature-superposition-principle (TTSP) and a torsion test. See Kenner, Knauss and Chai (1982) for a description of a simple torsiometer and the measurement of a master curve for a structural epoxy adhesive, FM-73, at 20.5° C.

The accurate measurement of broadband bulk modulus and Poisson's ratio presents greater difficulties. While the shear and tensile relaxation modulus vary by several orders of magnitude over many decades of time, the bulk modulus and Poisson's ratio vary very little over the same number of decades. To complicate matters more, in general, the variations of bulk relaxation modulus and Poisson's ratio are not synchronous with the shear and tensile relaxation modulus over the same time scale. To visualize the dilemma, consider that Poisson's ratio for most polymers is approximately 1/3 in the glassy range and approximately 1/2 in the rubbery range. Using the relation given in Chapter 2 for the bulk modulus in terms of Young's modulus and Poisson's ratio,

$$ K = \frac{E}{3(1-2v)} \qquad (9.63) $$

the bulk modulus is equal to the Young's modulus if $v = 1/3$ and is infinite if $v = 1/2$. In the glassy range of a polymer, measured values of the bulk modulus are nearly the same as measured values of extensional modulus (see, Arridge, (1974)). In the rubbery range, the bulk modulus is indeed large compared to the extensional modulus but certainly not infinite and it is doubtful that Poisson's ratio ever becomes exactly 1/2. Experiments do tend to verify that variations in bulk modulus from the glassy to rubbery range are small compared to either extensional or shear moduli (see Ferry, (1980) and Tschoegl, et al., (2002)). Methods to measure bulk modulus have been proposed by Arridge (1974), Duran and McKenna (1990), Sane and Knauss (2001), and Park, et al. (2004). Emri and Prodan (2006) have proposed a single apparatus to measure both the bulk and shear modulus. However discussion of the optimal procedures and accuracy required to attain true values of the bulk modulus or Poisson ratio's over time is still ongoing.

The contradictions posed by use of a viscoelastic Poisson's ratio have been discussed by Flugge, (1975), Shames and Cozzarelli, (1992), Lake, (1998) and Tschoegl, et al., (2002). Some issues that arise are non-physical values for Poisson's ratio when simple mechanical models are used (e.g. Maxwell or Kelvin), and even the fundamental definition of a time-dependent Poisson ratio. Various definitions of Poisson's ratio and its measurement are discussed in depth by Tschoegl, et al., (2002). Therein the suggestion is given that the ratio is best measured in relaxation and that extreme four-digit accuracy is required which is presently not found in the literature.

As a result, at the present time, use of one or more of the assumptions provided earlier in the chapter, together with broadband data for shear or extensional modulus, represent the most fruitful approach to the solution of viscoelastic boundary value problems.

9.8 Review Questions

9.1   Investigate and discuss the peculiar nature of Poisson's ratio for viscoelastic materials. (Hint: See Flugge (1974), Shames and Cozzarelli (1992).

9.2   Why are viscoelastic constitutive equations normally written using bulk and shear properties?

9.3   Describe three assumptions that are often made for viscoelastic stress analysis.

9.4   Describe three methods for solving viscoelastic boundary value problems.

9.5   Describe the three types of boundary value problems encountered in solid mechanics.

9.6   Which type of boundary value problem cannot be solved using the standard (or Alfrey) correspondence principle?

9.7   Name two frequently encountered viscoelastic boundary value problems in solid mechanics that cannot be solved with the standard correspondence prionciple.\

9.9 Problems

9.1. Develop an expression for the shear relaxation modulus assuming the tensile relaxation modulus can be represented by a Maxwell fluid. Hint: Use **Table 2.1**.

9.1b. Starting from **Eqs. 9.10**, derive a relationship between the total stress and total strain tensors, σ_{ij} and ε_{ij}. In the process, find the expression Lame's constant in the transform domain in terms of the transformed shear and bulk moduli.

9.1c. Show that the synchronous moduli assumption, **Eq. 9.22**, results in a Poisson's ratio being constant.

9.2. Find a solution for the radial displacement in a thin-wall cylindrical pressure vessel with closed (spherical) ends. Assume a step input in internal pressure, the cylinder is made of a polymer whose properties are elastic in bulk and Maxwellian in shear.

9.3. Obtain the solution for a reinforced thick wall cylinder similar the one of **Fig. 9.2**. Assume the shell is rigid and that the propellant can be represented by a Kelvin Material.

9.4. Determine the stresses in a thick wall cylinder similar the one of **Fig. 9.1** using the integral equation solution given by **Eq. 9.58**. Use the assumption of a step input in pressure as well as elastic response in bulk and Maxwellian in shear as in the earlier example. Compare your solution to that obtained using the correspondence principle.

9.5. Solve each of the problems below using the correspondence principle. The elastic solutions can be found from elementary books on solid mechanics (such as Timoshenko), an elasticity book (such as Timoshenko and Goodier) or from fundamental principles of either.

a. Find the solution of for the radial stress in a rotating disk. Assume steady state conditions, i.e., the disk is rotating at a constant angular velocity. Assume the disk is made of a polymer whose properties are elastic in bulk and Maxwellian in shear.

b. A polymer bar of circular cross section is compressed within a steel die whose internal diameter is exactly the same as the rubber bar. Assume that the axial compression load on the bar is $P(t) = P_0 Ht)$ and the die to be rigid. Also assume no friction and the properties of the polymer are elastic in bulk and Maxwellian in shear. Determine an expression for the pressure between the bar and the die. (see Fig. below).

c. A solution of a problem not listed above that can be obtained us-
    ing one of the methods of solutions discussed in class. (Here, it
    would be best to verify with the instructor that the problem you
    select is of similar level of difficulty as those above.)

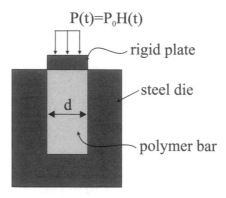

Schematic of polymer compressed in a rigid die for Problem 9.5b.

10. Nonlinear Viscoelasticity

Because Young's modulus of most polymers is relatively low compared to other structural materials such as metals, concrete, ceramics, etc., strains and deformations may be relatively large. A casual glance at the stress-strain response of polycarbonate given in **Fig. 3.7** indicates that the strain at yield is about 5% and at failure is more than 60%. Further, examination of the creep response of polycarbonate (Brinson, (1973)) as discussed in Chapter 11 indicates inear behavior for strains larger than about 3% and the material begins to neck or yield (Luder's bands form) for strains larger than about 5%. Obviously, polycarbonate as well as other polymers with similar behavior cannot be considered to be linear for such circumstances. For these reasons, it is appropriate to have basic understanding of nonlinear processes in order to be able to design structures made of polymeric materials. The intent here is to give basic definitions that will assist in identifying nonlinear effects when they occur and to review several nonlinear approaches.

As many nonlinear approaches are beyond the intended level and scope of this text, the focus will be on single integral mathematical models which are an outgrowth of linear viscoelastic hereditary integrals and lead to an extended superposition principle that can be used to evaluate nonlinear polymers. The emphasis will be on one-dimensional methods but these can be readily extended to three dimensions using deviatoric and dilatational stresses and strains as was the case for linear viscoelastic stress analysis as discussed in Chapters 2 and 9.

10.1. Types of Nonlinearities

The two types of nonlinearities that are most often encountered in practice and in the literature are identified as being either *material nonlinearities* or *geometric nonlinearities*. Material nonlinearities refer to nonlinear stress-strain response that occurs due to the inherent constitutive response of the material, while geometric nonlinearities refer to mathematical issues that arise when displacements and strains become large and the linearized definitions of stress and strain become inadequate. In this chapter we are concerned with material nonlinearities. Two complications should be mentioned at the outset. First, material nonlinearities typically become apparent in material response as the strain level increases. However, only a few percent strain is often sufficient for material nonlinearities to become important and at that level of strain, the linearized definitions of the stress and strain tensors are mathematically sufficient. Second, when dealing with metals, it is common to plot the stress-strain curve for a constant strain rate test and regard any deviation of that curve from linearity as an indication of the onset of material nonlinearity. As mentioned in Chapter 3, because of the dependence of viscoelastic material response on time, the stress-strain curve from constant a constant strain rate test for linear viscoelastic materials is not linear. As time increases during a test, relaxation occurs simultaneously with increasing strains. Thus, one must examine other methods to establish linearity for polymers, such as isochronous stress-strain plots at different times or modulus plots at different stress levels.

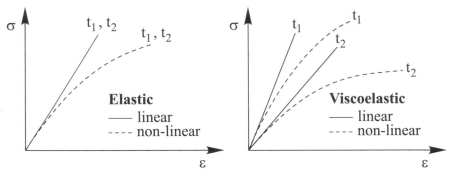

Fig. 10.1   Typical isochronous stress-strain diagrams of elastic and viscoelastic materials for two values of lapsed time.

Typical examples of tensile (isochronous) linear and nonlinear stress-strain diagrams for elastic and viscoelastic materials are shown in **Fig. 10.1**. For elastic materials, the response is time independent, so there is a single curve for multiple times and the nonlinearity is apparent as a deviation of the stress-strain response from linear. For linear viscoelastic materials, the isochronous response is linear, but the effective modulus decreases with time so that the stress-strain curves at different times are separated from one another. When a viscoelastic material behaves nonlinearly, the isochronous stress-strain curves begin to deviate from linearity at a certain stress level. **Fig. 10.2** shows creep compliance data for an epoxy adhesive as a function of stress level for various time intervals after initial loading.

Such a plot is sometimes more illuminating than the usual isochronous plot. For linear viscoelasticity, the compliance is independent of stress level and the isochronous compliance as a function of stress must be constant at each time instant (a horizontal line on **Fig. 10.2**). As the material enters the nonlinear range, the compliance begins to exhibit a dependence on stress and the isochronous curve starts to deviate from horizontal. For the material shown in **Fig. 10.2**, notice that the demarcation between linear and nonlinear behavior appears to be a function of time after initial loading (incubation time). The data suggests that if tests were conducted for a sufficient length of time, the material might appear to be nonlinear from initial loading. This result simply indicates that it is not always easy to tell from short time tests if a material is linear or nonlinear over a longer time period.

For some rubbery materials, stress may be linearly related to strain for strains as large as 20% to 50% or more. Such a case gives rise to "geometric" nonlinear behavior in which strains of higher order must be included in analyses. Here it should be noted that only odd order terms are considered in order to avoid negative values of stored energy. Of course, for large strains both material and geometric nonlinearities may occur simultaneously. Indeed for rubbers, where strains can easily reach 500%, a great deal of work has been devoted to development of accurate nonlinear elasticity models where both material and geometric nonlinearities are accommodated. See for example work by Arruda and Boyce (1993) where an 8-chain model is developed to represent the macromolecular deformations, and interaction effects of non-uniaxial loading are accounted for by limits on chain extension.

Extension ratios are most often used in cases of large strains or large deformations and can be found by examining the basic tensile strain definition as follows,

$$\varepsilon = \frac{\ell - \ell_0}{\ell_0} = \frac{\ell}{\ell_0} - 1 \qquad\qquad \textbf{(10.1a)}$$

or

$$\lambda = \frac{\ell}{\ell_0} = \varepsilon + 1 = \text{extension ratio} \qquad\qquad \textbf{(10.1b)}$$

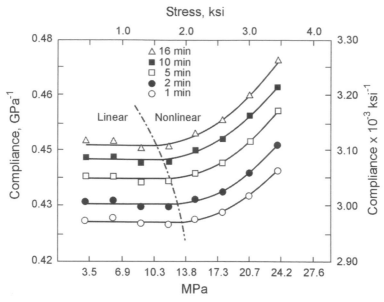

Fig. 10.2 Isochronous creep compliance of an adhesive (FM-300) at 60°C. (Data from Hiel, (1984). See also Brinson, (1985).)

As nonlinear elasticity constitutive models must include higher order terms the amount of error involved between linear and nonlinear formulations can easily be seen by comparing λ^2, λ^3, etc. to λ. It is easy to see that for strains at yield in most metals (0.2% or 0.002 in/in) higher order terms (λ^2, λ^3, etc.) lead to an error of about 0.4% and can most often be neglected. Even for strains at yield for polycarbonate (~5% or 0.05) higher order terms lead to only an error of about 10 % over linear theory.

Other less well-known types of nonlinearities include "interaction" and "intermode". In the former, stress-strain response for a fundamental load component (e.g. shear) in a multi-axial stress state is not equivalent to the stress-strain response in simple one component load test (e.g. simple shear). For example, **Fig. 10.3** shows that the stress-strain curve under pure shear loading of a composite specimen varies considerably from the shear stress-strain curve obtained from an off-axis specimen. In this type of test, a unidirectional laminate is tested in uniaxial tension where the fiber axis runs 15° to the tensile loading axis. A 90° strain gage rosette is applied to the specimen oriented to the fiber direction and normal to the fiber direction and thus obtain the strain components in the fiber coordinate system. Using simple coordinate transformations, the shear response of the unidirectional composite can be found (Daniel, 1993, Hyer, 1998). At small strains in the linear range, the shear response from the two tests coincide.

However, the difference observed at high strain levels is postulated to be due to the effect of stress normal to the fiber in addition to the shear stress along the fiber in the off-axis loading test. Similarly, tests of a neat polymer in simple torsion (shear) or tension-torsion/compression-torsion (multi-axial load) in **Fig. 10.4** also demonstrate that the shear compliance extracted from the test data differs in each case. This data shows an interaction nonlinearity and the authors postulate its origin lies in the coupling of dilatation and free volume for the polymer. Schapery (1969) and Lou, Y.C. and Schapery (1971) have shown that the invariant associated with the octahedral shear stress can be used as a normalizing parameter to account for such differences in PVC and in glass and carbon polymer matrix composites. A more recent paper by Knauss and Zsu (2002) investigated the nonlinear behavior of polycarbonate under multi-axial loading and concluded that the octahedral shear stress is nearly constant for all combinations of shear and normal stress investigated.

Even less well known are intermode nonlinearities that occur when several different mechanisms contribute to the deformation process simultaneously such as yielding and buckling. In this text only material nonlinearities will be considered with strains and deformations being small.

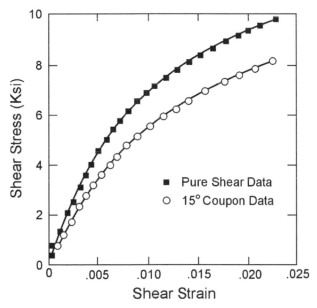

Fig. 10.3 Extent of stress interaction in off-axis unidirectional boron-epoxy coupons. (After Cole and Pipes, (1974); see Hiel et al., (1984))

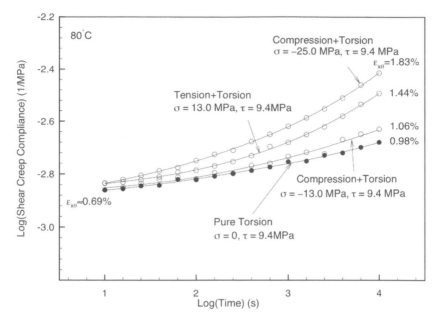

Fig. 10.4 Shear creep compliance curves for pure torsion and torsion with super-imposed tension or compression for PMMA. Change in creep compliance with multiaxial load test illustrates interaction nonlinearity (Lu and Knauss, (1999); reprinted with kind permission from Springer Science and Business Media).

10.2. Approaches to Nonlinear Viscoelastic Behavior

As mentioned earlier, there have been many attempts to develop mathematical models that would accurately represent the nonlinear stress-strain behavior of viscoelastic materials. This section will review a few of these but it is appropriate to note that those discussed are not all inclusive. For example, numerical approaches are most often the method of choice for all nonlinear problems involving viscoelastic materials but these are beyond the scope of this text. In addition, this chapter does not include circumstances of nonlinear behavior involving gross yielding such as the Luder's bands seen in polycarbonate in **Fig. 3.7**. An effort is made in Chapter 11 to discuss such cases in connection with viscoelastic-plasticity and/or viscoplasticity effects. The nonlinear models discussed here are restricted to a subset of small strain approaches, with an emphasis on the single integral approach developed by Schapery.

Nonlinear Mechanical Models: It is possible to represent nonlinear behavior by introducing nonlinear spring and damper elements into the derivation of differential stress-strain relations. For example, for the four-parameter fluid shown in **Fig. 10.5**, the spring moduli, damper viscosities and relaxation times are functions of stress, i.e.,

$$E_i = E_i(\sigma), \quad \mu_i = \mu_i(\sigma), \quad \tau_i = \tau_i(\sigma), \quad i = 1,2,...$$

If only the spring moduli are nonlinear, a nonlinear generalized Kelvin model can be represented by,

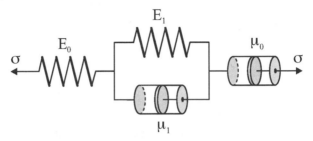

Fig. 10.5 Nonlinear four parameter fluid.

$$D(t) = \sum_{i=1}^{n} \frac{f_i(\sigma)}{E_i}(1 - e^{-t/\tau_i}) \tag{10.2}$$

where $f_i(\sigma)$ is a nonlinearizing function of stress. Obviously, three parameters must now be determined for each Kelvin element one of which is nonlinear with stress and if a large number of elements are needed the difficulty in determining properties from experimental data is increased considerably over the use of a linear model. If the nonlinearity can be modeled to affect all the springs in the same manner, $f_i(\sigma)$ becomes simply $f(\sigma)$ and the complexity is reduced.

Nonlinear Creep Power Law: It has been empirically observed that the creep of metals and other materials can be approximated using a creep power law of the form:

$$\varepsilon(t) = \varepsilon_0 + mt^n \tag{10.3}$$

For steady state (or secondary) creep of both metals and polymers it is often assumed that n = 1.0. In this form ε_0 is a fitting parameter and is found by extrapolation of the linear (with time) secondary creep portion of the curve to zero time (Dillard, (1981)). Another form used by Findley (1976) is

$$\varepsilon(t) = \sigma_0 D(t) = \sigma_0 \left(A + Bt^n \right) \tag{10.4a}$$

or using standard viscoelastic notation

$$\varepsilon(t) = \sigma_0 D(t) = \sigma_0 \left(D_0 + D_1 t^n \right) \tag{10.4b}$$

Where $A = D_0$, $B = D_1$, and n are material constants.

Early stress dependent power law models were developed for the creep of metals, mostly associated with the dependence of secondary or steady state creep on stress level, i.e. $d\varepsilon/dt = k\sigma^p$ (Findley et al., (1976)). The nonlinear form given in **Eq. 10.5** is sometimes called the Findley creep power law,

$$\varepsilon(t) = \varepsilon_0 \sinh\left(\frac{\sigma}{\sigma_0}\right) + mt^n \sinh\left(\frac{\sigma}{\sigma_n}\right) \tag{10.5}$$

where ε_0, σ_0, **m**, σ_n and are constants. However, many have contributed to the various forms of these equations including Andrade, Ludwik, Nadai, and Prandtl (see Findley et al., (1976), p. 8-21, for an excellent discussion and review of early efforts; also, see Dillard (1981) for a thorough description of the use of the power law as well as potential difficulties).

The fundamental idea behind the power law is to have a simple form, using few parameters, that will give a broadband approximation to a master curve rather than the more accurate generalized Maxwell or Kelvin models. The power law sacrifices accuracy at any one time to obtain a reasonable representation over the entire time scale from short term (glassy) behavior to long time (rubbery) behavior. For example, **Fig. 10.6** shows the compliance data for an epoxy from Chapter 7 along with several power law fits to the data. The first two fits use only data points from the earlier times and thus under predict at long times. When the longer time compliance slope is fit (the "fitlong" curve), the quality of the fit at shorter times is sacrificed. In addition, note that the power law is unable to fit the long time rubbery plateau of the material response as the mathematical form ensures ever increasing compliance values with time. Note that the Prony series fit also shown overlays the data exactly. Hiel et al. (1984) shows that the parameter **n** is most sensitive to experimental error. An example of the variation of **n** with the length of the creep tests used to collect data is shown in **Fig. 10.7a** for an epoxy resin often used in composites. The exponent **n** varies with the length of the test but becomes stable after relatively large creep times. He also shows that the exponent **n** becomes stable in a shorter time if it is determined from a creep recovery test as illustrated in **Fig. 10.7b**, where the stabilized value from **Fig. 10.7a** is obtained in tests an order of magnitude shorter in time. (The creep recovery test will be covered in detail later when discussing Schapery's method.)

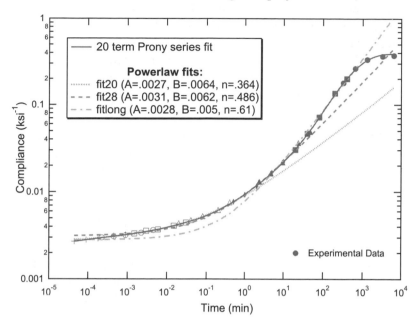

Fig. 10.6   Power law fits to compliance data for epoxy (Hysol 4290). Symbols show compliance data from **Fig. 7.3**.

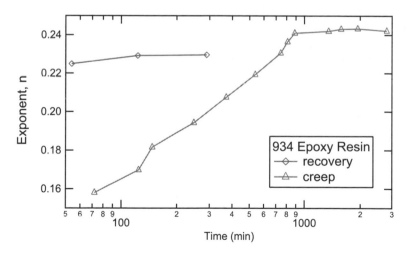

Fig. 10.7   Variation of power law exponent for an epoxy resin used in composites with length of creep test and length of recovery test. (Data from Hiel et al., (1984))

Multiple-Integral Approaches: The two preceding methods of representing nonlinear viscoelastic response and many others found in the literature were developed for the purpose of fitting one step uniaxial creep or constant strain-rate data. The major difficulty with nonlinear behavior is that superposition of the effects of multiple step stresses, a continuously varying stress or multiaxial stresses via the Boltzman superposition principle is no longer allowed. For this reason, the multiple and single integral approaches discussed in this section include modified superposition concepts that are necessary for successful stress analysis.

As noted by Robotnov (1980), the earliest description of a method to mathematically model nonlinear viscoelastic behavior was accomplished by Volterra using an earlier representation developed by Frechet in the early 1900's. The Volterra-Frechet equation for one dimension cited by Robotnov is,

$$\varepsilon(t) = \int_{-\infty}^{t} D_1(t-\tau_1)d\sigma(\tau_1) + \int_{-\infty}^{t}\int_{-\infty}^{t} D_2(t-\tau_1,t-\tau_2)d\sigma(\tau_1,\tau_2) + ... \quad (10.6)$$

According to Robotnov, this method was forgotten until the procedure was generalized to three dimensions by Rivlin and Green in 1954 (see Robotnov for reference). The multiple integral approach has been explored by many and an excellent description of various efforts are given by Robotnov (1980), Findley, et al. (1976) and Hiel, et al. (1984). Findley has perhaps documented the technique more fully than others both theoretically and experimentally in his 1976 book and in numerous journal articles cited therein. The following third order approximation is developed by Findley (1976) using a less rigorous approach than the functional analysis method given by others (see appendix A2 of his book for a derivation involving functional analysis),

$$\varepsilon(t) = \int_{-\infty}^{t} D_1(t-\tau_1)\frac{d\sigma(\tau_1)}{dt}d\tau$$

$$+ \int_{-\infty}^{t}\int_{-\infty}^{t} D_2(t-\tau_1,t-\tau_2)\frac{d\sigma(\tau_1)}{dt}\frac{d\sigma(\tau_2)}{dt}d\tau_1 d\tau_2 \quad (10.7)$$

$$+ \int_{-\infty}^{t}\int_{-\infty}^{t}\int_{-\infty}^{t} D_3(t-\tau_1,t-\tau_2,t-\tau_3)\frac{d\sigma(\tau_1)}{dt}\frac{d\sigma(\tau_2)}{dt}\frac{d\sigma(\tau_3)}{dt}d\tau_1 d\tau_2 d\tau_3$$

where the three kernel functions must be found from a three-step creep test. In an earlier paper, Findley, et al., (1965) gives a fourth order expansion that requires the determination of 14 kernel functions. Because the of the difficulty in experimentally evaluating a large number of functions and because of stability problems Robotnov (1980) suggests limiting studies to only a third order expansion. Both Findley and Robotnov give several approximate methods of evaluating kernel functions as well as experimental data and analysis.

Single Integral Approaches: Leadermann (1948) recognized the nonlinear nature of polymers and suggested an approach based on a linear hereditary integral given by,

$$\varepsilon(t,\sigma) = \int_{-\infty}^{t} D(t-\tau)\frac{d}{d\tau}\psi(\tau,\sigma)d\tau \qquad (10.8)$$

where nonlinear effects are incorporated in the stress measure, $\psi(t, \sigma)$. In his 1980 book Rabotnov[*] describes a similar equation for the one-dimensional behavior of metals (from his 1948 paper) that may be written as (Hiel, 1984),

$$\phi(t,\varepsilon) = \int_{-\infty}^{t} D(t-\tau)\frac{d\sigma(\tau)}{d\tau}d\tau \qquad (10.9)$$

except the nonlinearization was through the strain measure $\phi(t, \varepsilon)$. Hiel (1984) also reports that Koltunov[*] used a combination of the equations proposed by Leadermann and Rabotnov to obtain

$$\phi(t,\varepsilon) = \int_{-\omega}^{t} D(t-\tau)\frac{d}{d\tau}\psi(t,\sigma)d\tau \qquad (10.10)$$

which includes both a nonlinear strain and stress measure.

Two general methods for the development of single integral nonlinear constitutive equations that have been used are the rational (functional) thermodynamic approach and the state variable approach (or irreversible thermodynamic approach), each of which are described in a well-documented survey by K. Hutter (1977). In rational thermodynamics, the free energy is represented as a function of strain (or stress), temperature, etc, and then constitutive equations are formed by taking appropriate derivatives of the free energy. The state variable approach includes certain internal variables in order to represent the internal state of a material. Constitutive equations which describe the evolution of the internal state variables are included as a part of the theory. Onsager introduced the concept of internal variables in thermodynamics and this formalism was later used

---

[*] See Hiel for references. The 1948 Rabotnov paper and the Kotunov paper are in Russian journals and are not available to the authors.

by Biot in the derivation of constitutive equations of linear viscoelasticity. Schapery (1964, 1965) used this method to develop a modified linear hereditary integral approach to nonlinear viscoelastic materials. Knauss and Emri (1981) also used a single integral method to associate the nonlinearizing parameters to free volume and in this manner allowed the inclusion of stress-induced dilatation, moisture or other diffusion parameters in the theory.

10.3.   The Schapery Single-Integral Nonlinear Model

The Schapery single integral approach (1964, 1969) is an outgrowth of the irreversible thermodynamic procedures developed by Biot, and others and is likely the most widely used technique to represent the nonlinear time-dependent behavior of polymers. The thermodynamic derivation of the fundamental equations needed to represent data is beyond the scope of this text but an excellent description of the original derivation is given by Hiel, et al. (1984). Schapery in 1997 also provides an updated mathematical approach that includes viscoplasticity effects. The purpose here is to introduce the method as a means of representing polymer data and provide a basic understanding of how to obtain the necessary material parameters from experiments. The development of equations here closely follows the description of Schapery (1969) and Lou and Schapery (1971).

It is important to point out that the reason to develop a relatively simple and easy to use single integral method is not only to determine the necessary material parameters more easily, but to have a method that can be used with more ease and confidence in solving nonlinear boundary value problems to obtain stress, strain and displacement distributions for engineering design. This of necessity entails having a modified superposition approach as well as use of the time-shift principles discussed in Chapter 7.

10.3.1. Preliminary Considerations

In Chapter 6 it was shown that linear viscoelastic materials could be represented by the hereditary convolution integrals,

$$\varepsilon(t) = \int_{-\infty}^{t} D(t - \tau) \frac{d\sigma(\tau)}{d\tau} d\tau \qquad (10.11)$$

or

$$\sigma(t) = \int\limits_{-\infty}^{t} E(t-\tau)\frac{d\varepsilon(\tau)}{d\tau}d\tau \qquad (10.12)$$

where the lower limit is such that all previous history of loading is included. If the material is initially "dead" or has no previous history prior to time zero, the lower limit was $t = 0^-$ and the equations are,

$$\varepsilon(t) = \int\limits_{0^-}^{t} D(t-\tau)\frac{d\sigma(\tau)}{d\tau}d\tau \qquad (10.13)$$

or

$$\sigma(t) = \int\limits_{0^-}^{t} E(t-\tau)\frac{d\varepsilon(\tau)}{d\tau}d\tau \qquad (10.14)$$

Also as explained in Chapter 6, **Eqs. 10.13** and **10.14** can be written as,

$$\varepsilon(t) = \sigma_0 D(t)H(t) + \int\limits_{0^+}^{t} D(t-\tau)\frac{d\sigma(\tau)}{d\tau}d\tau \qquad (10.15)$$

or

$$\sigma(t) = \sigma_0 E(t)H(t) + \int\limits_{0^+}^{t} E(t-\tau)\frac{d\varepsilon(\tau)}{d\tau}d\tau \qquad (10.16)$$

when a step input occurs at $t = 0$. An additional form can be obtained by separating the creep compliance into instantaneous and transient terms such that,

$$D(t) = D_0 + \tilde{D}(t) \qquad (10.17)$$

or the relaxation modulus into equilibrium and transient terms such that,

$$E(t) = E_\infty + \tilde{E}(t) \qquad (10.18)$$

As an example the creep compliance of a three parameter solid may be written as,

$$D(t) = \left[\frac{1}{E_0} + \frac{1}{E_1}(1-e^{-t/\tau})\right] = D_0 + \tilde{D}(t) \qquad (10.19)$$

where

$$D_0 = \frac{1}{E_0} \quad \text{and} \quad \tilde{D}(t) = \frac{1}{E_1}\left(1-e^{-t/\tau}\right) \qquad (10.20)$$

Similarly, the relaxation modulus for a three parameter solid (from **Table 5.1**) is,

$$E(t) = q_0 + (\frac{q_1}{p_1} - q_0)e^{-t/p_1} \qquad (10.21)$$

where the coefficients p_1, q_0, and q_1 are given in Chapter 5 and,

$$E_\infty = q_0 = \frac{E_0 E_1}{E_0 + E_1} \quad \text{and} \quad \tilde{E}(t) = (\frac{q_1}{p_1} - q_0)e^{-t/p_1} \qquad (10.22)$$

Using the separated forms of the creep compliance or relaxation modulus, the linear viscoelastic constitutive laws 10.13 and 10.14 may be rewritten as

$$\varepsilon(t) = D_0\sigma(t)H(t) + \int_{0^-}^{t} \tilde{D}(t-\tau)\frac{d\sigma(\tau)}{d\tau}d\tau \qquad (10.23)$$

$$\sigma(t) = E_0\sigma(t)H(t) + \int_{0^-}^{t} \tilde{E}(t-\tau)\frac{d\varepsilon(\tau)}{d\tau}d\tau \qquad (10.24)$$

which are used as the base forms for the Schapery nonlinear model.

10.3.2. The Schapery Equation

Using irreversible thermodynamic (or energy) descriptions of the state of a viscoelastic material subjected to external loads, R. A. Schapery (1964, 1966) developed the following single-integral representation for strains due to a variable stress input,

$$\varepsilon(t,\sigma) = g_0 D_0\sigma(t)H(t) + g_1 \int_{0^-}^{t} \tilde{D}(\psi - \psi')\left\{\frac{d[g_2 \cdot \sigma(\tau)H(\tau)]}{d\tau}\right\}d\tau \qquad (10.25)$$

where g_0, g_1, g_2, a_σ are material parameters which are dependent on stress. The parameter a_σ is a shift factor which modulates the time scale much in the same way that the temperature dependent shift factor, a_T, modulates the time scale for temperature effects. The shifted stress dependent time scale is given by,

$$\psi(t,\sigma) = \int_{0}^{t} \frac{dt}{a_\sigma(t)} \quad \text{and} \quad \psi'(\tau,\sigma) = \int_{0}^{\tau} \frac{d\tau}{a_\sigma(\tau)} \qquad (10.26)$$

The Schapery method given by **Eqs. 10.25** and **10.26** is a mathematical definition of a time-stress-superposition-principle or TSSP that is analogous to the TTSP. Later it will be shown how to obtain stress dependent compliance and modulus master curves from experimental data using TSSP much in the same manner as temperature dependent master curves were determined from experimental data using the TTSP.

An analogous equation for stress under a variable input of strain was also developed by Schapery and is given by,

$$\sigma(t,\varepsilon) = h_\infty E_\infty \sigma(t) H(t) + h_1 \int_{0^-}^{t} \tilde{E}(\psi - \psi') \left\{ \frac{d[h_2 \cdot \varepsilon(\tau) H(\tau)]}{d\tau} \right\} d\tau \quad \textbf{(10.27)}$$

and

$$\psi(t,\varepsilon) = \int_0^t \frac{dt}{a_\varepsilon(t)} \quad \text{and} \quad \psi'(\tau,\varepsilon) = \int_0^\tau \frac{d\tau}{a_\varepsilon(\tau)} \quad \textbf{(10.28)}$$

In **Eqs. 10.25** and **10.27**, the parameters g_0, g_1, g_2 arise from third and higher order dependence of stress on the Gibbs* free energy while h_∞, h_1, h_2 arise from third and higher order dependence of the strain on the Hemholtz* free energy. The Hemholtz and Gibbs free energies are given by the energy balance equations,

$$\text{Hemholtz free energy}: A = U \cdot T \cdot S$$
$$\text{Gibbs free energy}: G = F = (U + PV) - T \cdot S \quad \textbf{(10.29)}$$

where **U** is the internal energy, **T** is the temperature, **S** is the entropy, **P** is the pressure (hydrostatic), **V** is the volume and **H=(U+PV)** is the enthalpy. (See also Chapter 7, section on rubber elasticity for additional discussion of thermodynamics.)

It should be noted that the Boltzman superposition integral for linear viscoelasticity is recovered in **Eq. 10.25** if the nonlinear parameters are each identically equal to one, i.e.,

$$g_0 = g_1 = g_2 = a_\sigma = 1 \quad \textbf{(10.30)}$$

Further, if all parameters except a_σ are unity Knauss's free volume model (Knauss and Emri, 1981) is recovered in which,

$$g_0 = g_1 = g_2 = 1 \quad \textbf{(10.31a)}$$

---

* See Chapter 7 for a brief thermodynamic description of Gibbs and Helmholz free energies.

$$a_\sigma \neq 1 \tag{10.31b}$$

$$a_\sigma \sim \text{Free Vol. (f)} \tag{10.31c}$$

$$f = f_0 + \alpha\,\Delta T + B\,\Delta\sigma + \gamma\,\Delta C \tag{10.31d}$$

where α is the coefficient of thermal expansion, **B** is a parameter relating stress to the amount of free volume and γ relates moisture concentration to free volume.

The Schapery Equation for a Two Step Stress Input: Determination of the material parameters necessary for the application of the Schapery Equation are best done by using creep-recovery data and will be demonstrated in a later section. Toward that end, we develop the specific form for the Schapery equation with a simple two-step load. In this section, we assume a general two step stress distribution such that,

$$\sigma(t) = \sigma_a H(t) + \left(\sigma_b - \sigma_a\right)H(t - t_a) \tag{10.32a}$$

or

$$\sigma(t) = \begin{cases} \sigma_a, & 0 \le t \le t_a \\ \sigma_b, & t > t_a \end{cases} \tag{10.32b}$$

also illustrated in graphical form in **Fig. 10.8**. To obtain the stress loading profile for creep-recovery, the second step is simply negative of the first step and this will be calculated explicitly in the next section.

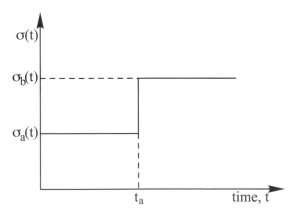

Fig. 10.8 Two step creep load.

Since stress is constant in each of the two time regions (below and above t_a), the nonlinear parameters are also constant in each of those regions. We refer to $\mathbf{g_i}$ and $\mathbf{a_\sigma}$ for $\mathbf{t \le t_a}$ as g_i^a and a_σ^a; similarly for $\mathbf{t > t_a}$ $\mathbf{g_i}$ and $\mathbf{a_\sigma}$ are referred to as g_i^b and a_σ^b. In application to Schapery's **Eq. 10.25**, the superscript for the $\mathbf{a_\sigma}$ and the $\mathbf{g_1}$ coefficients correspond to the time interval of evaluation of the term. The superscript for the $\mathbf{g_0}$ and $\mathbf{g_2}$ coefficients are affiliated with the stress value that each coefficient modifies. Using these parameters, Schapery's **Eq. 10.25** becomes

$$\varepsilon(t) = g_0^a \sigma_a D_0 H(t) + \left(g_0^b \sigma_b - g_0^a \sigma_a\right) D_0 H(t - t_a)$$

$$+ g_1 \int_{0^-}^{t} \tilde{D}(\psi - \psi') \frac{d}{d\tau}\left(g_2^a \sigma_a H(\tau) + \left(g_2^b \sigma_b - g_2^a \sigma_a\right) H(\tau - t_a)\right) d\tau \qquad (10.33)$$

The integral can be broken up into two integrals, before and after t_a and noting the derivative of the step function as the dirac delta function

$$\varepsilon(t) = g_0^a \sigma_a D_0 H(t) + \left(g_0^b \sigma_b - g_0^a \sigma_a\right) D_0 H(t - t_a)$$

$$+ g_1 \int_{0^-}^{t_a^-} \tilde{D}(\psi - \psi')\left(g_2^a \sigma_a \delta(\tau) + \left(g_2^b \sigma_b - g_2^a \sigma_a\right)\underbrace{\delta(\tau - t_a)}_{\text{zero}}\right) d\tau \qquad (10.34)$$

$$+ g_1^b \int_{t_a^-}^{t} \tilde{D}(\psi - \psi')\left(g_2^a \sigma_a \underbrace{\delta(\tau)}_{\text{zero}} + \left(g_2^b \sigma_b - g_2^a \sigma_a\right)\delta(\tau - t_a)\right) d\tau$$

Note the superscript on the $\mathbf{g_1}$ term remains unspecified for the first integral and will depend upon the time period of evaluation. The terms indicated with zero provide no contribution to the integral in which they appear and thus, the expression becomes

$$\varepsilon(t) = g_0^a \sigma_a D_0 H(t) + \left(g_0^b \sigma_b - g_0^a \sigma_a\right) D_0 H(t - t_a)$$

$$+ g_1 \int_{0^-}^{t_a^-} \tilde{D}(\psi - \psi')\left(g_2^a \sigma_a \delta(\tau)\right) d\tau \qquad (10.35)$$

$$+ g_1^b \int_{t_a^-}^{t} \tilde{D}(\psi - \psi')\left(\left(g_2^b \sigma_b - g_2^a \sigma_a\right)\delta(\tau - t_a)\right) d\tau$$

For $\mathbf{0 \le t \le t_a}$, the expression simplifies to,

$$\varepsilon(t) = g_0^a D_0 \sigma_a H(t) + g_1^a \int_{0^-}^{t} \tilde{D}(\psi - \psi')\left(g_2^a \sigma_a \delta(\tau)\right) d\tau \qquad (10.36)$$

where the integrand must be evaluated at $\tau = 0$. As a result, the effective times may be calculated as

$$\psi = \int_{0^-}^{t} \frac{dt}{a_\sigma^a} = \frac{t}{a_\sigma^a} \tag{10.37}$$

and

$$\psi' = \int_{0^-}^{t} \frac{dt'}{a_\sigma^a} = \frac{\tau}{a_\sigma^a} = 0 \tag{10.38}$$

With these considerations, **Eq. 10.36** becomes,

$$\varepsilon(t) = \left[g_0^a D_0 + g_1^a g_2^a \tilde{D}\left(\frac{t}{a_\sigma^a} \right) \right] \sigma_a H(t) , \qquad 0 \leq t \leq t_a \tag{10.39}$$

For the interval, $t > t_a$, the strain is given by,

$$\varepsilon(t) = g_0^b D_0 \sigma_b H(t-t_a) + g_1^b \int_{0^-}^{t_a^-} \tilde{D}(\psi - \psi') \left(g_2^a \sigma_a \delta(\tau) \right) d\tau +$$

$$+ g_1^b \int_{t_a^-}^{t} \tilde{D}(\psi - \psi') \left(g_2^b \sigma_b - g_2^a \sigma_a \right) \delta(\tau - t_a) d\tau \tag{10.40}$$

The first term is the effect of the step input of the stress, $\sigma_b = \sigma_a + (\sigma_b - \sigma_a)$, at $t = t_a$ which includes the effect of the step stress of σ_a at $t = 0$. The second term is the transient portion of the step input of stress, σ_a, at $t = 0$ whose effect continues beyond $t = t_a$ and the third term is the transient portion of the step input of the stress, $\sigma_b - \sigma_a$, at $t = t_a$.

The first integral must be evaluated at $\tau = 0$ and the second at $\tau = t_a$ to determine the effective times. For the first integral,

$$\psi = \int_{0^-}^{t} \frac{dt}{a_\sigma} = \int_{0^-}^{t_a} \frac{dt}{a_\sigma^a} + \int_{t_a}^{t} \frac{dt}{a_\sigma^b} \qquad \text{and} \qquad \psi' = \int_{0^-}^{\tau} \frac{d\tau}{a_\sigma} \tag{10.41}$$

or

$$\psi = \frac{t_a}{a_\sigma^a} + \frac{t-t_a}{a_\sigma^b} \qquad \text{and} \qquad \psi' = 0 \tag{10.42}$$

For the second integral,

$$\psi = \int\limits_{t_a}^{t} \frac{dt}{a_\sigma^b} \qquad\qquad \psi' = \int\limits_{\tau_a}^{\tau} \frac{d\tau}{a_\sigma^b} \qquad\qquad (10.43)$$

or

$$\psi = \frac{t - t_a}{a_\sigma^b} \qquad \text{and} \qquad \psi' = 0 \qquad\qquad (10.44)$$

As a result, the final equation for creep strain for $t > t_a$ is,

$$\varepsilon(t) = g_0^b D_0 \sigma_b + g_1^b g_2^a \tilde{D} \left(\frac{t_a}{a_\sigma^a} + \frac{t - t_a}{a_\sigma^b} \right) \sigma_a + g_1^b \left[\left(g_2^b \sigma_b - g_2^a \sigma_a \right) \tilde{D} \left(\frac{t - t_a}{a_\sigma^b} \right) \right] \quad (10.45)$$

Schapery Equation for a Creep and Creep Recovery Test: Schapery suggested using a creep and creep recovery test (as shown in **Fig. 10.9**) to determine the stress dependent parameters g_0, g_1, g_2, a_σ. This condition is a special case of the two step loading of **Eq. 10.32a** in which stresses $\sigma_a = \sigma_0$ and $\sigma_b = 0$ and thus

$$\sigma(t) = \sigma_0 H(t) - \sigma_0 H(t - t_1)$$

$$\text{or} \qquad\qquad\qquad (10.46)$$

$$\sigma(t) = \begin{cases} \sigma_0, & 0 \le t \le t_1 \\ 0, & t > t_1 \end{cases}$$

and the parameters associated with σ_b are,

$$g_0^b = g_1^b = g_2^b = a_\sigma^b = 1 \qquad\qquad (10.47)$$

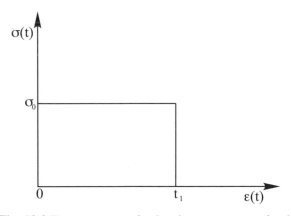

Fig. 10.9 Two-step creep load and creep recovery load.

It is necessary to specify $\tilde{D}(t)$ in order to evaluate the parameters and for that Schapery suggested using a power law given by,

$$D(t) = D_0 + D_1 t^n \qquad (10.48)$$

where D_0, D_1 and n are constants and $\tilde{D}(t)$ is given by,

$$\tilde{D}(t) = D_1 t^n \qquad (10.49)$$

The creep strain for the interval $0 < t < t_1$ from **Eq. 10.39** becomes after substituting the conditions given in **Eqs. 10.46 –10.49**,

$$\varepsilon(t) = \left[g_0 D_0 + g_1 g_2 D_1 \left(\frac{t}{a_\sigma} \right)^n \right] \sigma_0 H(t) , \qquad 0 \le t \le t_1 \qquad (10.50)$$

The recovery strain for $t > t_1$ is found from **Eq. 10.45** after substituting the same conditions given in **Eqs. 10.44** and **10.46 – 10.49** is,

$$\varepsilon_r(t) = \left[D_1 \left(\frac{t_1}{a_\sigma} + t - t_1 \right)^n - D_1 \left(t - t_1 \right)^n \right] g_2 \sigma_0 \qquad (10.51)$$

The constants D_0, D_1, and n as well as the stress dependent parameters g_0, g_1, g_2, and a_σ in **Eqs. 10.50** and **10.51** must be found or, in other words, seven material properties are needed to represent nonlinear uniaxial creep and creep recovery behavior. While this may seem excessive, it actually represents quite a large economy over the multiple integral form represented by **Eq. 10.6**. (See Findley (1976) and Rabotnov (1980) for examples of the use of multiple integrals.) For creep alone, only 5 parameters are needed if g_1 and g_2 are combined with D_1. However, the recovery strain is necessary to separate g_1 and g_2. Also, because of the sensitivity of n to the length of either a creep or creep recovery test as presented in the previous section on the power law, Schapery, et al. (1971) suggests using data from recovery to determine n. As will be seen in the following, the values of the strain jumps at the initial load and unloading arise naturally in determining the parameters. However, it is important to note that when dealing with experimental data both $\varepsilon(t = 0^+)$ and $\varepsilon(t = t_1^+)$ are ill-defined quantities because, as noted in Chapter 5, creep or recovery stresses (and hence strains) are not instantaneously applied in order to avoid dynamic effects. It is also important to note that even if the jump stresses are instantaneous at $t = 0^+$ and at $t = t_1$, the theory indicates the instantaneous jump strains are not equal in magnitude. For example, consider **Eq. 10.51** rewritten as,

$$\varepsilon_r(t) = g_2 D_1 \left(\frac{t_1}{a_\sigma}\right)^n \left\{ \left[1 + \left(\frac{a_\sigma}{t_1}\right)(t - t_1)\right]^n - \left(\frac{a_\sigma}{t_1}\right)^n (t - t_1)^n \right\} \sigma_0 \qquad (10.52)$$

or

$$\varepsilon_r(t) = g_2 D_1 \left(\frac{t_1}{a_\sigma}\right)^n \left\{ \left[1 + a_\sigma \lambda\right]^n - \left(a_\sigma \lambda\right)^n \right\} \sigma_0 \qquad (10.53)$$

where

$$\lambda = \frac{t - t_1}{t_1} \qquad (10.54)$$

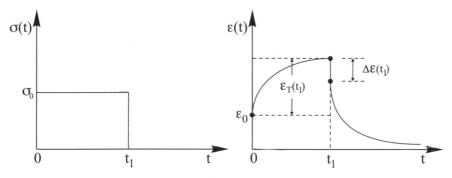

Fig. 10.10 Creep and Creep Recovery: Applied stress (left) and material response (right). Instantaneous strains at t=0 and t=t$_1$ are denoted as ε_0 and $\Delta\varepsilon(t_1)$. Transient strain refers to the time dependent strain and the magnitude of the transient strain at t=t$_1$ is depicted.

The creep and creep recovery data will appear as shown in **Fig. 10.10**. The magnitude of the instantaneous creep recovery, $\Delta\varepsilon(t_1)$ is given by,

$$\Delta\varepsilon(t_1) = \varepsilon(t_1^+) - \varepsilon(t_1^-) \qquad (10.55)$$

From **Eqs. 10.50** and **10.51**,

$$\varepsilon(t_1^-) = \left[g_0 D_0 + g_1 g_2 D_1 \left(\frac{t_1}{a_\sigma}\right)^n \right] \sigma_0 \qquad (10.56)$$

and

$$\varepsilon_r(t_1^+) = g_2 D_1 \left(\frac{t_1}{a_\sigma}\right)^n \sigma_0 \qquad (10.57)$$

or

$$\Delta\varepsilon(t_1) = \varepsilon_r(t_1^+) - \varepsilon(t_1^-) = -g_0 D_0 \sigma_0 + g_2 D_1 \left(\frac{t_1}{a_\sigma}\right)^n (1 - g_1)\sigma_0 \qquad (10.58)$$

Note that the jump in strain when the creep load is first applied at $t = 0$ is,

$$\varepsilon(t = 0) = g_0 D_0 \sigma_0 \qquad (10.59)$$

Comparison of **Eqs. 10.58** and **10.59**, shows that the strain jump discontinuity at $t=0$ and $t=t_1$ are not the same, even though the stress change is identical in magnitude. This dependence on load history is due to the nonlinearity of the material. In particular, the parameter g_1 can be identified as the source of this difference, since for a linear material $g_1=1$ and **Eq. 10.58** reduces to the negative of **Eq. 10.59** as is expected for a linear material.

10.3.3. Determining Material Parameters from a Creep and Creep Recovery Test

Currently the approach most often found in the literature to determine the seven material parameters to represent nonlinear viscoelastic behavior using the Schapery procedure is via numerical fitting of experimental data both in the linear range and in the nonlinear range at different stress levels. Examples of such a numerical approach can be found in work by Peretz and Weitsman (1982), Rochefort (1983), or Tuttle (1985). The latter two used a commercially available least-squares fitting program on a mainframe computer. Now it is easy to do the numerical curve fitting to a power law using a personal computer and programs such as Math Cad or MatLab (for example see Wing et al. (1995)). Schapery and co-workers have also described a method to determine the parameters based on DMA testing (Golden et al. (1999)), where the strains are first separated into oscillatory and transient components, then further dissected and the linear and nonlinear coefficients are determined directly or by integration of several expressions. In this section, however, we describe the approach originally set forth by Schapery (1969) and Lou and Schapery (1971), which although a bit cumbersome in its description, provides insight into the meaning and origin of the nonlinear parameters. Using this original approach, determination of the seven material parameters require creep and creep recovery tests to be performed at several stress levels. –

Finding the Material Constants D_0, D_1 and n: These linear material constants need to be determined from the experimental data in the linear stress range before the nonlinear parameters can be properly determined.

For linear viscoelastic response, when the transient strain is large compared to the initial step input, the strain vs. time on a log-log plot is a straight line at long times and the slope of the line is the power law exponent, n, (see **Fig. 10.6**). Additionally, the initial strain jump ε_0 equals $g_0 D_0 \sigma_0$, which could then provide the value of D_0, since g_0 is unity in the linear range. However, these approaches to determine n and ε_0 are inaccurate due to the inability to apply a truly instantaneous stress jump as mentioned earlier. Therefore, in general, ε_0 must be considered a fitting parameter that must be found in addition to the seven other parameters. The original approach set forth by Lou and Schapery (1971) circumvents these difficulties and provides a semi-graphical approach that allows all seven parameters to be found. Further, his approach is insightful in relating the nonlinear parameters to various portions of the creep and creep recovery process. Equally important, the approach allows for a convenient way for students to demonstrate their understanding of the analysis and parameters without recourse to numerical curve fitting packages – see Homework problem 10.5.) This approach is outlined in the following discussion.

From **Eq. 10.50** the transient creep strain at $t = t_1$ (depicted in **Fig. 10.10**) is,

$$\varepsilon_T(t_1) = g_1 g_2 D_1 \left(\frac{t_1}{a_\sigma}\right)^n \sigma_0 \tag{10.60}$$

The creep and creep recovery strain, **Eqs. 10.56** and **10.53**, can now be written as,

$$\varepsilon(t_1^-) = \left[g_0 D_0 + \varepsilon_T(t_1)\right]\sigma_0 \tag{10.61}$$

$$\varepsilon_r(t) = \frac{\varepsilon_T(t_1)}{g_1}\left\{\left[1 + a_\sigma \lambda\right]^n - \left(a_\sigma \lambda\right)^n\right\} \tag{10.62}$$

where

$$\lambda = \frac{t - t_1}{t_1} \tag{10.63}$$

Eq. 10.62 can be used to find n and $\varepsilon_T(t_1)$ for a creep and creep recovery stress in the linear range by noting that all nonlinear terms are unity, i.e., $g_0 = g_1 = g_2 = a_\sigma = 1$, and thus

$$\frac{\varepsilon_r(t)}{\varepsilon_T(t_1)} = \left[\left(1 + \lambda\right)^n - \lambda^n\right] \tag{10.64}$$

The numerator on the left hand side (LHS) of **Eq. 10.64** represents data and the right hand side (RHS) represents a mathematical representation of the data from which **n** and $\varepsilon_T(t_1)$ can be found. The RHS can be plotted as parametric family of curves with respect to **n** as shown on **Fig. 10.11** by the solid lines. The numerator on the LHS is known creep recovery data for a stress level in the linear range and is shown by square symbols in **Fig. 10.11**. The denominator represents the amount, $\varepsilon_T(t_1)$, the linear recovery data must be shifted downward on a log scale to match the curve with the proper exponent and is equivalent to the transient creep strain at t_1 for the same stress level in the linear range. The **x** symbol shows that the recovery data when shifted does not match the exponent **n = 0.25**. The diamond symbol shows that the recovery strain when shifted downward by the correct amount does fit the exponent **n = 0.15**. Thus the power law exponent is found as well as the transient creep strain, $\varepsilon_T(t_1)$, for the particular stress level used in the linear range.

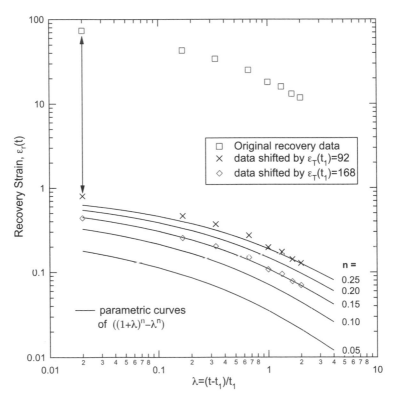

Fig. 10.11 Procedure for finding **n**. See also **Fig. 10.12**. (Data on FM-73 from Rochfort (1983) for a creep stress of 3.5 MPa (493psi).)

Alternately, **n** and $\varepsilon_T(t_1)$ could be found by simply solving **Eq. 10.64** with data taken for two different values of λ. The value of **n** and $\varepsilon_T(t_1)$ so determined for various values of λ should be the same but, due to experimental error, a small variation may be obtained using different points and an average should be used. At this point D_1 could be found from **Eq. (10.60)** since the nonlinear parameters are unity and $\varepsilon_T(t_1)$ and **n** have been determined. However the original approach differs slightly as follows.

The equation for creep strain (**Eq. 10.50**) for a linear viscoelastic material can be written as,

$$\varepsilon(t) = \left[D_0 + D_1 t^n\right]\sigma_0 = \varepsilon_0 + \varepsilon_1 t^n \tag{10.65}$$

where ε_0 is the initial strain and ε_1 is the transient strain coefficient. Selecting two values of transient strain at two values of time for a stress level in the linear range provides two equations from which ε_0 and ε_1 can be found and hence the coefficients D_0 and D_1 can be found. Note that the strains selected should be more than five times the amount of time required for the initial stress to be applied (See Lou and Schapery (1971)). Knowing ε_0 and ε_1 for a stress in the linear range allows the determination of $\varepsilon_T(t_1)$ for the same stress level and this should match the amount determined by shifting the linear data in **Fig. 10.11**. However, due to experimental error a small difference may be found.

Finding the quantities g_0 and $\dfrac{g_1 g_2}{a_\sigma^n}$: Using the creep strain **Eq. 10.50**,

$$\varepsilon(t) = \left[g_0 D_0 + g_1 g_2 D_1 \left(\frac{t}{a_\sigma}\right)^n\right]\sigma_0 \tag{10.66}$$

and two values of measured strain for two time values (again more than five times the initial loading time) will allow the determination of g_0 and $\dfrac{g_1 g_2}{a_\sigma^n}$ at each nonlinear stress level. This also allows the determination of the initial strain, ε_0, ε_1 and the transient strain, $\varepsilon_T(t_1)$, for each stress level in the nonlinear range.

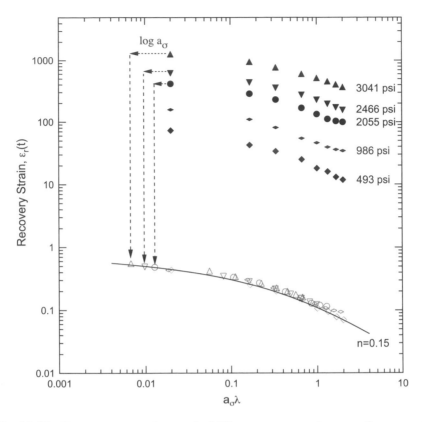

Fig. 10.12 Creep-recovery data and shifting process to form to for a master curve. (Data from Rochfort, (1983)).

Determination of g_1, g_2 and a_σ: Consider the recovery data shown in **Fig. 10.12** and the recovery strain given by **Eq. 10.62** written as,

$$\frac{\varepsilon_r(t)}{\varepsilon_T(t_1)/g_1} = \left[1 + a_\sigma \lambda\right]^n - \left(a_\sigma \lambda\right)^n \tag{10.67}$$

The data for all stress levels can be shifted to coincide with the linear viscoelastic data represented in **Fig. 10.11** by the power law exponent **n = 0.15** by moving each curve downward by the amount $\dfrac{\varepsilon_T(t_1)}{g_1}$ and to the left by the amount a_σ as shown in **Fig. 10.12**. This process forms a master curve as shown and is similar to time-temperature master curves discussed in Chapter 7. As a result this procedure is sometimes referred to as the analytical basis for the time- stress-superposition-principle (TSSP), which is discussed in the next section. As the transient strain, $\varepsilon_T(t_1)$, was previously determined for each stress level in the nonlinear range, g_1 can be determined. Thus, a_σ and g_1 are now known. The parameter g_2 can be found, as all other quantities in the expression $\varepsilon_1 = \dfrac{g_1 g_2}{a_\sigma^n} D_1 \sigma_0$ are known.

All parameters are now known that are needed to predict the response of a nonlinear viscoelastic material using the Schapery technique. Lou and Schapery (1971) used this semi-graphical procedure to characterize a glass epoxy composite and showed good correlation between creep and creep recovery experimental data and their analytical representations using **Eqs. 10.50** and **10.51**. Cartner used this approach to determine all the necessary Shapery parameters for a chopped glass fiber composite, SMC-25, and a structural adhesive, Metalbond 1113-2. He showed excellent correlation between creep data and theory for the SMC. The comparison was very good for the adhesive at low to moderate stress levels but diverged considerably at the higher stress level. Peretz and Weitsman (1982) used a computer-aided numerical least squares curve fitting approach to find all the parameters needed in the Schapery model to represent the structural adhesive FM-73. They showed a good correlation between data and theory. Rochefort (1983) used a similar computer-aided numerical least squares curve fitting approach to find the necessary parameters for FM-73 and his comparison between creep and creep recovery data is shown in **Fig. 10.13**. It is interesting to note that parameters in the two separate studies on FM-73 are similar even though performed in separate laboratories with materials made by different groups. For example Peretz and Weitsman found n = 0.12 and Rochefort found n = 0.15. Considerable differences were found in some of the nonlinear parameters, however.

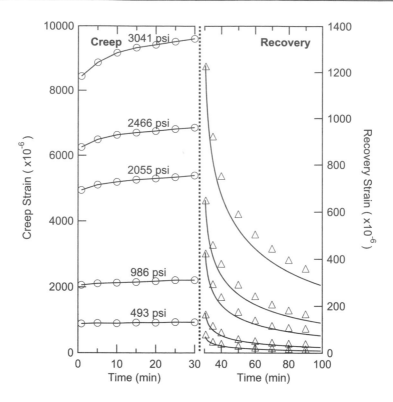

Fig. 10.13 Creep and recovery of FM-73 data and comparison to Schapery equation representation (Data from Rochefort, (1983)).

Experimental Procedures: Obtaining good experimental data from any creep and creep recovery testing program is difficult for high performance materials such as fiber reinforced polymer matrix composites and thin film adhesives and a few comments about procedures are necessary. Strain is typically measured using strain gages. As a result, it is important to recall the brief comments given in Chapter 3 about the possible reinforcement effects of strain gages and the possibility of strain gage heating effects. Indeed, Lou and Schapery (1971) estimated the effect of strain gage reinforcement to be about 2% for the glass-epoxy specimens they tested and considered this to be low enough to be neglected. However, without careful consideration, the error could be much larger especially for very thin specimens. Strain gage heating effects should be evaluated and can be minimized by limiting the amount of current used or by pulsing the current to gage only when a measurement is taken.

For materials such as the continuous fiber or chopped glass fiber composites as well as film type adhesives, it may be necessary to mechanically condition specimens prior to performing creep and creep recovery tests. The rationale is that numerous small and unstable flaws are created during processing and that when first tested these flaws grow by a small amount at relatively low stress and strain levels to a stable configuration. Therefore without mechanical conditioning prior to a creep test, an unknown portion of the initial strain and transient strain may be due to the accumulation of deformation associated with these flaws. Lou and Schapery, 1971 and Cartner, 1978 used cyclic constant strain rate tests to about 50% of ultimate to condition each test specimen. In Cartner's case, no change was found after the 20th cycle. For SMC materials the fibers flow in a random manner during the cure process and so it is necessary to take specimen from the central portion of the panel in the same direction to minimize potential scatter. Rochefort (1983) used a short creep and creep recovery test to mechanical condition specimens of FM-73. The test was repeated to convergence, where no change was found from the previous test.

Peretz and Weitsman (1982) describe an excellent test program to obtain the best results for an epoxy film adhesive (FM-73). Each test was performed in triplicate and each was repeated twice yielding six creep and creep recovery sets of data for each stress level that was then averaged. In this manner experimental scatter was minimized to be less than 2.5%.

All the testing programs described were for relatively short periods (less than one hour of creep and less than two hours of recovery). In reality, most structures made from the material used are designed to last days, months or years. As a result, a relative question to ask is: "how reliable would the use of predictive equations whose parameters were obtained from such short-term tests be in the design of structure for much longer periods of time"? In attempt to answer this question, Tuttle (1985) performed short-term tests on 90° and 10° unidirectional graphite epoxy specimens (creep of 480 minutes, recovery of 120 minutes) to obtain the necessary seven parameters. He then used the data in the Schapery model in conjunction with a lamination theory analysis to predict the long-term creep response of a symmetric composite laminate in a matrix-dominated direction. The result was compared with experimental data from independent long-term creep tests on the appropriate laminate. The analysis underpredicted the response at 10^5 minutes by about 8%. However, for several years, the error would be much greater. Upon performing a sensitivity analysis on all fitting parameters he found that the power law exponent was the largest contributor to error. As seen earlier, the power law does not provide the best fit to general viscoelastic creep compliance data, likely leading to this result. Thus while the small number of parameters makes the power law amenable to the semi-graphical approach to finding the Schapery model parameters, the numerical methods mentioned earlier can be successfully applied using a Prony series expansion for the creep compliance (Tuttle, (1995)) and may provide better results.

A TSSP recovery master curve such as the one shown in **Fig. 10.12** is not very useful for long-term predictions. However, creep data (or its analytical representation) as shown in **Fig. 10.13** can be converted into creep compliance data as a function of time for various stress levels. The data can then be shifted to form a master curve for any one of the individual stress levels similar to the process of shifting data at different temperatures to achieve a TTSP master curve as described in Chapter 7. By shifting data to the lowest stress level (analogous to shifting to the lowest temperature), a master curve extending to the longest times can be found. This approach to form a stress dependent master curve is discussed and demonstrated in the next section.

Before closing this section it is important to point out that the Schapery nonlinear characterization approach is best used for materials that do not have residual permanent deformation when the stress is removed. Therefore, the technique is best used for cross-linked or thermosetting polymers and not for thermoplastics or those referred to as linear polymers. It has been demonstrated that the Schapery technique may be used if the residual permanent deformation is subtracted from the creep and creep recovery response. However, the amount of permanent deformation needs to be small and a means to pro-rate the total amount over the total time scale is necessary. Tuttle et al. (1995) and coworkers (see also Pasricha et al. 1995 and Wing et al. 1995) have developed procedures to include parameters in the Schapery method that allow permanent deformation to be a part of the analysis. They use a nonlinear viscoplastic functional employed by Zapas and Crissman together with the Shapery model to find all material parameters for a graphite-bismaleimide composite and then use the results to pre dict the response of a laminate using classical laminated plate theory. Popelar, et al. (1990) has developed a nonlinear model that incorporates permanent deformation into the analysis and prediction of properties of polyethylene pipe. In addition, in recent years analytical models have been developed for nonlinear viscoelastic materials including the growth of damage and associated permanent deformation. For example see Weitsman (1988) and Ha and Schapery (1998). Segard et al. (2002) have used a procedure similar to that of Tuttle to model the behavior of a chopped glass fiber polypropylene composite with linear and nonlinear viscoelastic regions without damage and a nonlinear viscoelastic region with damage.

10.4. Empirical Approach To Time-Stress-Superposition (TSSP)

The fundamental concept behind both the TTSP and the TSSP is that the deformation mechanisms associated with time-dependent response at one temperature or stress (strain) level are the same as those at another level except that the time scale of the sequence of events is longer at a lower stress (strain) level or the time scale is shorter for a higher stress (strain) level. Basically, this allows for the deceleration or acceleration of the mechanisms of deformation and allows for either decelerated or accelerated predictions of response. Obviously, the latter is the most useful as by performing a test at a higher temperature or higher stress level, the collection of essential material data (parameters) can be shortened. For example, it is often necessary to design engineering structures for a life of 20 to 50 years. It is impossible to run tests for that duration to understand how polymer properties change over that time scale prior to making material decisions and building a structure. As a result, it is critical to have a process such as the TTSP and the TSSP to allow the determination of time dependent properties that may occur over a long time from tests that take place only over a short time.

The development of master curves using a semi-empirical TTSP approach was discussed in Chapter 7. A similar semi-empirical TSSP is can also be used to obtain a master curve valid over a long time at one stress level by shifting and superposing creep compliance (or relaxation modulus) data obtained at other stress levels in a short term test. This principle is illustrated in **Fig. 10.14**.

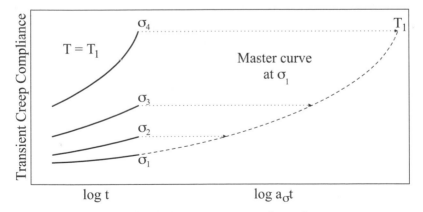

Fig. 10.14 TSSP master curve formation.

Hiel has used both the TSSP and the Schapery procedure to produce master curves of the shear behavior of a carbon epoxy composite. An example of his results is given in **Fig. 10.15**. The shear creep compliance of a carbon epoxy composite is shown for various stress levels. The data were shifted horizontally to form a smooth master curve for the lowest stress level as illustrated by the open symbols. The Schapery procedure was also used independently on this data set and the resulting master curve prediction is indicated by the solid line. For more details, see the cited reference in **Fig. 10.15**.

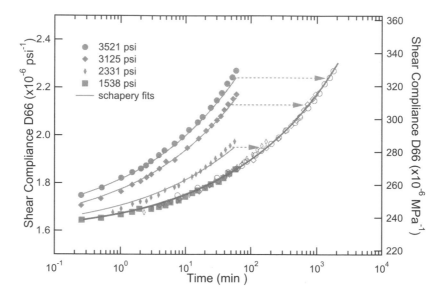

Fig. 10.15   Shear compliance of a carbon/epoxy composite at 320° F. (Data from Hiel et al., (1984)). Master curve shown with open symbols shifted by a_σ and Schapery master curve fit (thicker solid line); Schapery parameters retrofitted to the individual creep curves also shown.

The TTSP and the TSSP can be combined to produce a master curve that can be shifted both as a function of temperature and stress. The shift factors are therefore multiplicative or additive on a logarithmic time scale. This process is shown in **Fig. 10.16** where two paths are indicated to find the final master curve. In both cases, creep curves at different stress and temperature levels are found experimentally. Following the left path, the family of curves for each stress level is assembled on one graph and TTSP used to obtain TTSP master curves of the response at a reference temperature; one master curve for each stress level is obtained. Subsequently,

TSSP is performed on the master curves in (c) to obtain the grand Time-Temperature-Stress-Superposition master curve in (e). The right path is performed similarly, but starting with TSSP and then applying TTSP to the stress based master curves in (d). The same grand master curve should be obtained via either path. A detailed discussion of the limitations of the process can be found in Griffith (1980). An example for the formation of a master curve using TSSP is shown in **Fig. 10.17** and **Fig. 10.18** for the shear compliance of a carbon epoxy composite at 320° F.

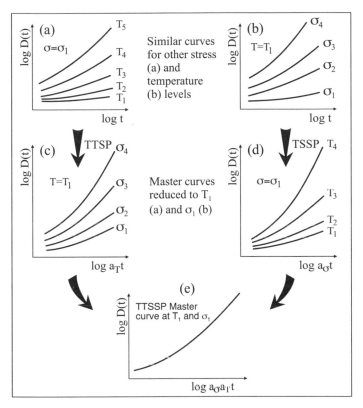

Fig. 10.16 Combination of TTSP and TSSP to form a TTSSP master curve.

TTSP was also used to form a master curve for the same material and the combined shift function surface is shown in **Fig. 10.19**.

From this discussion as well as the information in Chapter 7 on time-temperature relationships and time-aging time relationships, it is clear that there are a variety of environmental factors that affect the long-term response of polymers and their composites. These parameters include temperature, stress, moisture*, physical aging, chemical aging, and others. In many of these cases, as has been shown here and in Chapter 7 with temperature, physical aging and stress, it is possible to consider the effects of the variables individually over short time periods and to represent long time behavior via superposition principles. These superposition principles are such that shifting experimental data on the log time scale produces shift factors which are used as a multiplicative factor on the time in expressions for material properties and in constitutive equations. And it is in this manner that long-term predictions can be made for material response. Ideally, it is desirable to find a convenient way by which all the relevant environmental factors could be combined into a single procedure to make lifetime predictions. This concept is expressed in the shift factor surface as shown in **Fig. 10.19** for the two parameters of stress and temperature, and can in general be thought of as a multidimensional surface for all factors similar to that proposed by Landel and Fedders (1964). However, in the examples shown here, each effect is probed individually to produce the shift factors, while when the environmental conditions occur simultaneously there are nonlinear interaction effects that prevents application of such a simple concept universally. The nonlinear model pioneered by Knauss and Emri (1981) mentioned earlier is one approach to attempt to address these coupled effects theoretically in a single complex shift factor function. Recently, Popelar and Leichti (2003) have extended this approach to incorporate distortional changes into the Knauss free volume model that is largely related to dilatational effects. This area of long-term predictions of polymer behavior considering multiple coupled environmental variables is still an active area of research where continued effort is needed (*Going to Extremes*, National Academy Press, 2006).

---

*   A time-moisture-superposition-principle is discussed, for example, by Crossman and Flaggs, 1978. See also, Flaggs, D. L. and Crossman, F. W., 1981.

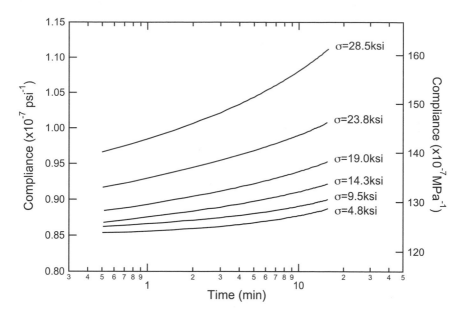

Fig. 10.17 Shear compliance of a carbon/epoxy composite at 320°F at various stresses. (Data from Griffith et al., (1980)).

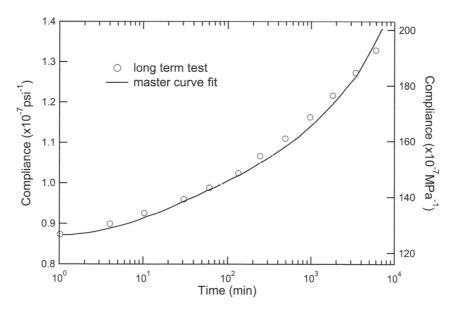

Fig. 10.18 Shear compliance master curve of a 10° carbon/epoxy composite developed from TSSP at 320°F. (Data from Griffith et al., (1980)).

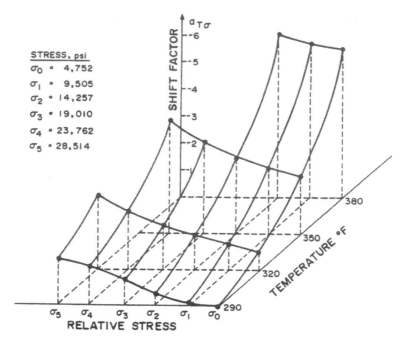

Fig. 10.19 Shift functions surface for a combined compliance master curve for the same material given in **Fig. 10.18**. (From Griffith et al., (1980)).

10.5. Review Questions

10.1. Name and describe the two types of nonlinearities most frequently used in engineering analysis.

10.2. Name and briefly describe two other less well-known types of nonlinearities.

10.3. Define extension ratio. In which type of nonlinearity is the extension ratio typically used?

10.4. Why is creep recovery data often used to determine the exponent, n, in the creep power law?

10.5. What is the TSSP? Describe the experimental process by which it is used.

10.6. What analytical approach is a mathematical statement of the TSSP?

10.7. Is it possible to combine TTSP and TSSP? If so describe the process.

10.8.   Is there evidence of a time- moisture-superposition principle?

10.9.   Why are concepts such as the TTSP and TSSP important? Give examples.

10.10. It is suggested that the reader perform additional reading on the subjects of the theory of rubber elasticity with especial emphasis on the thermodynamic approach to the theory.

10.6. Problems

10.1.   Discuss the power law in its various forms and compare it to the use of a Prony series to represent viscoelastic data.

10.2.   Discuss various methods to find the creep exponent in the power law. Is it ever a function of stress? Temperature? Explain.

10.3.   Perform a literature search for various forms of the power law with special attention to a modified form that allows a better representation of long time behavior. The latter is sometimes referred to as the modified or generalized power law. (Three sources are Hiel, et al., (1984), Halpin, (1967) and Landel and Fedders, (1964).)

10.4.   Determine the power law parameters for the following creep data.
$\sigma_0 = 6.19$ MPa (1ksi)

Time (min)	Strain(%)
1	0.324
5	0.335
10	0.340
20	0.348
30	0.352
40	0.355
50	0.359

(Hint: There are several methods used to find power law parameters but sometimes the fastest is trail and error. However, the reader would benefit from research for various possible procedures as suggested in problems 10.1 – 10.3.)

10.5.   Consider a nonlinear viscoelastic material which is well modeled by the Schapery approach. Would it be possible to determine all seven (7) material parameters only using creep tests? That is, not using recovery (unloading) data or a multiple steps in stress? Give a detailed explanation for your answer.

10.6. Using the nonlinear creep and creep recovery data given below, find the seven material parameters (D_0, D_1, n, g_0, g_1, g_2 a_σ) needed for representation by the Schapery equations.

Creep

Stress	1ksi	2ksi	3ksi	4ksi	5ksi
Time (min)	$\varepsilon\ 10^{+2}$ μ in/in	$\varepsilon\ 10^{+2}$ μ in/in	$\varepsilon\ 10^{+2}$ μ in/in	$\varepsilon\ 10^{+2}$ μ in/in	$\varepsilon\ 10^{+2}$ μ in/in
0.0	3.00	6.30	9.90	13.80	18.0
0.5	3.63	7.84	12.48	17.71	23.46
1.0	3.75	8.13	12.96	18.44	24.50
2.0	3.89	8.48	13.54	19.32	25.73
5.0	4.12	9.04	14.48	20.75	27.72
10.0	4.33	9.55	15.35	22.06	29.56
20.0	4.59	10.17	16.38	23.62	31.74
30.0	4.76	10.58	17.07	24.67	33.20

Recovery

Stress	1ksi	2ksi	3ksi	4ksi	5ksi
Time (min)	$\varepsilon\ 10^{+2}$ μ in/in	$\varepsilon\ 10^{+2}$ μ in/in	$\varepsilon\ 10^{+2}$ μ in/in	$\varepsilon\ 10^{+2}$ μ in/in	$\varepsilon\ 10^{+2}$ μ in/in
30.0	1.52	3.38	5.44	7.94	10.72
30.5	1.13	2.55	4.14	6.11	8.35
31.0	1.02	2.30	3.76	5.57	7.65
32.0	0.89	2.02	3.32	4.95	6.85
35.0	0.70	1.61	2.67	4.02	5.62
40.0	0.55	1.28	2.13	3.25	4.60
50.0	0.41	0.95	1.61	2.48	3.56
60.0	0.33	0.78	1.33	2.06	2.98

11. Rate and Time-Dependent Failure: Mechanisms and Predictive Models

No text on polymer science and viscoelasticity is complete without a discussion of time-dependent failure and just as with other structural materials, failure must be defined. In this chapter, only failure by a creep to yield or a creep to rupture (separation) will be considered. We will address both the mechanisms of deformation that often precede these types of failures as well as modeling to describe this behavior. The primary focus will be on one-dimensional models but many of the models discussed have been or can be extended to three-dimensions. The procedures to be discussed are not new and are relatively easy to use by the design engineer to make estimates of the time for either yielding or rupture to occur. While no discussion of either viscoelastic fracture mechanics or fatigue crack growth will be given these are very important topics and the reader is referred to Knauss (1973, 2003) for the former and to Kinloch and Young (1983) for the latter for an in-depth discussion of these topics. Fracture based approaches for prediction of time to failure work best when a crack of a known size exists. The same is true for fatigue as a relation between crack growth rates and time to failure can be established. Newer approaches provided by damage mechanics (Krajcinovic, (1983)) and viscoplasticity (Lubliner, (1990)) provide a more rational but highly mathematical approach to damage and/or failure evolution for three-dimensional stress states and are perhaps best suited for numerical procedures such as the finite element method. Here we restrict ourselves to simpler, analytic approaches to introduce the fundamental issues.

Failure is a defined quantity that must be established in the initial design stages. Typically, structural failure is defined as excessive deflection, yielding, or rupture. Excessive deflection may occur while materials of a structure are linear elastic or viscoelastic without yielding and for such circumstances can be predicted and prevented by elastic or viscoelastic stress, strain and deflection analysis as described in earlier chapters. The focus in this chapter will be on excessive deformation due to time-dependent yielding and/or progressive damage accumulation leading to rupture.

For metals, concrete and other usual building materials various design criteria have emerged to avoid failure by either yielding or rupture. These are often called "theories of strength" and date back to discussions by Galileo in 1638 (see Sandhu, (1972)). Three such strength theories discussed previously in Chapter 2 are the maximum normal stress theory, the maximum shear stress theory and the maximum distortion energy theory. While these are the most used theories, in reality there are hundreds more that have been proposed since the days of Galileo. Sandu (1972) describes more than 30 theories of failure for isotropic and anisotropic materials, most of which have been used for fiber-reinforced laminated (polymer and metal matrix) composites but none include a creep to yield or creep to rupture process. A review article by Yu (2002) entitled "Advances in strength theories for materials under complex stress state in the 20th Century" cites more than a 1000 references but only about ten are related to polymers and these do not explicitly speak to a time dependent failure process.

A major difficulty with predicting any type of failure including those for time dependent materials is that our analytical foundation for stress, strain and deformation analysis is based upon continuum mechanics that assumes that the material is continuous without flaws down to infinitesimal dimensions. Certainly such an assumption is not true for any realistic structural material including polymers. Therefore a method is needed to include a distribution of defects into continuum models which is what time dependent versions of plasticity theories, fracture mechanics and damage mechanics attempt to do. Herein some of the earlier approaches for the prediction of time dependent failures will be presented, several of which provide explicit elementary equations that can be used to predict the onset of time dependent yield and/or rupture. Two of these approaches (Nagdi and Murch, (1963) and Reiner, (1939 and 1964)) are unique in that the viscoelastic constitutive model for the material is contained in the failure law and for that reason are sometimes called a "unified models". With these introductory approaches and accompanying data, we demonstrate the fundamental issues of creep yielding and creep rupture in polymers along with simple tools to describe such behavior. Armed with this knowledge, the interested reader can delve into more advanced treatises on viscoelastic fracture and damage accumulation mentioned earlier.

11.1   Failure Mechanisms in Polymers

Before entering into mathematical descriptions of creep yielding and rupture, it is instructive to describe several of the physical deformation mechanisms most often considered to lead to polymer failure. These mechanisms involve large-scale irreversible molecular changes with the most common optically visible result of shear banding or crazing as described below.

11.1.1 Atomic Bond Separation Mechanisms

As discussed in Chapter 2, the rupture or fracture of materials must, at the atomic and molecular scale, involve the separation of individual atoms and molecules. A long-standing interpretation of interatomic forces and the resulting energy necessary for equilibrium is as given in **Fig. 2.22(d)**. In order to break the bond between two atoms the applied external force must generate internal forces which exceed the maximum. Equivalently, the amount of energy created by external forces must be larger than the bond energy, $E_B = D$, at the equilibrium spacing r_0. From **Fig. 2.22(a)** the addition of the attractive force and the repulsive force gives total force and is,

$$F = \frac{n\alpha}{r^{n+1}} - \frac{m\beta}{r^{m+1}}$$

(11.1a)

and the total energy is,

$$E = \frac{\beta}{r^m} - \frac{\alpha}{r^n}$$

(11.1b)

where the variables are all as defined in Chapter 2. These equations are for only two atoms. However, expressions for groups of atoms are similar to the above equations with summation over the group.

As discussed in Chapter 4, atoms are in a constant state of motion with the frequency and amplitude being related to the temperature. In a polymer, the motion is related to the amount of free volume and is small below the glass transition temperature and increases dramatically as the temperature is increased above the glass transition temperature. At the T_g the free volume in many polymers is approximately 1/40 or 2.5 % of the total volume.

In thermoplastic polymers the bonds between individual chains are secondary and the amount of free volume is sufficient for local chain motion. In thermosetting polymers interchain interactions between cross-linked sites are also secondary bonds and motion of these segments is similar. A mechanism of "switching" often used to describe the nature of motion in a viscous liquid is sometimes used to describe these local atomic movements in polymers. As illustrated in **Fig. 11.1(a)**, the atoms in a liquid can change positions through a rotational jump. (See McClintock and Argon, (1966); Courtney, (1990); and Shames and Cozzarelli, (1992) for more detailed discussion of this process.)

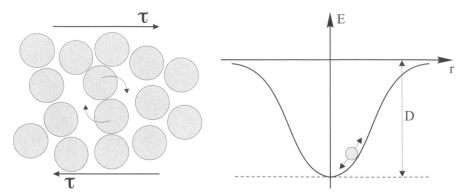

(a) Rotational jump in a liquid.   (b) Atom at temperature T vibrating in an energy well.

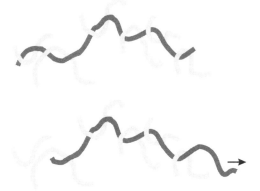

(c) Schematic of polymer chain movement through its neighbors by reptation.

Fig. 11.1 Molecular mechanism for flow of liquids and creep of solids.

In a liquid with a low viscosity or in a gas, the switching can take place spontaneously without the application of stress. For solid thermoplastic polymers, a first approximation is to assume that the application of external forces creates an internal shear stress sufficient to cause an atom to escape the energy well shown in **Fig. 2.22(e)** and **Fig. 11.1(b)** (**D** is the dissociation energy) and thus enable switching.

While many have used the above analogy for metals and for polymers, the nature of the switching phenomena is quite different in a polymer than in a simple liquid. In a polymer the beads in **Fig. 11.1(a)** must represent atoms along the backbone chain and, hence, it is difficult to visualize how the switching can take place without affecting the primary bonds along the backbone chain. A newer approach that treats the motion of polymer molecules in terms of reptation or a worm-like creeping motion of one polymer molecule through the matrix formed by its neighbors is likely a better visualization for polymers (**Fig. 11.1(c)**; see also Aklonis, et al., (1983) for a brief description).

It is possible to show that the binding or disassociation energy that must be overcome is (see; Shames and Cozzarelli, (1992)),

$$D = \frac{\alpha}{r_0^n}(1 - \frac{n}{m})$$

(11.2)

Clearly the energy needed to escape the energy well varies with temperature and since all properties of polymers are both time and temperature dependent, it is reasonable to assume that the disassociation energy, D, for polymers is also a function of time and rate. Often this time and temperature dependence is modeled by the Ahrrenius reaction rate equation,

$$\Phi = Ae^{-\frac{E_a}{kT}}$$

(11.3)

where Φ is the rate of a process, **A** is a constant, E_a is the activation energy, **k** is Boltzman's constant and **T** is the absolute temperature. This equation was developed by Ahrrenius presumably for the purpose of explaining the rate of chemical reactions but is also widely used to model the rate of many processes. Undoubtedly, Ahrrenius was strongly influenced by Boltzman (he was an associate of Boltzman's for a time) and his equation has a strong resemblance to the well-known Maxwell-Boltzman equation. The Maxwell-Boltzman equation was developed for the gaseous state and defines the probability that a molecule will have a particular energy state among all the energy states of the total number of molecules in the volume (for a discussion, see; Freudenthal, (1950) and Glasstone, S, Laidler, K.J. and Iyring, H., (1941)).

Iyring has suggested that the Ahrrenius equation is inadequate in many instances and that an equation with two reaction rates is more appropriate. His equation can be written as,

$$\Phi = \Phi_0 e^{nS} = aT^w \exp(\frac{-E_a}{kT})\exp\left[S\left(c + \frac{-E_{a2}}{kT}\right)\right]$$

(11.4)

where **a, c, w, E$_a$, E$_{a2}$** and **k** are constants. In **Eq. 11.4 E$_a$,** and **E$_{a2}$** are the two activation energies while **S** is a stress dependent function and $n = c + \dfrac{-E_{a_2}}{kT}$ (for a description of this approach, see: Carfagno and Gibson, (1980); or Ward and Hadley, (1993)). Others have also noted that the activation energy is not a constant and have suggested that the activation should be represented as a variable function of temperature and stress.

The activation energy approach has been used to develop both time dependent yield and time dependent rupture models, wherein a critical activation energy is defined and the expressions can be used to determine the time to yield or rupture under given static loading conditions. A few of these approaches will be discussed briefly in later sections.

Before closing this section it should be mentioned that efforts have been made to directly calculate the failure or fracture (separation) strength of a solid using the atomic bonding model. It is relatively easy to show that, based on **Fig. 2.22** and **Eq. 11.1,** the theoretical strength of a perfectly arrayed crystalline solid should be on the order of the elastic modulus. For example, since the modulus of mild steel is 206×10^6 MPa (30×10^6 psi) the strength should be of similar magnitude. Since the tensile strength of mild steel is only 206×10^3 MPa (30×10^3 psi) there is obviously something significant missing from the strict atomic bond separation prediction for strength. A similar argument can be made for any solid polymer. The answer is, of course, that both types of solids have many inherent flaws due to production processes that drastically lower the tensile strength. Some of these flaws and imperfections that lead to lower strength in polymers will be discussed in succeeding sections as well as potential mechanisms that lead to lower strength. It is known, however, that the strength of ether metals or polymers can be drastically improved if the production process is better controlled to avoid flaws and imperfections. Further, by creating more perfect crystalline structures strength properties can be greatly improved.

11.1.2 Shear Bands

Shear bands develop in polymers due to large-scale movement of molecular chains and usually initiate at a site of higher stress than the surrounding region or a point of stress concentration. An example of a shear band formation in a uniaxial tensile test of a thin specimen of polycarbonate is shown in **Fig. 11.2**. The Luder's band begins to form at a point of high stress (likely due to an edge defect created during machining) such as that indicated by the high concentration of isochromatic fringes at the beginning of the tapered region of a tensile specimen of polycarbonate shown in (a). The yield region grows into a V-shaped band as shown in (b) where the photo is taken after removal of the load. The existence of the birefringence fringes after unloading indicates residual permanent strains remain after unloading. A neck forms with continued loading beyond the point of Luder's band initiation as shown by the unloaded specimen in (c).

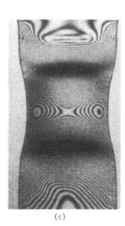

(a)   (b)   (c)

Fig. 11.2 Typical isochromatic fringe patterns[1] showing stress contours in polycarbonate: (From: Brinson, (1973)) (a) at incipient yielding. (b) unloaded specimen after Luder's band formation. (c) unloaded specimen after neck formation.

Another example of a Luder's band in polycarbonate is shown in **Fig. 11.3(a)**. Also shown in **Fig. 11.3(b)** are micro-shear bands which form in polystyrene. In each case, birefringence photos of an unloaded specimen show the residual plastic deformation remaining after load removal. In polycarbonate, yielding initially produces a single slip (shear) band at a 54.7° angle with the long axis of the specimen as seen in **Fig 11.3(a)**. With propagation and depending on the specimen thickness, the shear band can

---

[1] The birefringence photos here and elsewhere were taken by viewing a specimen using polarizing filters such that stress or strain induced birefringence could be viewed. The fringes are termed isochromatics because if they are viewed with polychromatic light they will appear in various colors dependent upon the stress field. The fringes are black here as the specimen is illuminated with monochromatic light. In this text the isochromatics are not being used for stress or strain analysis but simply to enhance the ability to view the shear band region. For more information, see Optical Methods in engineering Analysis by G. Cloud, Cambridge University Press.

develop a V-shape as shown in **Fig. 11.2**. Very thick specimens will from a band (or neck) perpendicular to long axis of the specimen. Interestingly, the angle of the slip band does not coincide with the direction of maximum shear stress of 45° as might be expected. Rather, the angle conforms to the direction associated with the maximum distortion energy or the octahedral angle as described in Chapter 2. (See Nadai, A., (1950) for a discussion of Luder's bands in metals and Hetenyi, (1952) for a discussion of Luder's bands in nylon 66.) However, the angles of the slip bands will depend upon the ductility of the material. Polycarbonate exhibits nearly perfectly plastic flow past the yield point and a distortion energy failure law is reasonable.

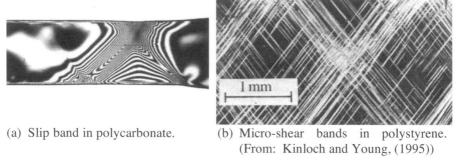

(a) Slip band in polycarbonate.

(b) Micro-shear bands in polystyrene. (From: Kinloch and Young, (1995))

Fig. 11.3 Shear (slip) bands in two polymers.

Molecular mechanisms associated with shear band formation are indicated schematically in **Fig. 11.4**. Shear bands form due to the orientation of molecules in regions of high stress. Initially, deformation in glassy polymers is associated with stretching bond angles and small conformation changes due to bond rotations. However, as the external loads increase, the internal stresses on the molecular scale increase and the level of molecular energy nears the disassociation energy for secondary bonds, large movement of the molecules can occur. Some motion comes from the relaxation of kinks in the structure during polymerization and some comes from conformation changes. That is, the molecules tend to become unentangled and they begin to orient in the direction of the local maximum octahedral shear stress (or the direction of maximum distortion energy). Eventually, the molecules will tend to orient with the direction of maximum external load as shown in **Fig. 11.4**.

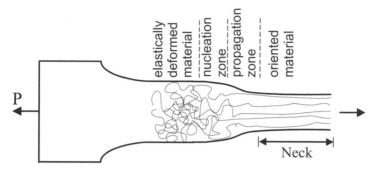

Fig. 11.4 Mechanisms for the formation of slip bands and a neck in a ductile polymer.

11.1.3 Crazing

Crazing is another deformation mechanism for glassy polymers but unlike shear bands crazes form perpendicular to the maximum normal stress. Crazes are micro-cracks that occur due to the formation of micro-voids at points of high stress concentration such as surface scratches, particulate inclusions such as dust particles or even small voids occurring during processing. Crazes in a modified (rubber toughened) epoxy are shown in **Fig. 11.5**. Here the crazes are quite small but many of them join together to produce white striations across the specimen. For transparent polymers, a milky appearance or translucency may occur while in an opaque polymer, as in **Fig. 11.5**, the crazes appear as white regions often called stress whitening. (This is the principle behind plastic labeling tape). Generally, crazes are caused by dilatational stresses while shear bands are caused by deviatoric stresses.

Crazes occur in both brittle and ductile polymers but they are often very hard to see with the naked eye. For example, they can be seen in tensile tests of thin polycarbonate fracture specimens if viewed from the correct angle and with lighting such that the edge of the tiny cracks are positioned to reflect light back to the viewer. A single edge notched tensile specimen is shown in **Fig. 11.6** after plastic zone growth and load removal. The crack is at the extreme left and the plastic zone is the long, horizontal flame shaped region in the center; the sudden reduction in thickness of the specimen at the edges of the plastic zone changes the refraction of light from the specimen allowing its visualization. Just ahead of the plastic zone is a region of crazes that formed while loading and due to the residual permanent deformation remain visible after unloading. The craze zone at the end of the plastic zone is nearly circular and represents the intense energy region often discussed in regard to the stress field in front of a crack in a brittle material that arises using the theory of linear elastic fracture mechanics. The distortion of the circular shape is due to the angle at which the specimen was photographed. Linear elastic fracture mechanics forms the basis for the analysis of cracks in ductile materials via the Dugdale model. (See Brinson, (1969)).

Several mechanisms have been suggested to explain the formation of crazes in polymers. One approach suggests that crazes initiate either on the surface of a polymer at imperfections such as small flaws or scratches, or at internal defects such as air bubbles dust particles, etc. The mechanism for the crazes shown in **Fig. 11.5** is likely the inclusion of rubber particles as discussed in Chapter 3 (see **Fig. 3.2**). The crazes shown in **Fig. 11.6** may be both due to small surface or internal cracks occurring in the intense stress region at the crack or plastic zone tip. While a craze may start at an imperfection such as a dust particle or small void, a mechanism for craze growth is needed to account for the multiplicity of crazes at the plastic zone tip in polycarbonate seen in **Fig. 11.6** which, of necessity, must be different than that due to the inclusion of rubber toughened particles as in shown in **Fig. 11.5**. One explanation is that the triaxial stress field stress field in the region ahead of the tip of a micro-crack must be sufficient to cause a new crack to nucleate immediately ahead of the old crack while leaving a small ligament in between. The nucleation process is repeated until a number of the micro-cracks coalesce to form a larger visible crack. Numerous such visible cracks are formed and eventually one will dominate and lead to eventual failure. This mechanism of craze growth and others are described in detail by Kinloch and Young, (1995) and Courtney, (1990).

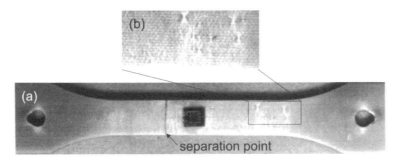

Fig. 11.5 Crazes (whitened regions) in a modified epoxy Metalbond 1113-2 (Renieri, (1976)). (a) Failed tensile specimen. (b) Enlargement of a central crazed area.

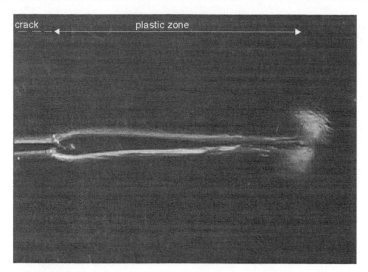

Fig. 11.6 Craze region at tip of plastic yield zone ahead of an edge crack in thin sheet of polycarbonate. (The white region along the crack and the long plastic zone is reflected light due to the oblique angle of exposure (out of the plane of the page).

11.2 Rate Dependent Yielding

First it is important to place in perspective the concept of yield behavior. For most metals yielding is defined as the point on the tensile stress strain diagram found in a constant strain-rate test after which a permanent deformation will exist on unloading. As it is experimentally difficult to determine this point, often a 0.2% offset method is used as described in Chapter 2. However, in the case of mild steel it is customary to define yielding as having occurred when the load in a constant strain rate test decreases while the strain continues to increase as shown in **Fig. 2.8**. In this manner, both upper and lower yield points are identified. Typically the lower yield point associated with the plateau region is defined as the correct one to use in analysis. In Chapter 3, **Fig. 3.7** shows that polycarbonate has a stress strain behavior similar to that of mild steel and again it is appropriate to define yielding at the lower yield point. However, for polycarbonate, if true stress and strain are used no stress decrease occurs and yielding may be considered as the beginning of the plateau region. Such a description agrees with the use of Considere's definition of yielding given in **Fig. 3.5**. It should be noted that many different yield criteria have been used for polymers and no single definitive definition is available which is suitable for all polymers.

For mild steel the tensile stress strain behavior in a constant strain-rate test is often approximated by two straight lines as shown earlier in **Fig. 2.9(b)** and here in **Fig. 11.7**. Here the permanent strain, ε_p, after unloading is indicated and the total strain, ε_{tot}, at some arbitrary point is given by,

$$\varepsilon_{tot} = \varepsilon_e + \varepsilon_p \qquad (11.5)$$

where ε_e is the elastic strain and ε_p is the plastic strain or the amount of strain past the yield point. This type of stress strain diagram forms the basis for classical plasticity theory. The stress strain law for the elastic portion of the diagram is given by $\sigma = E\varepsilon$ and for the yielded portion by $\sigma = \mu\dot{\varepsilon}$. That is, the elastic portion is represented by Hooke's law and the yielded portion is represented by the Newtonian law for viscous liquids. It might appear that the yielded portion is rate dependent but that is not the case as in classical plasticity theory the yield stress is assumed to be rate independent. Newton's law of viscosity simply provides a convenient way to calculate the deformation past the yield point inasmuch as a constant strain rate test is being used.

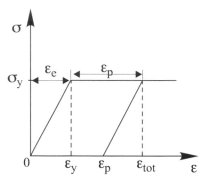

Fig. 11.7. Idealized stress-strain diagram for mild steel.

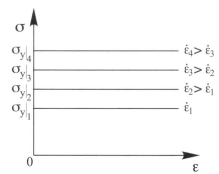

Fig. 11.8. Rate dependent ideal rigid-plastic stress-strain response.

It is well known that many materials have yield points that vary with strain rate. Notably mild steel has a significant variation in the yield stress with strain rate at high temperature, as do other metals. As a result, various rate dependent plasticity theories have been developed for metals and some of these have been extended to polymers. Early approaches used idealized stress-strain response such as that shown in **Fig. 11.8** and **Fig. 11.10**. In **Fig. 11.8** the material is assumed to be rigid but with a rate dependent yield point. Such an assumption can be reasonable in cases where the component for the elastic strain is very small compared to the plastic strain component and for practical purposes can sometimes assumed to be zero. A more realistic ideal elastic-plastic rate dependent material is shown in **Fig. 11.10.**

A mechanical model to represent a rigid-plastic material is shown in **Fig. 11.9(a).** The model is simply a friction element that moves or slides only when the frictional resistance is overcome. Thus the constitutive equation for the friction element is

$$\begin{cases} \sigma_f = \sigma \\ \varepsilon = 0 \end{cases} \quad \text{for } \sigma < \sigma_y$$

$$\sigma_f = \sigma_y \quad \text{for } \sigma \geq \sigma_y$$

(11.6)

Note that when the stress applied to the friction element reaches σ_y, the strain increase is not defined and thus this element must be used in conjunction with other elements to define the strain (rate) changes.

A rigid-elastic element is shown in **Fig. 11.9(b)** that moves linear elastically after friction is overcome. Stress-strain diagrams of the two materials are also shown in **Fig. 11.9**.

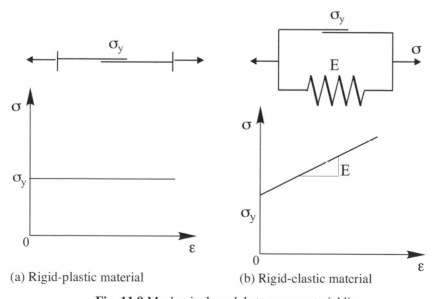

(a) Rigid-plastic material   (b) Rigid-elastic material

Fig. 11.9 Mechanical models to represent yielding.

In studies of viscous fluids (such as paint), Bingham (1922) suggested that some fluids do indeed have a yield point and suggested the model shown in **Fig. 11.10(a)**. In this model, a viscous element and a friction element are in parallel. Upon applying a stress, no movement occurs until the resistance of the friction element is overcome. The stress-strain response is that of a rigid-viscous material and is as shown in **Fig. 11.10(b)**.

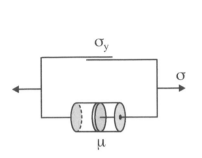
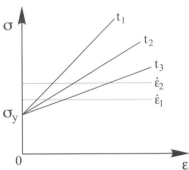

(a) Rigid-viscous model of a fluid with a yield point.
(b) Isochronous stress-strain behavior of a rigid viscous fluid.

Fig. 11.10 Bingham model for a rigid viscous-fluid with a yield point.

To account for rate effects after yielding in solids Ludwik, (1909) and Prandtl, (1928) observed that for some materials the yield stress in uniaxial tension was linearly related to the logarithm of strain rate and suggested use of the equation,

$$\sigma_y = \sigma_0 + \sigma_1 \log\left(\frac{\dot{\varepsilon}_p}{\dot{\varepsilon}_0}\right)$$

(11.7)

where σ_y is the applied tensile stress at yield for the strain rate $\dot{\varepsilon}_p$, σ_1 is a constant and σ_0 is the yield stress for the strain rate $\dot{\varepsilon}_0$. Constant strain (head) rate tests on polycarbonate shown in **Fig. 11.11** reveal the applicability of **Eq. 11.7** to polymers. Also, it should be noted, that similar results were obtained for a modified (or rubber toughened adhesive (Brinson, et al., (1975)). In **Fig. 11.11(a)** the term "initial" applied to the strain rate emphasizes that in reality the strain rate varies slightly in a constant head rate test especially for a viscoelastic polymer at stresses and strains near the yield point.

Equations to represent the rate dependence of polymers have been developed by Bauwens-Crowet, et al. (1969) using the Eyring activation energy method resulting in the following expression,

$$\sigma_y = \frac{E_a}{v} + \left(\frac{kT}{v}\right)\ln\left(\frac{\dot{\varepsilon}}{c}\right) \tag{11.8}$$

where v is an activation volume, c, is a constant and the other parameters are as described previously. A plot of the yield stress as a function of strain rate and temperature for polycarbonate is given in **Fig. 11.11(b)**. See, Miller, (1996) for a brief description of this model, as well as the original reference.

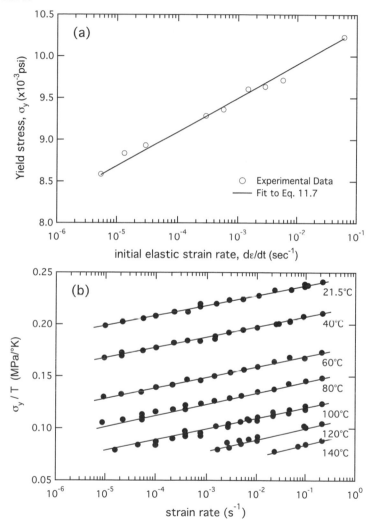

Fig. 11.11 Yield stress strain-rate behavior of (polycarbonate). (a) Room temperature data (Brinson, (1973)). (b) Variation with temperature (data from Bauwens-Crowet, (1969); see also Miller, (1996)).

Rearranging **Eq. 11.8** results in

$$\dot{\varepsilon}_p = \dot{\varepsilon}_0 \exp\left(\frac{\sigma_y - \sigma_0}{\sigma_1}\right) \tag{11.9}$$

where the conversion factor from base 10 logarithms to natural logarithms is contained in $\dot{\varepsilon}_0$.

Use of **Eq. 11.9** to describe rate effects is often called the over-stress model which was developed mostly for metals under high rates of loading and used more in dynamic circumstances of impact, ballistic penetration, wave propagation, etc. (See Christescu, (1967)). Krempl and his associates have used the over-stress technique for various polymers (e.g., Bordonaro and Krempl, (1992)).

Malvern, expressed stress as a function of plastic strain rate as,

$$\sigma = f(\varepsilon) + a \ln(1 + E\dot{\varepsilon}_p) \tag{11.10}$$

which can be expressed as,

$$\dot{\varepsilon}_p = \frac{1}{E}\left\{\exp\left[\frac{\sigma - f(\varepsilon)}{a}\right] - 1\right\} \tag{11.11a}$$

or

$$E\dot{\varepsilon}_p = F\big(\sigma - f(\varepsilon)\big) \tag{11.11b}$$

and since, $\dot{\varepsilon} = \dot{\varepsilon}_e + \dot{\varepsilon}_p$, then,

$$E\dot{\varepsilon} = \dot{\sigma} + F\big(\sigma - f(\varepsilon)\big) \tag{11.12}$$

Malvern also used the more general form,

$$E\dot{\varepsilon} = \dot{\sigma} + g(\sigma, \varepsilon) \tag{11.13}$$

Sokolovsky, suggested the equation,

$$\dot{\varepsilon} = \frac{\dot{\sigma}}{E}, \qquad \sigma < \sigma_y$$

$$\dot{\varepsilon} = \frac{\dot{\sigma}}{E} + g\left(\sigma - \sigma_y\right), \qquad \sigma > \sigma_y \tag{11.14}$$

Perzyna, generalized the overstress concept to obtain,

$$\dot{e}_{ij} = \frac{\dot{s}_{ij}}{2G} \, , \qquad f < 0$$

$$\dot{e}_{ij} = \frac{\dot{s}_{ij}}{2G} + \gamma\phi(f)\frac{\partial f}{\partial \sigma_{ij}} \, , \qquad f > 0$$

(11.15)

where $\gamma \sim \mu$ is a material constant.

The above mathematical models[2] (and later derivatives) define constitutive relationships for the plastic strain regime and they all assume a linear elastic behavior terminated by a yield point that is rate dependent. Hence the yield surface of the material is rate dependent. Since the purpose of these models are to develop methods to calculate deformations which are rate dependent beyond the yield point of a material they are often referred to by the term viscoplasticity. (see Perzyna, (1980), Christescu, (1982)). This practice is analogous to referring to methods to calculate deformation beyond the yield point of an ideal rate independent elastic-plastic material as classical plasticity. However, more general theories of viscoplasticity have been developed in some of which no yield stress is necessary. See Bodner, (1975) and Lubliner, (1990) for examples.

11.3   Delayed or Time Dependent Failure of Polymers

Due to their inherent viscoelastic behavior many polymers exhibit a time dependent failure process either by delayed yielding or rupture under conditions of constant load. Depending upon the type of structure and loading circumstance this may occur under either creep or relaxation conditions. The creep response of Polycarbonate is shown in **Fig. 11.12(a)** and indicates that creep to yield occurred in 5 minutes at a stress level 9,056 psi but took 40 hours for yielding to occur at 7,952 psi. In **Fig. 11.12(b)** a delayed rupture occurred in a $[\pm 45°]_{4s}$ graphite epoxy specimen containing a centrally located circular hole. Here several tensile specimens were ramp loaded in a closed loop hydraulic testing machine and at a certain point the machine was stopped and held in a fixed grip (or relaxation) mode. Depending on the aspect ratio of the hole diameter to specimen width, rupture (complete separation) of two specimens occurred at the times indicated. (The load was removed prior to failure for the specimen with the lowest

---

[2] For a review of early models, see E. Sancaktar, (1987). Also see Christescu, (1967, 1982) for the references cited as well as further discussion. For a more complete description of of plasticity and viscoplasticity see, I.H. Shames and F.A. Cozzarelli, (1992) as well as Bodner, (1975) and Lubliner, (1990).

aspect ratio.) The fact that the specimen failed while the load was decreasing significantly can be attributed to the viscoelastic behavior of the matrix which led to time delayed failure. Examination of the perimeter of the hole revealed small growing cracks in the outer plies in the +45° direction while other cracks on the interior plies in the -45° were growing in the opposite direction. For this reason, far field strains (and hence loads) were decreasing to compensate for the increased local strains in the cracked region in such a way that the overall (global) deformation could remain constant.

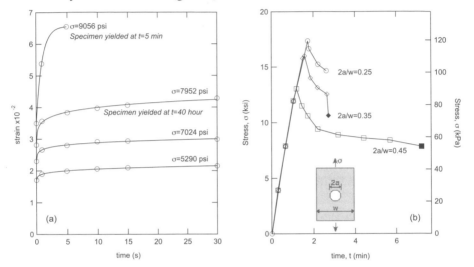

Fig. 11.12 **Delayed Failures in polymer based materials.** (a) Creep of polycarbonate (data from Brinson, (1973)). (b) Deformation control tensile test of a graphite/epoxy specimen; solid points represent rupture (data from Brinson, et al., (1981)).

Creep failures such as those illustrated in **Fig. 11.12** are often called static fatigue and are not uncommon in practical applications such as pressurized piping applications. Kinloch and Young (1983) gives data on the creep rupture of high-density polyethylene pipe (HDPE) and an excellent discussion of the mechanisms associated with static fatigue.

At this juncture it is appropriate to recall the failure envelope given in Chapter 2, **Fig. 2.21** that displays a comparison of failure stresses for both metals and polymers to the three failure theories mentioned therein. The data provided for both polymers and metals were developed without regard to possible rate and/or viscoelastic effects. In **Fig. 11.11** and in **Fig. 11.12** it has been demonstrated that yielding of polycarbonate is both rate and time dependent. The same is true for most ductile polymers and, as a result, the yield (or failure) surface for polymers should be understood to change with rate or time as depicted in **Fig. 11.13**. While not to be considered herein, strain hardening (indicated by κ) is also well known to change the yield surface for metals. However, little information on strain hardening often associated with the Bauschinger effect is available for polymers. (See the excellent article by Drucker, (1962) or Lubliner, (1990) for an introduction to plasticity and the Bauchinger effect in metals.) **Fig. 11.13** shows that the yield surface expands due to strain rate effects, shrinks due to time effects and expand due to strain hardening effects. Clearly yielding in particular and failure in general are very complicated aspects of the behavior of polymers. Obviously viscoelastic processes are involved in the delayed failure behavior of polymers and it would be desirable to have a delayed failure analytical model that combines the prediction of failure with viscoelastic analytical constitutive models such as those discussed in Chapters 5, 6 and 10. The next sections address this issue where the objective is to develop relatively simple closed form equations that would allow the prediction (and prevention) of time dependent failure by design engineers without recourse to extensive numerical procedures.

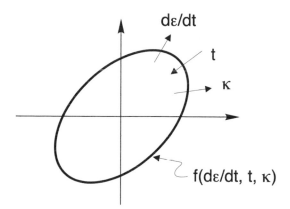

Fig. 11.13   von Mises yield surface displaying the effect of rate, time or strain hardening.

11.3.1 A Mathematical Model for Viscoelastic-Plastic Behavior

In 1963 Nagdi and Murch published a paper entitled "On the Mechanical Behavior of Viscoelastic-Plastic Solids" just before Schapery first published his efforts on a thermodynamic approach to nonlinear viscoelasticity and prior to the development of early viscoplasticity theories. Likely because both nonlinear viscoelasticity and/or viscoplasticity caught the attention of the technical community, Nagdi's work received little attention. However, if the tensile stress-strain diagrams of polycarbonate given in **Fig. 3.7** and **3.8** are combined, the result will be the diagram shown in **Fig. 11.14** where the stress-strain diagram has been extended past the yield point by the dashed lines. The dashed lines could be extended out to a strain of about 60% but have been abbreviated for clarity. Therefore, it is easy to think of polycarbonate as a viscoelastic perfectly plastic material in much the same way that mild steel is often considered to be a perfectly elastic-plastic material. For a tensile stress less than 4,000 psi, rate and time effects are quite small and the material is essentially elastic. For the rates used to produce **Fig. 11.14** rate and time effects are substantial above a tensile stress of 4,000 psi and cannot be neglected (note the creep data given in **Fig. 11.12(a)**). For these reasons polycarbonate can be considered to be nearly elastic below 4,000 psi and viscoelastic between 4,000 psi and 10,000 psi. Above the instability point of approximately 8,500 psi, depending upon the strain rate, the material exhibits a tensile instability. That is, Luder's bands form and the material begins to neck. As a result it is appropriate to think of polycarbonate as a material that has three regions of behavior, i.e., a linear elastic region for low stresses and strains, a viscoelastic region (portions of which may be linear and nonlinear) for intermediate stress levels and a plastic flow region when the stress is high. When a step stress is applied in the viscoelastic region, delayed yielding will occur after a sufficient incubation time. It is interesting to note that the yield strain (defined as peak of the stress-strain curve) increases with strain-rate as does the yield stress. The fact that yield strain increases with strain rate is, in fact, similar to the phenomena of increasing failure stress for increasing rates in polymers (in the rubbery region) and elastomers. (See, Smith, (1965) and Landel and Fedders, (1964)).

The Nagdi-Murch Model

For the purposes here, the behavior of polycarbonate appears to be a good candidate to explore the use of the Nagdi and Murch viscoelastic-plastic theory to determine if it is possible to develop, as suggested in the preceding section, a relatively simple closed form equation that would allow the prediction of time dependent yielding including viscoelastic effects without recourse to extensive numerical procedures.

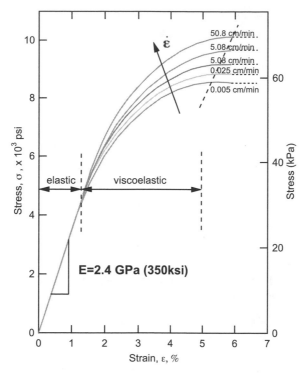

Fig. 11.14 Constant strain-rate behavior of polycarbonate. (Data from Brinson (1973).)

Here it is necessary for a few brief comments about plasticity and visco-plasticity theories without providing details needed to fully appreciate the mathematical procedures involved in their development. First, it is noted that failure (here referred to as yield) theories as discussed in Chapter 2 are mainly used to predict the onset of yielding and might be properly called theories of insipient yielding. Plasticity and viscoplasticity theories are essentially subsequent yielding theories developed for the purpose of determining the growth of yielded regions within load bearing structures. For this reason, plasticity or viscoplasticity constitutive equations are provided for regions of loading, neutral loading and for unloading. These include a flow rule as well as conditions for normality and convexity of the yield surface.

The von Mises theory for yielding given in Chapter 2 can be written in the form,

$$\left(\sigma_1 - \sigma_2\right)^2 + \left(\sigma_2 - \sigma_3\right)^2 + \left(\sigma_3 - \sigma_1\right)^2 - 2\sigma_y = 0 \qquad \textbf{(11.16)}$$

This and all yield (or failure) theories can be written in functional form as,

$$f(\sigma_1, \sigma_2, \sigma_3, \sigma_y) = 0 \qquad (11.17)$$

More general plasticity theories as well as the one of Nagdi and Murch write **Eq. 11.17** in the form,

$$f(\sigma_{ij}, \varepsilon_{ij}^p, \chi_{ij}, \kappa_{ij}) = 0 \qquad (11.18)$$

where σ_{ij} is the stress tensor, ε_{ij}^p is the plastic strain tensor, and κ_{ij} is a tensor representing strain hardening (which will not be used herein). In **Eq. 11.18** a time dependent factor, χ_{ij}, introduced by Nagdi and Murch is included to account for viscoelastic effects.

The Nagdi and Murch theory of viscoelastic-plasticity contains many of the same caveats as in plasticity theory and, in fact, reduces to the two limiting cases of plasticity for non-viscoelastic materials and linear viscoelasticity for non-yielded materials. The only portions needed here are the linear viscoelastic constitutive equations given in Chapter 6 and a generic failure law given by,

$$f(\sigma_{ij}, \varepsilon_{ij}^p, \chi_{ij}) = 0 \qquad (11.19)$$

The important feature is the form of the time dependent term which Nagdi and Murch assumed to be a function of the time dependent strains such that,

$$\chi_{ij} = \chi_{ij}(\varepsilon_{ij}^v - \varepsilon_{ij}^e) \qquad (11.20)$$

where ε_{ij}^v is the viscoelastic strain. Here the elastic strains are subtracted from the viscoelastic strains as in Chapter 10.

The Crochet Model Time Dependent Yielding Model

Later Crochet, (1966) (Nagdi's research assistant) assumed χ_{ij}, to have the specific form,

$$\chi = \left[\left(\varepsilon_{ij}^V - \varepsilon_{ij}^E\right)\left(\varepsilon_{ij}^V - \varepsilon_{ij}^E\right)\right]^{1/2} \qquad (11.21)$$

undoubtedly guided by the von Mises yield criteria written as,

$$f = j_2 = \frac{1}{2} s_{ij} s_{ij} - k^2 = 0 \qquad (11.22)$$

where the yield stress $k = \sigma_y = f(\chi)$. **Eq. 11.22** reduces to **Eq. 11.16** upon expansion. Crochet applied the Nagdi and Murch theory to obtain the solution for a viscoelatic-plastic cylinder under internal pressure in a state of plane strain. He also used the approach to address the solution of a viscoelastic-plastic cylinder under torsion.

Note that χ in the form given in **Eq. 11.21** is a scalar and when expanded for the case of uniaxial tension becomes,

$$\chi = \left[\left(\varepsilon_{11}^V - \varepsilon_{11}^E \right)^2 + \left(\varepsilon_{22}^V - \varepsilon_{22}^E \right)^2 + \left(\varepsilon_{33}^V - \varepsilon_{33}^E \right)^2 \right]^{1/2} \tag{11.23}$$

For uniaxial tension, Crochet assumed that the yield stress was related to the χ function as follows,

$$\sigma_y = A + B\left[\exp(-C\chi) \right] \tag{11.24}$$

where A, B, and C are constants. No rationale for this equation was given except to note that the assumption agrees with the fact that the yield stress is an increasing function of strain rate for many materials including metals and polymers as described by others. A careful examination of **Fig. 11.14** reveals that the yield strain actually increases with strain rate as does the yield stress in polycarbonate and this might lead one to question the assumption associated with **Eq. 11.24**. However, if the time for yielding calculated from the strain at yield in **Fig. 11.14** is divided by the strain rate for yield in **Fig. 11.11(a)**, it is clear that the time for yielding to occur in a constant strain rate test decreases with increasing strain rate. Therefore **Eq. 11.24** is a reasonable assumption based upon experimental evidence. Also, as discussed in the next section in regard to the Zhurkov (1965) theory for time dependent failure some consider the rate at which failure occurs to be an activated process and therefore in this light **Eq. 11.24** is quite reasonable.

Phenomenologically, the behavior of polycarbonate can be represented by a mechanical model containing a friction or "stick-slip" element to represent yielding. Such a model was first introduced by Bingham (See Bingham, (1922) and Reiner, (1971)) to explain the behavior of certain fluids such as paint and later adapted to explain yielding in various materials including polymers with various modifications as shown in **Fig. 11.15**.

The relation between stress and strain for the Modified Bingham model in **Fig. 11.15(b)** has three regions of behavior, linear elastic, linear viscoelastic and plastic flow. The relations between stress and strain for the first two regions in **Fig. 11.16(d)** can be described by,

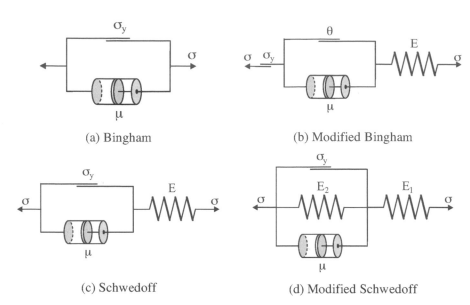

(a) Bingham   (b) Modified Bingham

(c) Schwedoff   (d) Modified Schwedoff

Fig. 11.15 Various mechanical models with a yielding (friction) element. (After Brinson and DasGupta, (1975)).

$$\varepsilon = \frac{\sigma}{E} , \quad \sigma \le \theta \tag{11.25}$$

$$\dot{\varepsilon} = \frac{\dot{\sigma}}{E} + \frac{\sigma - \theta}{\mu} , \quad \theta \le \sigma \le \sigma_y \tag{11.26}$$

where θ is defined as the linear elastic limit stress and σ_y is defined as the yield stress. Above the yield stress a suitable relation for plastic flow must be used. The stress-strain equations corresponding to a constant strain-rate test can be shown to be,

$$\sigma = E\varepsilon , \quad \sigma \le \theta$$

$$\sigma(t) = \theta + \tau ER\left[1 - e^{-(t-t_0)/\tau}\right] , \quad \theta \le \sigma \le \sigma_y \tag{11.27}$$

where the relaxtion time, τ, is given by $\tau = \mu/E$ and t_0 is the time at the elastic limit. The second equation can be written as,

$$\sigma(\varepsilon) = \theta + \tau ER\left[1 - e^{-(\varepsilon - \phi)/R\tau}\right] , \quad \theta \le \sigma \le \sigma_y \tag{11.28}$$

where $\mathbf{R}$ is the strain rate and ϕ is the linear elastic limit strain. The stress-strain response for a constant strain-rate test up to the yield point can be accurately model by **Eq. 11.28** for any particular strain rate as shown by Brinson and DasGupta, (1975). To model all rates with only one equation requires that the relaxation time vary with strain-rate which makes the representation physically inconsistent. That is, if the relaxation time varies with strain rate then the relaxation time is a function of either stress and/or strain and would mean that the material is nonlinear. However, as will be shown this simple model can be used with the viscoelastic-plastic theory of Nagdi, Murch and Crochet to allow the prediction of delayed failures for polycarbonate and other polymers.

The differential equation **Eq. 11.26** for the modified Bingham model of **Fig. 11.15(b)** can be solved for creep to give,

$$\varepsilon(t) = \frac{\sigma_0 - \theta}{\mu} t + \frac{\sigma_0}{E}, \qquad \theta < \sigma < \sigma_y \tag{11.29}$$

This result predicts a linear variation of strain with time and, therefore, does not well represent the case of creep for polycarbonate. Combining models b and d in **Fig. 11.15** would give the desired form of creep response. (See HW problem 11.4.) The modified Bingham model is, however, compatible with the experimental data given in **Fig. 11.14**, in the sense that there appears to be a stress below which creep or relaxation will not occur.

Using the modified Bingham model in the Nagdi-Murch analytical approach, the difference between viscoelastic and elastic strains in a creep test becomes,

$$\left(\varepsilon_{11}^V - \varepsilon_{11}^E\right)^2 = \left(\frac{\sigma_0 - \theta}{\mu} t\right)^2, \qquad \sigma \geq \theta \tag{11.30}$$

and the lateral strains become, upon assuming a constant Poisson's ratio,

$$\left(\varepsilon_{22}^V - \varepsilon_{22}^E\right)^2 = \left(\varepsilon_{33}^V - \varepsilon_{33}^E\right)^2 = v^2 \left(\frac{\sigma_0 - \theta}{\mu} t\right)^2, \qquad \sigma \geq \theta \tag{11.31}$$

The time factor, χ, in Crochet's time dependent yield criteria for uniaxial tension now becomes,

$$\chi = \left(\frac{\sigma_0 - \theta}{\mu} t\right)\left(1 + 2v^2\right)^{1/2}, \qquad \sigma \geq \theta \tag{11.32}$$

and the expression for the time dependent yield behavior of **Eq. 11.24** becomes,

$$\sigma_y(t) = A + B\exp\left[-K(\sigma_0 - \theta)t\right], \quad \sigma \geq \theta \tag{11.33a}$$

where the constant **K** is given by,

$$K = \frac{C(1 + 2v^2)^{1/2}}{\mu} \tag{11.33b}$$

In **Eq. 11.33a** σ_y and σ_0 are the same, as this is the representation for a creep test. Also, the time in **Eq. 11.33a** is the time for yielding to occur. As a result **Eq. 11.33a** is rewritten as,

$$\sigma_y(t) = A + B\exp\left[-K(\sigma_y - \theta)t_f\right], \quad \sigma \geq \theta \tag{11.34}$$

where t_f is the time to yield or time to rupture in the case of a more brittle material. An explicit expression for the time to failure can be obtained by rearranging **Eq. 11.34**,

$$t_f = \frac{1}{K(\sigma_y - \theta)}\ln\frac{B}{\sigma_y - A} \tag{11.35}$$

The yield stress (Luder's band formation) vs. creep to yield time from **Fig. 11.12** is shown in **Fig. 11.16** and compared to **Eq. 11.35**. The constants A, B and C in **Eq. 11.35** were determined from the creep to yield data. Poisson's ratio was assumed to be 0.4 and all other parameters were determined for the modified Bingham model. A similar procedure was used to obtain the creep to yield behavior of a rubber-toughened adhesive (Brinson, et al., (1975)).

The modified Schwedoff model as given in Fig. 11.15 could easily be changed to have a friction element in series with model which would become the yield point and change the friction element in parallel with the spring and damper to the elastic limit stress. (See HW problem 11.4.) Such a model would better represent the creep process but would be more cumbersome to use and it would not be possible to develop a closed form equation for the creep to yield time as given by Eq. 11.35. At the higher stresses polycarbonate is nonlinear and it would be best to use a nonlinear approach rather than mechanical models using friction elements. From Eqs. 10.47 and 11.23 and it is easily shown that the χ parameter including the Schapery parameters is given by (Carter, et al., (1978))

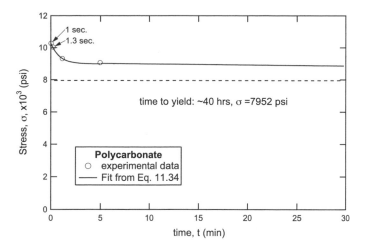

Fig. 11.16   Uniaxial creep to yield of polycarbonate (Data from Brinson, (1973)).

$$\chi = D_1 \frac{g_1 g_2}{a_\sigma^n} \sigma t^n \left(1 + 2v^2\right)^{1/2} \qquad (11.36)$$

which when substituted in **Eq. 11.24** results in the time to failure equation,

$$t_f = \left[\frac{1}{C\beta\sigma_f} \ln \frac{B}{\sigma_f - A}\right]^{1/n} \qquad (11.37)$$

where β is given by,

$$\beta = D_1 \frac{g_1 g_2}{a_\sigma^n} \left(1 + 2v^2\right)^{1/2} \qquad (11.38)$$

Here the symbols σ_f is used rather than σ_y to indicate that the process may be used for creep to rupture as well as creep to yield behavior. **Eq. 11.37** was used to represent time dependent failure data for a chopped fiber composite (SMC 25) and a modified adhesive (Metlbond 1113-2) as given in **Fig. 11.17**.

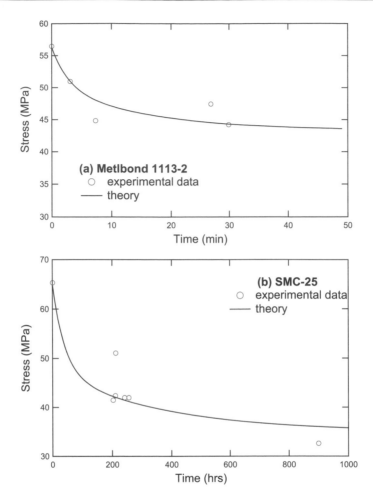

Fig. 11.17 Uniaxial creep to failure of (a) Metlbond 1113-2 and (b) SMC-25 (Data from Cartner, (1978)).

Long Term Delayed Yielding and Three-Dimensional Problems

The forgoing development only presents the framework for including the viscoelastic constitutive equation in developing a method and equations to predict the onset of delayed yielding due to the viscoelastic behavior of a polymer. Further, only one-dimensional examples have been given and only for relatively short times. Naturally, for realistic circumstances, any such approach needs to be modified to predict time dependent yielding in more complicated problems associated with real structures where the time scale may be on the order of years.

An illustration of such a realistic problem was the July 2006 failure of anchor bolts held in place by an epoxy adhesive that led to massive amounts of concrete to fall on motorists in the D Street portal of the Interstate 90 (I-90) connector tunnel in Boston, resulting in one death. The NTSB Highway Safety Board announced in July of 2007 that the cause of the failure was due to creep of the epoxy adhesive. The time from installation until failure was a number of years, perhaps as many as ten. Obtaining ten-year data in advance of a project is not realistic but perhaps well-defined creep to yield or rupture tests could have established a lower bound and given engineers a better understanding of how to make sure that such a failure would not occur.

One way to obtain long-term information is through the use of the time-temperature-superposition principle detailed in Chapter 7. Indeed, J. Lohr, (1965) (the California wine maker) while at the NASA Ames Research Center conducted constant strain rate tests from 0.003 to 300 min^{-1} and from 15° C above the glass transition temperature to 100° C below the glass transition temperature to produce yield stress master curves for poly(methyl methacrylate), polystyrene, polyvinyl chloride, and polyethylene terephthalate. It should not be surprising that time or rate dependent yield (rupture) stress master curves can be developed as yield (rupture) is a single point on a correctly determined isochronous stress-strain curve. Whether linear or nonlinear, the stress is related to the strain through a modulus function at the yield point (rupture) location. As a result, a time dependent master curve for yield, rupture, or other failure parameters should be possible in the same way that a master curve of modulus is possible as demonstrated in Chapter 7 and 10.

To avoid time dependent yielding in circumstances where a two or three-dimensional viscoelastic stress analysis is needed it would be necessary to define the lifetime of the structure. Then tensile creep tests are needed to determine a failure envelope for the defined lifetime similar to those depicted by **Fig. 2.20**. A loading for the structure would be selected such that the stresses determined in a viscoelastic stress analysis of the structure would be inside the envelope.

11.3.2 Analytical Approaches to Creep Rupture

In this section the concern is failure by rupture or separation rather than by yielding as in the previous section. For example, the toughened epoxy tensile specimen shown in **Fig. 11.5** failed by separating into two parts with the fracture surface being perpendicular to the specimen length. That is, of all the many visible crazes, eventually only those on the line of separation reached the critical state. To reiterate, for failures of this type in many materials including polymers, it is thought that the fracture process begins at small microscopic defects or flaws in the material and if an induced stress field around these flaws is sufficient, additional cracks will nucleate near the tip of the flaw. There will be many of these competing micro-cracks and eventually one dominant crack will prevail and a tensile specimen will fail as in **Fig. 11.5**. This has led some to consider the fracture process to be a stochastic event and to develop statistical tools for the evaluation of viscoelastic fracture processes. For a description, see Halpin and Polley, (1967). Indeed, they demonstrated that the breaking stress for an SBR Gum polymer at different temperatures could be shifted to form a breaking strength master curve that could be fit with a modified power law.

Knauss, (1963), suggested that that weak bonding areas in a polymer could also serve as the site for small cracks to nucleate. That is, during the polymerization process not all molecules are able to move freely to reach the optimum position for maximum bond strength. As a result, weakly bonded regions are distributed throughout the bulk polymer and serve as an ideal location for an induced stress field to create a small fissure. In developing a comprehensive molecular based approach to fracture, the Arrhenius rate law given earlier was used as a starting point.

In the following several time dependent failure laws will be considered that can be used by the design engineer to make estimates of the probable time for rupture failure in uniaxial tensile tests. The section will conclude with a brief discussion of how to apply these approaches to more complicated structures. While the following methods have been developed primarily for a creep to rupture phenomenon, they can potentially be used for creep to yield as well or even possibly as a means of determining the demarcation between linear and nonlinear viscoelastic regimes. Some of the examples included are applied in this manner.

Activation Energy Approach to Creep Rupture

The creep behavior of many materials including most metals and thermoplastic polymers is often described as given in **Fig. 11.18** and contains three stages; primary (transient), secondary (steady state) and tertiary portions. Ultimately, with sufficient loading and sufficient time the material will creep until rupture occurs. The time associated with transient and tertiary response is often very small compare to the time associated with secondary creep. As a result, an approximation for the time to creep to rupture, t_r, can be obtained by using the Ahrrenius reaction rate equation. Assuming that the secondary region can be extended to approximate the tertiary portion, the rate of strain can be calculated and equated to the reaction rate,

$$\frac{\varepsilon_r - \varepsilon_0'}{t_r} = \Phi = A e^{-\frac{E_a}{kT}} = \dot{\varepsilon} = \text{constant} \qquad (11.39)$$

where t_r is the time to creep to rupture, ε_r, ε_0' are defined in **Fig. 11.18** and all other parameters are previously defined in **Eq. 11.3**. Rearranging gives,

$$t_r = A' e^{\frac{E_a}{kT}} \qquad (11.40)$$

where **A'** is a new constant.

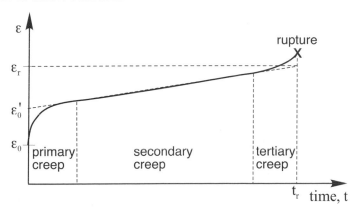

Fig. 11.18 Typical creep curve for metals and thermoplastic polymers.

Taking logarithms of both sides and converting to base 10 logarithms gives,

$$\log_{10} t_r = \log_{10} A' + M \frac{E_a}{kT} \qquad (11.41)$$

where **M = 0.4343** is a factor relating natural logarithms to base 10 logarithms. From **Eq. 11.39** both **A'** and E_a are functions of the (constant) stress level. However, the Larson-Miller Parameter method for determining creep rupture time of a material assumes E_a is a function of stress while $\mathbf{A}'$ is a constant. Based on this assumption and **Eq. 11.41** the Larson-Miller parameter is defined as,

$$LMP = T\left(\log_{10} t_r + C_{lm}\right) = f(\sigma) \qquad (11.42)$$

where $\mathbf{C_{lm}}$ is a new constant. The Sherby-Dorn method to determine the creep rupture time of a material assumes $\mathbf{A}'$ is a function of stress while $\mathbf{E_a}$ is a constant. Again with this assumption and **Eq. 11.41** the Sherby-Dorn parameter is defined as,

$$SDP = \log_{10} t_r - \frac{C_{sd}}{T} = f(\sigma) \qquad (11.43)$$

(For a more complete description of these time dependent failure approaches, see Dowling, (1993)). As an illustration of the use of the utility of Larson-Millard parameter, method data from Dillard (1981) for the uniaxial creep rupture of a [90/60/-60/90]$_{2S}$ graphite/epoxy composite at various temperatures is given in **Fig. 11.19**. The rupture stress is plotted against the LMP as shown and reduces failure stresses at three different temperatures to a single linear master curve. Such an approach is very convenient for a designer as it provides an upper bound for the failure stress as a function of both temperature and time that can be used for the engineering design of structures. If all stresses in a structure are kept below the data scatter, then failure is not likely to occur for any temperature and the line can be extrapolated so a design lifetime can be estimated.

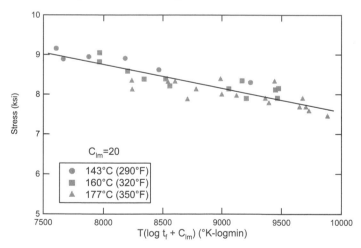

Fig. 11.19 Application of Larson-Miller parameter method to creep of a [90/60/-60/90]$_{2S}$ graphite/epoxy laminate. (Data from Dillard, (1981)). Line is a fit of the data using the Larson Miller equation.

The Larson-Miller and other similar methods have been widely used for metals but here it is important to note that difficulties arise for fiber reinforced composite laminates because the constants are only valid for one configuration of the plies and a more general approach is needed. Dillard (1981) developed an incremental viscoelastic time dependent lamination theory approach that included the Tsai-Hill failure law modified to account for delayed failures using the Zhurkov time dependent failure model that will be discussed in the next section. The advantage of the Dillard approach is that information on the viscoelastic behavior as well as the delayed failure behavior of 0°, 10° and 90° plies can be used to predict the behavior of general laminate configurations.

The Zhurkov Method

Another variant of the activation energy approach (**Eq. 11.40**) is the Zhurkov method, sometimes referred to as the kinetic rate theory, which is based upon tests on more than 50 different materials including both metals and polymers (see Zhurkov, (1965)) and results in an equation for the time to creep to rupture given by,

$$t_r = t_0 e^{\frac{E_a - \gamma\sigma}{kT}} \qquad (11.44)$$

where t_r is the time to creep to rupture in a uniaxial tensile test, t_0, and γ, are constants, E_a is a constant activation energy, σ, is the applied true stress, k, is Boltzman's constant and T is the absolute temperature. The parameter, t_0, is described by Zhurkov as the period of natural oscillation of the atomic structure and is said to be constant for all materials. In the original form, the activation energy was defined by a parameter, u_0, as an energy barrier that must be overcome to rupture the atomic structure and is therefore similar to the activation energy, E_a, in the Ahrrenius equation. Therefore, in **Eq. 11.44** the standard activation energy symbol has been used. This modified activation energy approach is quite similar to **Eq. 11.40**.

If the activation energy, E_a, is equal to the quantity, $\gamma\sigma$, the creep rupture time is independent of temperature and implies the existence of a common "pole" as shown in **Fig. 11.20**. That is,

$$\lim_{\gamma\sigma \to E_a} t_r = \lim_{\gamma\sigma \to E_a} t_0 \exp(\frac{E_a - \gamma\sigma}{kT}) = t_0 \qquad (11.45)$$

and data for different temperatures will intersect at the common pole as shown in **Fig. 11.20**. All the materials tested by Zhurkov displayed this behavior.

Various modifications of Zhurkov's equation have been suggested and a discussion of these can be found in Griffith, (1980). The Zurkov equation can be written in the form,

$$\ln t_r = A + \frac{B}{T} \qquad (11.46)$$

where B is a function of stress and $A = \ln t_0$.

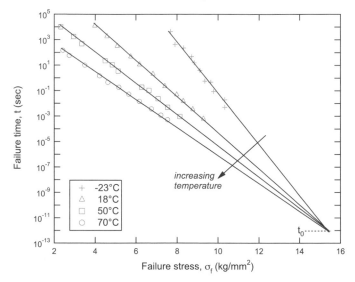

Fig. 11.20 Temperature dependence of the Zhurkov equation for creep rupture for PMMA. (Data from Zurkov, (1965))

Cumulative Creep Damage of Polymers

For multi-step creep loading or even single-step creep loading in inhomogenous materials, it is necessary to have a cumulative creep damage rule similar to Minor's rule for cumulative fatigue damage. Such a rule is essential for both creep to yield and creep to rupture theories and can be used in conjunction with a given failure theory. Note that for materials with several constituents such as a laminated polymer composite material it is possible for a single lamina to see a change in stress level even when the laminate is under a constant uniaxial creep loading. Often cumulative damage under multi-level creep loads is represented by an approach known as Robinson's life fraction rule shown in **Table 11.1** where t_i is the time at the creep stress level σ_i and t_{fi} is the time to creep rupture for the stress σ_i. Therefore, if the rule holds it is possible to predict the time to creep to failure under an arbitrary number of step stress inputs providing the creep rupture times are known for various single step input. While this theory suggest that damage is proportioned equally to the creep time for each stress level, actual data indicates that the right hand side of Robinson's equation varies between 0.3 and 2.0 for various particular polymers. These variations are not surprising, as Minor's rule is notorious for it's inability to properly predict the failure of a material with multiple stress steps under fatigue loading. For this reason a number of modifications to the Robinson rule have been suggested and are given in **Table 11.1**.

Table 11.1 Various Cumulative Damage Rules for Polymers

1.	Robinson's life fraction rule	$\sum_i \dfrac{t_i}{t_{fi}} = 1$
2.	Lieberman's creep strain-fraction rule	$\sum_i \dfrac{\varepsilon_i(t)}{\varepsilon_{fi}(t)} = 1$
3.	Oding and Burdusky proposed a rate of void production rule proportional to the secondary creep rate, $\dot{\varepsilon}$, and the rate of void accumulation, m.	$\sum_i \left(\dfrac{t_i}{t_{fi}}\right)^m = 1$
4.	Johnson proposed a rate of void production rule similar to 3 except that the rate of void accumulation was also related to primary creep, μ.	$\sum_i \left(\dfrac{t_i}{t_{fi}}\right)^{m+\mu} = 1$
5.	Freeman and Voorhees proposed a combination of equations of 1 and 2.	$\sum_i \left(\dfrac{t_i}{t_{fi}} + \dfrac{\varepsilon_i}{\varepsilon_{fi}}\right)^{1/2} = 1$
6.	Abo El Ata and Finnie proposed a combination of equations 1 and 2.	$K\sum_i \left(\dfrac{t_i}{t_{fi}}\right) + (1-K)\sum_i \left(\dfrac{\varepsilon_i}{\varepsilon_{fi}}\right) = 1$
7.	Kargin and Slonimsky proposed an integral approach for varying stresses and temperatures	$\displaystyle\int_0^\tau \dfrac{dt}{t_f\left[\sigma(t), T(t)\right]} = 1$

For a more complete discussion of the various cumulative creep damage rules as well as references to those given in **Table 11.1** see Zhang, et al., (1986) and Dillard, (1981).

Zhang conducted extensive creep and creep to failure tests of polycarbonate and polysulfone for both single creep loads and multiple step-up or step-down creep loads as illustrated in **Fig. 11.21** and developed an equation of the form,

$$\sum_{i=1}^{q} K_i \left[\frac{t_i'}{t_{fi}} \right]^n = 1 \tag{11.47}$$

where **n** is a constant, K_i is a constant for each creep time interval, t_i' is the time interval for stress σ_i and t_{fi} is the time for failure at a single step stress of σ_i. For a two step loading **Eq. 11.47** becomes,

$$K_1 \left[\frac{t_1'}{t_{f1}} \right]^n + K_2 \left[\frac{t_2'}{t_{f2}} \right]^n = 1 \tag{11.48}$$

In experiments for both polycarbonate and polysulfone Luder's bands formed. In polycarbonate the Luder's band was a precursor to yielding but in polysulfone the Luder's bands were a precursor to rupture that occurred almost simultaneous with formation.

Creep to yield times (Luder's band formation) for a single creep load are shown in **Fig. 11.22** for polycarbonate and vary linearly with log time. Creep to yield times for a single step-down or step-up loading for polycarbonate are shown in **Fig. 11.23** and **11.24** in non-dimensional form. The dashed line is Robinson's life fraction rule for a single step-down or step-up loading. Solid lines represents polycarbonate yield data for step-down or step-up loading fitted with **Eq. 11.48** where K_1, K_2 and **n** are given in the each figure. Each data point in **Fig. 11.22** - **Fig. 11.24** represents five independent tests.

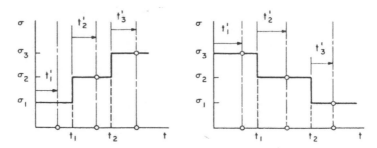

Fig. 11.21 Typical step-up and step-down creep tests. Zhang, et al. (1986). Reprinted with kind permission of Springer Science and Business Media.

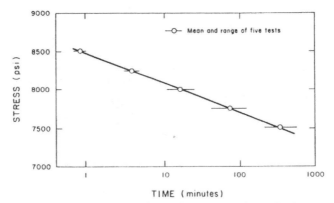

TIME (minutes)

Fig. 11.22 Creep to failure (rupture) for Polycarbonate for a single creep load. Zhang, et al. (1986). Reprinted with kind permission from Springer Science and Business Media.

Clearly Robinson's life fraction rule is invalid for polycarbonate and the same was true for polysulfone though not shown here. This is not surprising as the effects of memory are not included, illustrating once again that using analytical methods developed for metals are not usually viable for polymer based materials. Unlike Robinson's life fraction rule, Eq. 11.48 is nonlinear and fits the data for polycarbonate well though the equation is still empirical. The data for two different loading histories given in Fig. 11.24 demonstrates that the parameters in Eq. 11.48 vary with time and must be determined for each load history. Perhaps the Nagdi-Murch-Crochet approach could be used for a multi-step load though that awaits further study.

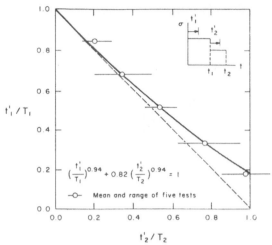

Fig. 11.23 Step down loading creep to failure (yield) for polycarbonate. Zhang, et al. (1986). Reprinted with kind permission from Springer Science and Business Media.

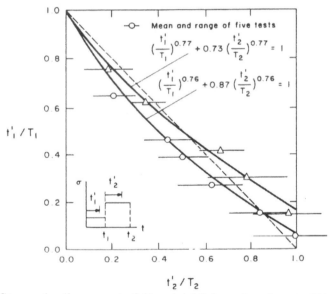

Fig. 11.24 Step-up loading creep to failure (yield) for polycarbonate. Triangles and circles are for different loading histories. Zhang, et al. (1986). Reprinted with kind permission from Springer Science and Business Media.

Reiner-Weissenberg Criteria for Failure

As discussed previously, failure is most often treated as a separate issue from the determination of modulus properties of materials. In fact, most failure laws are derived empirically from observations related to a catastrophic event such as yielding or rupture. As a result, a great deal of testing and data analysis is necessary to establish an appropriate law. On the other hand, modulus or constitutive laws are derived by more rational means of relating deformations to the forces that produce them. For this reason, often much less testing is necessary to define a constitutive law for a material especially if deformations do not depart from the linear elastic or reversible deformation range of a material.

Failure, however defined, should be a part of a complete constitutive description of a material as discussed in the previous sections. In other, words, the key to dealing effectively with the failure of time dependent or viscoelastic polymers lies in treating failure properties as a termination of a nonlinear viscoelastic process. Perhaps, for this reason, a number of investigators have suggested that modulus and strength laws should be related to each other for polymers (eg., Landel, (1964)).

The concept of distortional energy as a measure for the critical magnitude of the stress state a material can endure at a point in an elastic material (von Mises criterion) cannot be carried over directly to viscoelastic materials because viscoelastic deformation involves dissipative mechanisms. Thus, at any point in time the energy balance must be written as,

Total deformation energy = Stored (Free) energy + Dissipated Energy

Reiner and Weissenberg, (1939) suggested that the energy storage capacity of a material is responsible for the transition from viscoelastic response to yielding in ductile materials or to rupture (fracture) in brittle materials. They assume that a threshold value of the distortional free (or recoverable) energy, called the resilience of the material, is the quantity that governs failure. If the Reiner-Weissenberg (R-W) approach is applied to a material with zero dissipation (elastic material) it becomes identical to the von Mises failure law. When applied to a viscoelastic material, however, the free energy under constant load changes with time and the variation must be known. If the mechanisms through which total deformation energy is transformed into dissipated energy are activated such that no free energy can accumulate, there is practically no limit to the amount of deformation energy that can be applied without failure occurring. That is, forces up to a certain magnitude can be applied for any length of time without leading to rupture. If the material cannot accommodate this energy redistribution fast enough then the material will store energy until the critical value needed for failure is achieved. At that time failure will occur. The failure process is therefore delayed (or time dependent) which limits the life of a given structure to a finite value or defines the time required for failure to occur after initial loading. The instant of yielding or rupture is thus clearly dependent upon the final outcome of the connection between deviatoric free energy and deviatoric dissipated energy. Thus, the effect of the strain history on delayed yielding or rupture follows from this model in a natural way.

The advantage of this approach is that the onset of failure is defined by a single parameter, the distortional free energy, while that the former method of Nagdi and Murch and Crochet required the determination of three new parameters in addition to those needed to describe constitutive behavior.

The following description is a brief review of the Reiner-Weissenberg criterion that follows that given by Hiel, (1984) and Bruller, (1978, 1981). (See Hiel for additional Brueller references to his extensive investigations and application to polymeric materials.)

Free Energy Accumulation in a Three-Parameter Model Under Creep Loading: In order to apply the R-W criterion the stored (free) and dissipative energy must be calculated and, as an example, these will be determined for a three-parameter model with the notation given in **Fig. 5.1**. Recall from Chapter 5 that work or total energy is,

$$W = \int_0^{\sigma_0} \sigma \, d\varepsilon \qquad (11.49)$$

which for an elastic material reduces to

$$W = \frac{1}{2}\sigma^2 D \qquad (11.50)$$

where **D** is the compliance (inverse of modulus, **E**). Thus, the energy stored in the elastic spring under creep loading is,

$$w_0 = \frac{1}{2}\sigma_0^2 D_0 \qquad (11.51)$$

and the energy stored in the Kelvin element spring can be shown to be,

$$w_1 = \frac{1}{2}\sigma_0^2 D_1 (1 - e^{-t/\tau_1})^2 \qquad (11.52)$$

The total stored energy in the springs under creep is therefore,

$$W_{springs} = W_{stored} = \frac{1}{2}\sigma_0^2 D_0 + \frac{1}{2}\sigma_0^2 D_1(1-e^{-t/\tau_1})^2 \qquad (11.53)$$

The dissipated energy in a 3-parameter solid is due to the damper, μ_1, in the Kelvin element and can be calculated from,

$$W_{damper} = W_\mu = \int_0^t \sigma_\mu \frac{d\varepsilon_\mu}{dt} dt \qquad (11.54)$$

where

$$\sigma_\mu = \mu \frac{d\varepsilon_\mu}{dt}, \quad \varepsilon_\mu = \varepsilon_K = \sigma_0 D_1(1-e^{-t/\tau_1}) , \quad \frac{d\varepsilon_\mu}{dt} = \sigma_0 \frac{D_1}{\tau}e^{-t/\tau_1} \qquad (11.55)$$

where the subscript μ represents the damper and K represents the Kelvin element. Therefore, the dissipated energy is,

$$W_{damper} = W_\mu = \frac{1}{2}\sigma_0^2 D_1(1-e^{-2t/\tau_1}) \qquad (11.56)$$

The total energy in the three-parameter model under creep loading is found by adding **Eqs. 11.53** and **11.56** to obtain,

$$W_{total} = W_{springs} + W_{damper} = \sigma_0^2\left[\frac{D_0}{2} + D_1(1-\exp(-t/\tau_1))\right] \qquad (11.57)$$

The extension of **Eqs. 11.53**, **11.56** and **11.57** to an N-element Kelvin unit with a free spring is,

$$W_{stored} = W_{springs} = \sigma_0^2\left[\frac{D_0}{2} + \sum_{i=1}^{N}\frac{D_i}{2}(1-\exp(-t/\tau_i))^2\right] \qquad (11.58)$$

$$W_{dissipated} = W_{dampers} = \sigma_0^2\left[\sum_{i=1}^{N}\frac{D_i}{2}(1-\exp(-2t/\tau_i))\right] \qquad (11.59)$$

$$W_{total} = \sigma_0^2\left[\frac{D_0}{2} + \sum_{i=1}^{N}D_i(1-\exp(-t/\tau_i))\right] \qquad (11.60)$$

Taking the limits for $t \rightarrow \infty$ gives,

$$\left[W_{springs}\right]_{\lim t \rightarrow \infty} = \sigma_0^2\left[\frac{D_0}{2} + \sum_{i=1}^{N}\frac{D_i}{2}\right] \qquad (11.61)$$

$$\left[W_{dampers}\right]_{\lim t \to \infty} = \sigma_0^2 \sum_{i=1}^{N}\left[\frac{D_i}{2}\right] \tag{11.62}$$

$$\left[W_{total}\right]_{\lim t \to \infty} = \sigma_0^2 \left[\frac{D_0}{2} + \sum_{i=1}^{N} D_i\right] \tag{11.63}$$

Thus, half the work done by the external forces on the Kelvin elements goes to increase the free energy while the other half is dissipated.

If a torsion test is used, **Eqs. (11.58 – 11.60)** give the stored (free), dissipated and total deviatoric energy. However, if a uniaxial tensile test is used, **Eqs. (11.58 – 11.60)** give the total (shear and bulk) strain energy, each of which has a stored and dissipated energy component. However, assuming Poisson's ratio to be a constant, it can be shown that the deviatoric stored energy comprises 93% of the total stored energy in a uniaxial tension test. See HW 11.9.) The stored energy due to volume change is relatively small and as only unidirectional data will be considered here it is assumed that the deviatoric stored energy and total stored energy are the same.

The master curve for a $[90°]_{8s}$ graphite/epoxy composite in uniaxial tension using TTSP is shown in **Fig. 11.25**. The following six-term Prony series representation of the data is also shown in **Fig. 11.25** and as may be observed the agreement between the two is excellent.

$$D(t) = 0.71 + 0.02(1 - e^{-t/0.01}) + 0.02(1 - e^{-t/1}) + 0.04(1 - e^{-t/10})$$
$$+0.04(1 - e^{-t/100}) + 0.18(1 - e^{-t/1000}) + 0.25(1 - e^{-t/10000}) \times 10^{-10} \text{psi}^{-1} \tag{11.64}$$

Using expression **11.64** in **Eqs. 11.58 – 11.60**, the total energy, the stored (free) energy and the dissipated energy for creep was determined with the results shown in **Fig. 11.26** normalized with respect to the initial total energy. The total energy and the stored energy are the same initially and therefore, if sufficient creep stress were imposed, failure would occur upon loading. If a lower creep stress were imposed, dissipation would prevent failure until a critical stored energy is reached.

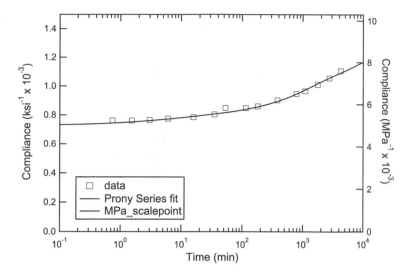

Fig. 11.25 Comparison between a TTSP master curve and a six term Prony series for a [90°] graphite/epoxy composite at 160° C (320° F) and a stress of 35.7 MPa (5.18 ksi). Data from Hiel, (1984).

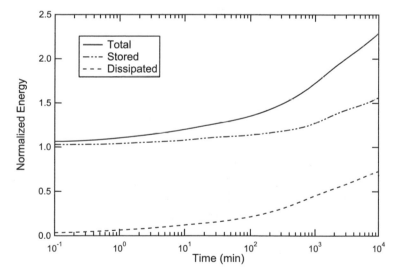

Fig. 11.26 Total stored and dissipated energy as calculated from **Eqs. 11.58-11.60** normalized with respect to the initial total energy (t = 0 min) using the Prony series representation of the master curve in **Fig. 11.25** Due to the log scale starting at t > 0, the normalized total energy is slightly larger than one at the left end of the plot.

Creep to rupture data for a $[90°]_{8S}$ graphite/epoxy composite in uniaxial tension is shown in **Fig. 11.27** (from Griffith, (1980)). The R-W theory suggests that the critical value of stored or free energy for failure at different creep stresses is a constant. Thus, the data for a single or a set of stress-time failure point(s) can be used with **Eq. 11.58** via an appropriate numerical fitting method to determine this material parameter, w_{crit}. Using this w_{crit}, the variation of rupture time with failure stress can be calculated. This was done for the material in **Fig. 11.27** in order to verify the theory and the result is shown superimposed upon the experimental data.

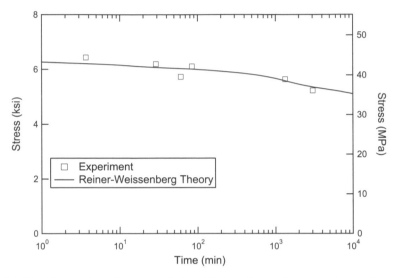

Fig. 11.27   Comparison between Reiner-Weissenberg theory and experimentally determined creep rupture times.

Power Law Approximation for Free Energy: The stored free energy expression in the form of **Eq. 11.58** requires the determination of **2N+1** material constants for D_1 and τ_i. (Thirteen material parameters were needed in the six term series used above.) This number can be reduced using the power law,

$$D(t) = D_0 + D_1 t^n \qquad (11.65)$$

with **Eq 11.49** to obtain the equivalent of **Eqs. 11.59** and **11.60** as,

$$\frac{W_{dissipated}}{\sigma_0^2} = \left[\sum_{i=1}^{N} \frac{D_i}{2} \left(1 - \exp(-2t/\tau_i)\right) \right] = \frac{1}{2} D_1 (2t)^n \qquad (11.66)$$

$$\frac{W_{total}}{\sigma_0^2} - \frac{D_0}{2} = \sum_{i=1}^{N} D_i \left(1 - \exp(-t/\tau_i)\right) = D_1 t^n \tag{11.67}$$

Subtracting **11.66** from **11.67** gives the free energy of the springs,

$$\frac{W_{stored}}{\sigma_0^2} = \left\{ \frac{D_0}{2} + D_1 \left[t^n - \frac{1}{2}(2t)^n \right] \right\} \tag{11.68}$$

Due to the decreased number of constants and the simple form of the power law, **Eq. 11.68** can be used to obtain a closed-form analytic expression for the stress σ_r required to rupture a polymeric material in uniaxial tension at the time t_r,

$$\sigma_r = \sigma_{failure} = \sigma_0 = \sqrt{\frac{W_{crit}}{\frac{D_0}{2} + D_1 \left[t^n - \frac{1}{2}(2t)^n \right]}} \tag{11.69}$$

where W_{crit} is the critical free energy for rupture. Conversely, the delay time for rupture t_r for a creep stress of σ_r can be determined by,

$$t_r = \left\{ \frac{1}{\beta \sigma^2} \left(W_{crit} - \frac{D_0}{2} \sigma^2 \right) \right\}^{\frac{1}{n}} \tag{11.70}$$

where

$$\beta = D\left(1 - 2^{n-1}\right) \tag{11.71}$$

This approach requires less parameters the Prony series method just discussed and also requires less parameters than the Nagdi-Murch-Crochet method described in a previous section. However it is dependent upon the power law providing a good representation of the master curve.

If the polymer is nonlinear viscoelastic and if the Schapery parameters g_0, g_1, g_2 and a_s are known, **Eqs. 11.69** and **11.70** would become (see homework problem 11.11),

$$\sigma_r^2 = \sigma_{failure}^2 = \sigma_0^2 = \frac{W_{crit}}{\left\{ g_0 \frac{D_0}{2} + g_1 g_2 D_1 \left[\left(\frac{t_r}{a_\sigma} \right)^n - \frac{1}{2} \left(\frac{2t_r}{a_\sigma} \right)^n \right] \right\}} \tag{11.72}$$

$$t_r = \left\{ \frac{1}{\beta\sigma^2} \left(W_{crit} - \frac{g_0 D_0}{2} \sigma^2 \right) \right\}^{\frac{1}{n}}$$ (11.73)

where

$$\beta = \frac{g_1 g_2}{a_\sigma^n} D_1 \left(1 - 2^{n-1} \right)$$ (11.74)

An attempt was made to use the power law expression defined by **Eq. 11.68** with coefficients appropriate for the master curve of **Fig. 11.25**. However, with this approach it was not possible to find a constant stored energy that would represent the creep rupture data of **Fig. 11.27**. A likely reason can be visualized by examination of **Fig. 11.28** where the six term Prony series given previously in **Fig. 11.25** is extended an additional two decades. It is to be noted that the series representation of the master curve reaches a plateau that is typical of all polymers. The power law can be made to fit a portion of the master curve but will not reach a plateau and as is shown schematically will continue to rise without bound. As a result, it would be surprising if the time-to-failure given by **Eq. 11.68** using the power law representation could adequately represent creep to rupture data over the glassy, transition and rubbery regions of a polymer. On the other hand, obviously from **Fig. 11.28** the Prony series can adequately represent all the regions of behavior of a polymer.

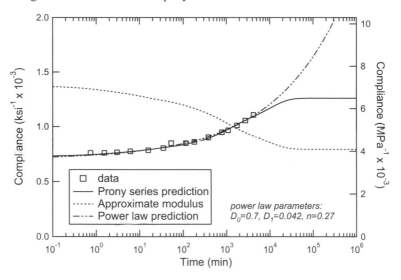

Fig. 11.28 Six term Prony series representation of the compliance master curve of **Fig. 11.25** extended to 10^6 minutes.

Shown in **Fig. 11.29** is the Reiner-Weissenberg predictions of creep rupture given in **Fig. 11.27** extended an additional two decades together with creep rupture data. Here it is seen that the failure stress prediction will also reach a plateau for a given constant stored energy at failure. Again, it is quite reasonable for the failure stress to become a constant in the rubbery plateau region of a polymer. The inverse of the Prony series representation of the compliance is given in **Fig. 11.25** to approximate schematically the change in relaxation modulus with time. It is interesting to note that variation in the measured and predicted failure stress with time has essentially the same shape as the variation of relaxation modulus with time. In other words, this seems to verify that a change in modulus does indeed represent a change in strength as is suggested by recent viscoplasticity or damage evolution theories.

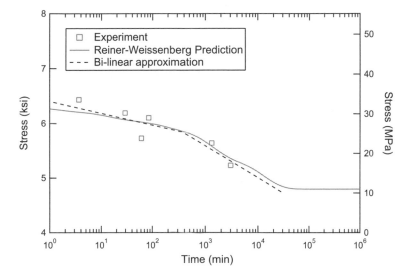

Fig. 11.29 R-W predictions of **Fig. 11.27** extended to 10^6 minutes using **Eq. 11.58** and Prony series compliance from **Fig. 11.28**. Also shown an empirical bi-linear approximation of creep rupture data.

Included in **Fig. 11.29** is an empirical bi-linear approximation to the creep rupture data. An examination of the R-W theory gives credence to the notion of "knee" in the creep to rupture life of a polymer as has been reported by others (see Kinloch and Young, (1983), p. 213). It is reasonable to assume that the so-called knee is just a manifestation of the change of material properties when moving from the glassy region to the transition region of the polymer. In fact, perhaps another "knee" would be found at longer times for a lower creep stress at the change of properties taking place in going from the transition to rubbery plateau region of a polymer.

Assuming a creep test was conducted at room temperature on the graphite/epoxy discussed here, the time to failure would be exceedingly long. The only reason that creep ruptures could be found within four decades of time was because the tests were conducted at 160° C (380° F) and a high level of stress. In this instance caution is suggested for the preceding interpretations as no data for the master curves shown in **Fig. 11.25** or the strength curve shown in **Fig. 11.29** was obtained for times longer than 4 decades and, therefore, the existence of a rubbery plateau was not verified for this composite.

The Riener-Weissenberg energy based time dependent failure criteria has been used by Brueller (1978) to identify the craze limit for PMMA defined as the first appearance of visible crazes which he argues is the beginning of the failure process. Together with Schmidt (1979) he has also used the R-W method to define the linear viscoelastic limit in PMMA. That such a limit exists and is identifiable is shown in **Fig. 10.2**. Arenz (1999) using an elegant torsion test attempted to use the R-W method to identify the linear viscoelastic limit in poly(vinyl acetate) without success and raises concerns about the applicability of the R-W technique for such use. Since the linear viscoelastic limit is very difficult to determine and may, in fact, be a function of measurement sensitivity, using the R-W approach or any other technique for this purpose may be quite subjective.

A very thorough investigation of the R-W method by Guedes (2004) compares this approach with several other time dependent failure theories in predicting the failure of two forms of nylon 66 and polycarbonate. These are a stress work theory, a maximum strain theory and a modified version of the R-W method he proposes. In the latter, instead of assuming there to be a single value of the free energy that causes failure, he assumes that the amount of free energy required for failure varies with stress intensity. Generally his approach appears to fit experimental data better than the original R-W technique but in one case the original R-W method gives the better fit.

The difficulty in using the R-W technique or any other method to predict time dependent failure is subject to the variability of experimental data. It is well known that large variability in creep rupture data is the norm. One obvious reason for the variability is that the slope of a stress vs. failure time plot of data is quite small and, for this reason, a small change in failure stress can translate into a large change in time. It is for this reason that log-log or semi-log plots are used such as the Larson-Miller parameter method.

Again, as for the section on time dependent yielding it is necessary to consider how to obtain data that can be useful in preventing failure over the lifetime of a structure that may be intended to last for 20 to 100 years. The only rational means for this is some type of accelerated testing such as that offered by the time-temperature or time-stress-superposition procedure. Also, it is reiterated that failure is a stochastic event and of necessity a reliable statistical analysis should be performed.

Finally, just as for yielding, it is necessary to incorporate the various creep rupture approaches presented here into a viscoelastic stress analysis of realistic structures. In the case of the R-W technique the stored distortional free energy would need to be calculated and compared to that in simple torsion. Such procedures can be used to design structures so that the probability of failure is very low and within prescribed safety standards.

11.4   Review Questions

11.1.   Why is the process of switching important in describing the deformation mechanisms in polymers?

11.2.   What is the mechanism of reptation and why is it needed for describing deformation mechanisms in polymers.

11.3.   Describe the deformation mechanisms associated with shear bands.

11.4.   Describe the method by which crazes form.

11.5.   What is a Luder's band and what stress is it's formation associated with?

11.6.   What is the purpose of the mathematical models rate dependent yield behavior.

11.7.   What is a condition of normality? Convexity?

11.5   Problems

11.1   Using the inter-atomic force model given by **Eq. 11.1** and shown in **Fig. 2.22** can be used to estimate the tensile strength of a material to be of the same order of magnitude as the elastic modulus.

11.2 Construct isochronous stress-strain curves for polycarbonate from the creep data for t = 1 min. and t = 5 min. in **Fig. 11.12**.

11.3 Develop the differential equation for modified Bingham model given by **Eq. 11.29**.

11.4 Develop the differential equation for the following version of the modified Schwedoff model.

11.5 The Crochet modification to the Nagdi and Murch χ_{ij} function is.

$$\chi = \left[\left(\varepsilon_{kl}^{V} - \varepsilon_{kl}^{E} \right) \left(\varepsilon_{kl}^{V} - \varepsilon_{kl}^{E} \right) \right]^{1/2}$$

11.6 Verify **Eq. 11.35** for the creep to yield time assuming use of the modified Bingham model and the Crochet equation $Y(t) = A + B \exp(-C\chi)$.

11.7 Verify that the χ function of **Eq. 11.36** for creep of a nonlinear viscoelastic material in simple tension that can be represented by the Schapery model using a power law for creep. (Assume a constant Poisson's ratio.)

11.8 Develop an expression for the yield stress as a function of time using the Crochet equation and the expression for creep for the model shown in problem 11.4.

11.9 Show that the deviatoric stored energy comprises 93% of the total stored energy in a uniaxial tension test.

11.10 Discuss the relative merits of the Larson Miller parameter method and the Zurkov method compared to the Reiner-Weissenberg criteria.

11.11 Verify **Eqs. 11.72** and **11.73.**

11.12 Perform a literature search for other time dependent failure models (both yield and rupture) and compare their relative merits compared to those given in this chapter.

Appendix A - Step and Singularity Functions

Unit (Heavyside) Step Function

The unit or Heavyside step function is defined as,

$$H(t) = \begin{Bmatrix} 1, & t \geq 0 \\ 0, & t \leq 0 \end{Bmatrix}$$

and can be represented graphically as,

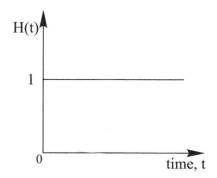

The unit step function is used to indicate a discontinuous change in another function at a particular point in time. If for example the function **F(t)** is given as,

$$F(t) = f(t)H(t - t_1)$$

It states that the function **f(t)** is zero before $t = t_1$ and is only defined for **t >** t_1 as illustrated graphically by,

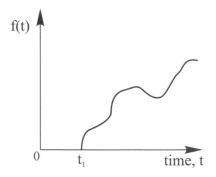

Differentiation of the unit step function yields the singularity, or Dirac delta, function which is defined as,

$$\delta(t) = \begin{cases} 0, & t \neq 0 \\ \infty, & t = 0 \end{cases}$$

and is represented graphically as,

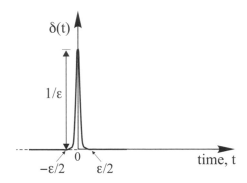

The function $\delta(t)$ becomes infinite in the limit as $\varepsilon \to 0$ and is defined such that the integral of the function yields unity

$$\int_{-\infty}^{+\infty} \delta(t)dt = 1$$

The singularity function results when the step function is differentiated, i.e.,

$$\frac{d[H(t)]}{dt} = \delta(t) = \begin{cases} 0, & t \neq 0 \\ \infty, & t = 0 \end{cases}$$

and

$$\frac{d\left[H(t-t_1)\right]}{dt} = \delta(t-t_1) = \begin{cases} 0, & t_1 \neq 0 \\ \infty, & t_1 = 0 \end{cases}$$

In the latter case,

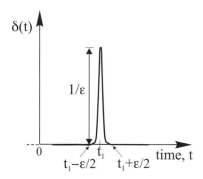

The singularity function has unique properties. For example, when a function such as **f(t)** is multiplied by a singularity function, $\delta(\mathbf{t})$, and then integrated, the result is the function evaluated at the location of the singularity function. That is,

$$\int_0^t f(t)\delta(t-t_1)dt = \int_0^t f(t)\frac{d\left[H(t-t_1)\right]}{dt}dt = f(t_1)$$

This is known as the sifting property of the Dirac delta function.

Appendix B – Transforms

Laplace Transforms

The Laplace transform is a linear operator which is defined as,

$$\mathcal{L}\{f(t)\} = \bar{f}(s) = \int_0^t f(t)e^{-st}dt$$

The above integral is easily evaluated for many simple functions which can be illustrated by finding the Laplace transform of an exponential as follows,

$$\mathcal{L}\{e^{at}\} = \int_0^t e^{at}e^{-st}dt = \int_0^t e^{-(s-a)t}dt = \frac{1}{s-a}$$

The Laplace transform of many simple trigonometric, exponential and other functions results in the given function being replaced by an algebraic function. For example, the Laplace transform of the derivative of a function is,

$$\mathcal{L}\{f'(t)\} = \int_0^t \frac{d[f(t)]}{dt}e^{-st}dt = s\bar{f}(s) - f(0)$$

Because the Laplace transform of a time derivative (of any order) is an algebraic function, differential equations involving time are converted to algebraic expressions by a Laplace transformation. Thus, differential equations involving time can often be solved in the transform domain by using usual algebraic techniques provided that the result can be inverted back to the time domain.

The Laplace transform of many functions have been evaluated and the results tabulated. Extensive tables may be found in many texts on the subject. The Laplace transform of a few functions is given in the table below,

Example of Function - Laplace Transform Pairs

$f(t)$	$\bar{f}(s)$
c	$\dfrac{c}{s}$
t	$\dfrac{1}{s^2}$
e^{at}	$\dfrac{1}{s-a}$
$\sin \omega t$	$\dfrac{\omega}{s^2 + \omega^2}$
$\cos \omega t$	$\dfrac{\omega}{s^2 + \omega^2}$
$f'(t)$	$s\bar{f}(s) - f(0)$
$f(t-a)H(t-a)$	$e^{-as}\bar{f}(s)$

Obviously, one approach to find the inverse Laplace transform is by using the known transforms that can be found in tables. Often, manipulation of an algebraic transform (such as breaking up expressions using partial fractions) can result in a form that is easily found from a table of known transforms. When such a simple procedure is not applicable, the inversion may be possible using the inversion integral,

$$f(t) = \mathcal{L}^{-1}\{\bar{f}(s)\} = \frac{1}{2\pi i} \int_{c-i\infty}^{c+i\infty} \bar{f}(s)e^{st}ds$$

This method involves contour integration in the complex plane and is beyond the scope of this text.

The convolution (Faltung) integral is defined as,

$$\int_0^t f(t)g(t-\tau)d\tau = \int_0^t g(t)f(t-\tau)d\tau$$

The Laplace transform of the convolution integral is,

$$\mathcal{L}\left\{ \int_0^t f(t)g(t-\tau)d\tau \right\} = \mathcal{L}\left\{ \int_0^t g(t)f(t-\tau)d\tau \right\} = \bar{f}(s)\bar{g}(s)$$

Due to the occurrence of convolution integrals naturally in the viscoelastic constitutve law, Laplace transforms can be quite useful.

Fourier Transform

A function, f(t), may be related to itself via the Fourier integral as

$$f(t) = \frac{1}{2\pi} \int\limits_{-\infty}^{+\infty} e^{i\omega t} \left\{ \int\limits_{-\infty}^{+\infty} f(\xi)e^{-i\omega\xi} d\xi \right\} d\omega$$

and is often represented by the Fourier transform pair,

$$F(\omega) = \int\limits_{-\infty}^{+\infty} f(\xi)e^{-i\omega\xi} d\xi$$

$$f(t) = \frac{1}{2\pi} \int\limits_{-\infty}^{+\infty} F(\omega)e^{i\omega t} d\omega$$

If the f(t) in the Fourier integral has no value for t < 0, the integral becomes,

$$f(t) = \frac{1}{2\pi} \int\limits_{-\infty}^{+\infty} e^{i\omega t} \left\{ \int\limits_{0}^{+\infty} f(\xi)e^{-i\omega\xi} d\xi \right\} d\omega$$

and can be shown to lead to the Laplace transform pair (Thompson, (1960),

$$f(s) = \int\limits_{0}^{+\infty} f(t)e^{-st} dt$$

$$\bar{f}(s) = \int\limits_{c-\infty}^{c+\infty} \bar{f}(s)e^{st} ds$$

Similar to Laplace transforms, Fourier transforms also have special properties under differentiation and integration making them a very effective method for solving differential equations. Due to the form of the constitutive laws in viscoelasticity, Fourier transforms are quite useful in analysis of viscoelastic problems.

References

Adamson, M.J., "A Conceptual Model of the Thermal Spike Mechanism in Graphite/Epoxy Laminates", *The Long Term Behavior of Composites*, (K. O'Brien, ed.), ASTM-STP 813, ASTM, Philadelphia, 1983, p. 179-191.

Aklonis, J.J. and McKnight, W.J., *Introduction to Polymer Viscoelasticity*, JW, NY, 1983.

Alfrey, T. and Gurne, E.F., *Organic Polymers*, Prentice-Hall, NJ, 1967.

Alfrey, T, "Non-homogeneous stresses in viscoelastic solids", *Q. Applied Math*. II (2), 1944, p, 113-119.

Arenz, R.J., "Nonlinear Shear Behavior of Poly(vinyl acetate) Material", *Mech. of Time-Dependent Materials* 2, 1999, p. 287–305.

Armistead, J. P., and Hoffman, J. D., "Direct Evidence of Regimes I, II, and III in Linear Polyethylene Fractions As Revealed by Spherulite Growth Rates", *Macromolecules*, 35 (10), 2002, p. 3895 -3913.

Arridge, R.G.C., *J. of Physics E: Scientific Instruments*, 1974 Volume 7, 1974, p.399-401.

Arridge, R.G.C., *Mechanics of Polymers*, Oxford, London, 1975.

Arridge, R.G.C., *An Introduction to Polymer Mechanics*, Oxford, London, 1985.

Arruda E.M., and Boyce M.C., "A Three-Dimensional Constitutive Model for the Large Stretch Behavior of Rubber Elastic Materials", *Journal of the Mechanics and Physics of Solids*, Vol. 41, pp. 389-412, 1993.

Baer, Eric (ed.), *Engineering Design for Plastics*, Reinhold, New York, 1964.

Bauwens-Crowet, C, Bauwens, J.C., and Holmes, G., *J. Polmer Sci.*, A2 (7), 1969, p. 735.

Beers, K. L., Douglas, J. F., Amis, E. J. and Karim, A., "Combinatorial Measurements of Crystallization Growth Rate and Morphology in Thin Films of Isotactic Polystyrene", *Langmuir*, 19, 2003, p. 3935-3940.

Brandrup, J., Immergut, E. H., Grulke, E. A., Akihiro, A., Bloch, D. R., (Ed'.s} *Polymer Handbook* (4th Edition), J. Wiley & Sons, 2005

Billmeyer, F.W., *Textbook of Polymer Science*, 3rd ed. JW, NY, 1984.

Bingham, E.C., *Fluidity and Plasticity*, McGraw Hill, NY, 1922.

Bodner, S.R. and Partom, Y., "Constitutive Equations for Elastic-Viscoplastic Strain-Hardening Materials", *J. of Applied Mechanics*, June, 1975, p. 385-389.

Bordonaro, C.M. and Krempl, E., "The Effect of Strain Rate on the Deformation and Relaxation Behavior of Nylon 6/6 at Room Temperature", *Polymer Engineering and Science*, V. 32, No. 16, Aug. 1992, p. 1066-1072.

Boresi, A.P. and Schmidt, R.J., and Sidebottom, O.M., *Advanced Mechanics of Materials*, 5th ed., J. Wiley, NY, 1993.

Bradshaw, R., Brinson, L.C., "A Sign Control Method for Fitting and Interconverting Material Functions for Linearly Viscoelastic Solids" *Mech. of Time-Dependent Materials,* Vol. 1, No. 1 / March, 1997, p. 85-108.

Brill, W.A., *Basic Studies in Photoplasticity*, Ph.D. Dissertation, Stanford University, 1965.

Brinson, H. F., *Studies in Photoviscoelasticity*, Ph.D. Dissertation, Stanford University, 1965.

Brinson, H. F., "Mechanical and Optical Viscoelastic Characterization of Hysol 4290," *Experimental Mechanics,* Dec. 1968, p. 561-566.

Brinson, H. F., "The Ductile Fracture of Polycarbonate," *Experimental Mechanics,* Feb. 1969, p. 72-77.

Brinson, H. F., "The Viscoelastic Behavior of a Ductile Polymer," *Deformation and Fracture of High Polymers,* (H. Kausch, et al., Ed.'s.) Plenum Press, NY, 1973, p. 397-416.

Brinson, H. F., and Das Gupta, A., "The Strain Rate Behavior of Ductile Polymers," *Experimental Mechanics,* Dec. 1975, p. 458-463.

Brinson, H. F., "The Behavior of Photoelastic Polyurethane at Elevated Temperatures," *Experimental Mechanics,* Jan. 1976, p. 1-4.

Brinson, H. F., Renieri, M. P., and Herakovich, C. T., "Rate and Time Dependent Failure of Structural Adhesives," *Fracture Mechanics of Composites,* STP 593, ASTM, Phil. PA, 1975, p. 177-199.

Brinson HF, Griffith WI, Morris DH. Creep rupture of polymer matrix composites. *Experimental Mechanics*, 1981; 21(9): 329-336

Brinson, H.F., "Durability Predictions of Adhesively Bonded Composite Structures Using Accelerated Characterization Methods," *Composite Structures*, I.H. Marshall, ed., Elvesier, 1985, p.1-18.

Brinson, H.F. "Matrix Dominated Time Dependent Failure Predictions in Polymer Matrix Composites", *J. of Composite Structures*, 47 (1999), p. 445-456.

Brinson, L.C. and Gates, T., "Viscoelasticity and Aging of Polymer Matrix Composites", *Comprehensive Composite Materials,* (A. Kelly and C. Zweben, Eds.), Vol. 2.10, *Polymer Matrix Composites*, (R. Talreja, Ed.), Pergamon, Oxford, 2000, p. 333-368.

Bruller, O.S., "On the Damage Energy of Polymers in Creep", *Polymer Engineering and Science*, 18, 1, Jan. 1978, p. 42-44.

Bruller, O.S., Potente, H. and Menges, G., "Energiebetrachtungen Kriechverhalten von Polymers", *Rheol. Acta*, 16, 1977, p. 282-290,.

Bruller, O.S., "Crazing Limit of polymers in creep and relaxation", *Polymer*, vol. 19, Oct. 1978, p. 1195-1198.

Brueller, O.S. and Schmidt, H.H., 'On the linear viscoelastic limit of polymers – Exemplified on poly(methyl methacrylate)', *Polymer Engineering and Science* 19, 1979, p, 883–887.

Bruller, O.S., "Energy-related failure criteria of thermoplastics", *Polymer Engineering and Science*, 23, 3, Feb. 1981, p. 145-150.

Bruller O.S., *Phenomenological characterization of the nonlinear viscoelastic behavior of plastics*, Private Communication, Technical University of Munich, 1989.

Bucknall, C. B., Ayre, D. S. and Dijkstra, D. J. "Detection of rubber particle cavitation in toughened plastics using thermal contraction tests", *Polymer*, v. 41, issue 15, 2000, p. 5937-5947.

Callister, W.D., Jr., *Materials Science and Engineering* (3rd ed.), JW, 1994.

Carfagno, S.P. and Gibson, R.J., *A Review of Equipment Aging Theory and Technology*, EPRI Report, NP-1558, Sept. 1980

Cartner, J.S., *The Nonlinear Viscoleastic Behavior of Adhesives and Chopped Fiber Composites*, M.S. Thesis, VPI&SU, 1978.

Cartner, J.S. and Brinson, H.F., "The Nonlinear Viscoleastic Behavior of Adhesives and Chopped Fiber Composites", Virginia Tech. Report, *VPI-E-78-21*, Aug. 1978.

Cartner, J.S., Griffith, W. I., and Brinson, H. F., "The Viscoelastic Behavior of Composite Materials for Automotive Applications," *Composite Materials in the Automotive Industry*, ASME, NY, 1978, pp. 159-169.

Catsiff, E. and Tobolsky, A.V., 'Stress-relaxation of polyisobutylene in the transition region (1, 2)', Journal of Colloidal and Interface Science 10, 1955,375–392

Christensen, R.M., *Theory of viscoelasticity*, 2nd ed., Academic Press, 1982.

Clegg, D.W. and Collyer, A.A., *The Structure and Prop. of Polymeric Mat'ls*, Inst. of Materials, London, 1993.

Cloud, G., *Optical Methods in Engineering Analysis,* Cambridge University Press, 1995.

Coker, E.G. and Filon, L.N.G., *A Treatise on Photoelasticity*, Cambridge, 1931.

Cole, B. and Pipes, R., "Filamentary Composite Laminates Subjected to Biaxial Stress Fields", AFFDL-TR-73-115, 1974.

Collins, E.A., Bares, J., Billmeyer, F.W., *Experiments in Polymer Science*, JW, NY, 1973.

Cost, T.L. and Becker, E.B., "A multidata method of approximate Laplace transform inversion", *Int.l J.l for Numerical Methods in Engineering 2*, 1970, p. 207-219.

Courtney, T.H., *Mechanical Behavior of Materials*, McGraw-Hill, 1990.

Crawford, R.J., *Plastics Engineering*, Pergamon, NY, 1992.

Crissman, J.M. and McKenna, "Physical and Chemical Aging in PMMA and Their Effects on Creep and Creep Rupture Behavior", *J. Poly. Sc., Part B: Polymer Physics*, Vol. 28, 1990, p. 1463-1473.

Cristescu, N., *Dynamic Plasticity*, N. Holland, 1967.

Cristescu, N. and Suliciu, I., *Viscoplasticity*, Kluwer, Boston, 1982.

Crochet, M.J., "Symmetric Deformations of Viscoelastic-Plastic Cylinders", *J. of Applied Mechanics*, 33, 1966, p. 321.

Crossman, F. W and Flaggs, D. L., *Viscoelastic analysis Hygrothermally Altered Laminate Stresses and Dimensions*, LMSC Report LMSC-D633086, 1978,

Daniel, I.M., and Ishai, O., *Engineering Mechanics of Composite Materials*, Oxford, NY, 2005.

Dillard, D.A., *Creep and Creep Rupture of Laminated Graphite/Epoxy Composites*, Ph.D. Thesis, Virginia Tech, 1981.

Dillard, D. A., Morris, D. H., and Brinson, H. F., "Creep and Creep Rupture of Laminated Graphite/Epoxy Composites," *VPI-E-81-3*, March 1981.

Dowling, N. Mechanical Behavior of Materials. Englewood Cliffs, New Jersey: Prentice Hall, (1993).

Drucker, D.C., "Basic Concepts of Plasticity and Viscoelasticity", in *Handbook of Engineering Mechanics* (W. Flugge, Ed.), McGraw Hill, NY, 1962.

Duran, R.S. and McKenna, G.B.: A torsional dilatometer for volume change measurements on deformed glasses: Instrument description and measurements on equilibrated glasses, *J. Rheol.* 34, 1990, p. 813-839.

Durelli, A.J., Phillips, E.A. and Tsao, C.H., *Introduction to the Theoretical and Experimental Analysis of Stress and Strain*, McGraw-Hill, NY, 1958.

Emri, I and T. Prodan: A Measuring System for Bulk and Shear Characterization of Polymers, *Experimental Mechanics*, V., 46, No. 4, August 2006, p. 429-439.

Emri, I. and Tschoegl, N.W., "Generating line spectra from experimental responses. Part I: Relaxation modulus and creep compliance", *Rheologica Acta* 32, 1993, p. 311-321.

Farris, R.J. and Adams, G.W., "Latent energy of deformation of amorphous polymers: 1. Deformation calorimetry", *Polymer*, V. 30, Issue 10, 1989, p. 1824-1828.

Farris, J. and Adams, G.W., "Latent energy of deformation of amorphous polymers: 2. Thermomechanical and dynamic properties", *Polymer*, V. 30, Issue 10, 1989, p. 1824-1828.

Ferry, J.D.,*Viscoelastic Properties of Polymers*, 3rd ed., JW, NY, 1980.

Findley, W. N., *Creep and Relaxation of Non-Linear Viscoelastic Materials*, N. Holland, 1976.

Flaggs, D. L. and Crossman, F. W., "Analysis of viscoelastic response of composite laminates during hygrothermal exposure," *J. of Composite Materials* 15:21-40, 1981.

Flory, P.J., *Principles of Polymer Chemistry*, Cornell, Ithaca, 1953.

Flügge, Wilhelm, *Tensor analysis and continuum mechanics*, Springer-Verlag NY, 1972.

Flügge, W., *Viscoelasticity*, Springer-Verlag, NY, 1974.

Fredrickson, A.G., *Principles and Applications of Rheology*, Prentice Hall, NJ, 1964.

Fried, J. R., *Polymer Science and Technology*, Prentice Hall PTR, NJ, 1995.

Freudenthal, A.M., *The Inelastic Behavior of Engineering Materials and Structures*, JW, 1950.

Fung, Y.C., *Foundations of Solid Mechanics*, Prentice Hall, NY, 1965.

Geil, P.H., "Nylon Single Crystals", J. Polymer Science, 44, 1960, p. 449-458.

Gittus, John, *Creep, Viscoelasticity and Creep Fracture of Solids,* JW, NY, 1975.

Glasstone, S, Laidler, K.J. and Iyring, H., *The Theory of Rate Processes*, McGraw Hill, 1941.

Golden, H.J., Strganac, T.W., Schapery, R.A., "An Approach to Characterize Nonlinear Viscoelastic Material Behavior using Dynamic Mechanical Tests and Analyses", *Journal of Applied Mechanics*, Vol. 66, pp. 872-878, 1999.

Graham, G.A.C., "The Correspondence Principle of Linear Viscoelasticity Theory for Mixed Boundary Value Problems Involving Time-

Dependent Boundary Regions", *Quart. Appl. Math.* 26, 1968, p. 167-174.

Green, A. E. and Rivilin, R.S., "The Mechanics of Non-Linear Materials with Memory, I," *Arch. Ration. Mech. Anal.* I. 1, 1957.

Griffith, W.I., *Accelerated Characterization of Graphite/Epoxy Composites*, Ph.D. Thesis, VPI&SU, 1980.

Griffith, W. I., Morris, D. H., and Brinson, H. F., "Accelerated Characterization of Graphite/Epoxy Composites," *VPI-E-80-27*, Sept. 1980.

Gross, Bernard, *Mathematical Structure of the Theories of Viscoelasticity*, Herman, Paris, 1953.

Guedes, R.M., "Mathematical Analysis of Energies for Viscoelastic Materials and Energy Based Failure Criteria for Creep Loading", Mech. of Time-Dependent Materials 8: 169–192, 2004.

Guo, Y. and Bradshaw, R.D. "Isothermal physical aging characterization of polyether-ether-ketone (PEEK) and polyphenylene sulfide (PPS) films by creep and stress relaxation," Mechanics of Time-Dependent Materials, 2007, 11(1), p. 61-89.

Ha, K. and Schapery RA. A three-dimensional viscoelastic constitutive model for particulate composites with growing damage and its experimental verification. Int. J. of Solids Structures, 35(26-27), 1998, p 3479-3517.

Halpin, J.C. and Polley, H.W., "Observations on the Fracture of Viscoelastic Bodies", J. Composite Materials, Vol. 1, 1967, p. 64-81.

Hertzberg, R.W., *Deformation and Fracture Mechanics of Engineering Materials*, JW, NY, 1989.

Hetenyi, N., The Fundamentals of Three-Dimensional Photoelasticity, *J. Applied Mechanics*, p. 149, 1938.

Hetenyi, N. (Ed.), *Handbook of Experimental Stress Analysis*, JW, 1950.

Hetenyi, N. "A Study in Photoplasticity", Proc. First U.S. National Congr. Appl. Mech, 1952.

Hiel, C., *The Nonlinear Viscoelastic Response of Resin Matrix Composite Laminates*, Ph.D. Thesis, Free University of Brussels (VUB), 1984.

Hiel, C., A.H. Cardon, and H. F. Brinson, "The Nonlinear Viscoelastic Response of Resin Matrix Composite Laminates", *NASA CR 3772*, July 1984. (See also, Hiel, C., Ph.D.

Hiel, C., Cardon, A. and Brinson, H. F.., Proceed. of the V Int. Cong. on Exp. Mech., Montreal 1984, p. 263-267.

Hilton, Harry, "Viscoelastic Analysis", (E. Baer, ed.*), Engineering Design for Plastics*, Reinhold, New York, 1964.

Hobbs, J.K., Binger, D. R., Keller, A., and Barham, P.J., "Spiralling Optical Morphologies in Spherulites of Poly(hydroxybutyrate)", *J. of*

Polymer Science: Part B: Polymer Physics, vol. 38, 1575–1583, (2000)

Houwink, R., *Elasticity, Plasticity and the Structure of Matter*, Harren, 1953. (Also, University Press, Cambridge , England, 1937.)

Hyer, M.W., *Stress Analysis of Fiber-Reinforced Composite Materials*, McGraw Hill, NY, 1998.

Hutter, K., "The Foundations of Thermodynamics, Its Basic Postulates and Implications. A Review of Modern Thermodynamics", *Acta Mechanica 27*, 1-54 (1977).

Kachanov, L.M., *Introduction to Continuum Mechanics*, Kluer, Boston, 1986.

Kaelble, D.H., *Computer Aided Design of Polymers and Composites*, Marcel Decker, NY, 1985.

Kennedy, J., *Creep and Fatigue of Metals*, JW.

Kenner, V. H., Knauss, W. G., Chai, H., "A Simple Creep Torsiometer and its Use in the Thermorheological Characterization of a Structural Adhesive", Experimental Mechanics, Feb., 1982, p. 75-80.

Kinloch, A.J., and Young, R.J., *Fracture Behavior of Polymers*, Applied Science, London, 1983; 2nd ed., 1995.

Kinloch, A.J., *Adhesion and Adhesives, Science and Technology.*, Chapman and Hall, London, 1990.

Kishore K. Indukuri and Alan J. Lesser," Comparative deformational characteristics of poly(styrene-b-ethylene-co-butylene-b-styrene) thermoplastic elastomers and vulcanized natural rubber", *Polymer*, 46 (18) 2005, 7218-7229.

Knauss, W.G., "Rupture Phenomena In Viscoelastic Materials", Ph. D. Thesis, California Institute of Technology, 1963.

Knauss, The mechanics of polymer fracture, Lead article in *Apl. Mech. Rev.*, Jan. 26, 1973.

Knauss, W.G. and Emri, I.J., "Nonlinear Viscoelasticity Based on Free Volume Consideration", *Computers and Structures*, 13, p. 123-128, 1981.

Knauss, W.G., "Viscoelasticity and the time-dependent fracture of polymers", in Vol. 2 of *Comprehensive Structural Integrity*, I. Milne, R.O. Ritchie and B. Karihaloo, (Eds.), Elsevier, 2003.

Knauss, W.G. and Zhu W. "Nonlinearly Viscoelastic Behavior of Polycarbonate. II. The Role of Volumetric Strain, *Mech. of Time-Dependent Materials* 6: 301–322, 2002.

Kobayashi, A., *Handbook of Experimental Mechanics*, 1985.

Kolsky, H., *Stress Waves in Solids*, Dover, NY, p. 91. , 1963

Krajcinovic, Dusan, "Creep of structures - A continuous damage mechanics approach," *J. Structural Mechanics*, Vol. 11, no. 1, pp. 1-11. 1983.

Krempl, E., "Models of Viscoplasticity-Some Comments on Equilibrium (Back) Stress and Drag Stress", Acta Mech. vol. 69, p. 25-42, 1987.

van Krevelen, D.W., et. al., *Prop. of Polymers, Correlation w. Chem. Struc.* Elsevier, London, 1972.

Kumar, Anil and Gupta, R.K., *Fundamentals of Polymers*, McGraw-Hill, NY, 1998.

Kuhn, H.H., Skontrop, A. and Wang, S.S., "High Temperature Physical and Chemical Aging in Carbon Fiber Polyimide Composites: Experiment and Theory" in *Recent Advances in Composite Materials,* (S.R. White, H.T. Hahn, and W.T. Jones, Ed.'s), ASME MD Vol. 56, 1995, p. 193-202.

Lakes, R. S., *Viscoelastic solids* , CRC Press, Boca Raton, FL, (1998).

Landel, R.F., and Fedders, R.F., "Rupture of Amorphous Unfilled Polymers", in *Fracture Processes in Polymeric Solids* (B. Rosen, ed.), 1964, p. 361-485.

Lazan, B.J., "*Damping of Materials and Members in Structural Mechanics*", Pergammon, 1968

Leaderman, H., *Elastic and Creep Properties of Filamentous Materials and Other High Polymers*, The Textile Foundation, Washington, D.C., 1943.

Lekhnitskii, S.G., *Theory of an Anisotropic Elastic Body*, Holden Day, SF, 1963.

Lee, E.H., "Stress analysis in viscoelastic bodies", *Q. Appl.* 13, 183-190, 1955.

Lee, E. H., "Stress Analysis of Viscoelastic Bodies", in *Viscoelasticity: Phenomenological Aspects* (J.T. Bergen, Ed.), Wiley, 1960.

Lee, E.H., and T.G.Rogers, "Solution of viscoelastic stress analysis problems using measured creep or relaxation functions," *J. Appl. Mech.* 30, 127–133 (1963).

Lesser, A.J., and Indukuri, K., "Comparative deformational characteristics of SEBS TPEs and vulcanized natural rubber," Society of Plastics Engineers ANTEC, Boston, May 2-5 (2005).

Lubliner, J. *Plasticity Theory*, Macmillan, NY, 1990.

Lohr, J.J., "Yield Stress Master Curves for Various Polymers below Their Glass Transition Temperatures", J. of Rheology, vol. 9, Issue 1, March 1965, pp. 65-81.

Lou, Y.C. and Schapery, R.A., "Viscoelastic Characterization of a Nonlinear Fiber-Reinforced Plastic", *J. Comp. Mat'ls.* Vol. 5, April 1971, p. 208-234.

Lu, H. and Knauss, W.G.,"The Role of Dilatation in the Nonlinearly Viscoelastic Behavior of PMMA under Multiaxial Stress States,", *Mechanics of Time Dependent Materials*, Vol. 2, No. 4, 1998, pp. 307-334.

Lubliner, Jacob, *Plasticity Theory*, Macmillan, NY, 1990.

Ludwik, P., *Phys. Z.*, 10, 1909, p. 411-417.

McClintock, F.A. and Argon, A., *Mechanical Behavior of Materials*, Addisson-Wesley, 1966.

McCrum, N.G., Buckley, C.P., and Bucknall, C.B., *Principles of Polymer Engineering*, 2nd ed. Oxford University Press, Oxford, 1997.

McKenna, G.B., "Glass Formation and Glassy Behavior", Comprehensive *Polymer Science,* (G. Allen and J.C. Bevington, eds.), Polymer Properties, Vol. 2, p. 311-362, Pergamon Press, 1989.

McKenna, G.B., "On the Physics Required for the Prediction of Long Term Performance of Polymers and Their Composites", *J. of NIST*, Vol. 99, No. 2, March-April, 1994, p. 169-189.

Malkin, A.Y., et. al., *Experimental Methods of Polymer Physics*, Prentice Hall, NJ, 1983.

Mark, J.E. and Erman, B., *Rubberlike Elasticity*, A Molecular Primer, JW, NY, 1988.

Markovitz, Hershel, "Superposition in Rheology", *J. Polymer Sci.*: Symposium No. 50, P. 431-456, 1975.

Markovitz, Hershel, "Boltzmann and the Beginnings of Rheology", *Transactions of the Society of Rheology*, 21:3, P. 381-398, 1977.

Menard, K.P., *Dynamic Mechanical Properties: A Practical Introduction*, CRC Press, NY, 1999.

Miller, E., *Introduction to Plastics and Composites*, Marcel Dekker, NY, 1996.

Mueller, F.H., "Thermodynamics of Deformation Calorimetric Investigations of Deformations Processes", in *Rheology Theory and Applications* (F.R. Eirich, Ed.), Vol. 5, Academic Press, NY, 1969, p. 417-489,

More, C.R., *Properties and Processing of Polymers for Engineers*, SPI, NJ, 1984.

Nadai, A., *Theory of Flow and Fracture of Solids*, McGraw-Hill, 1950.

Nagdi, P.M. and Murch, S.A., On the Mechanical Behavior of Viscoelastic-Plastic Solids" *J. of Applied Mechanics*, 30, 1963, p. 321.

National Research Council: Committee on Durability and Life Prediction of Polymer Matrix Composites in Extreme Environments, *Going to Extremes: Meeting the Emerging Demand for Durable Polymer Matrix Composites*, The National Academies Press, 2005.

Nielsen, L.E., *The Mechanical Properties of Polymers*, Reinhold, NY, 1965.

Nielsen, L.E. and Landel, R.F., *The Mechanical Properties of Polymers and Composites*, 2nd ed. M. Decker, NY, 1994.

Painter, P.C. and Coleman, M.M., *Fundamentals of Polymer Science*, Technomic, Lancaster, 1994.

Park, S.J., Liechti, K.M. and Roy, S.: Simplified bulk experiments and hygrothermal nonlinear viscoelasticty, *Mech. Time-Depend Mat.* **8**, 303-344 (2004).

Pasricha, A., Tuttle, M., and Emery, A., "Time Dependent Response of IM7/5260 Composites Subjected to Cyclic Thermo-Mechanical Loading", *Composite Science and Technology*, Vol 55 (1), pp 49-56 (1995).

Passaglia, E. and Knox, J.R., "Viscoelastic Behavior and Time-Temperature Relationships", (E. Baer, ed.), *Engineering Design for Plastics*, Reinhold, New York, 1964.

Peretz, D. and Weitsman, Y., "Nonlinear Viscoelastic Characterization of FM-73 Adhesive" J. of Rheology, 26(3),245-261, 1982.

Perzyna, P., "Modified Theory of Viscoplasticity: Applications to Advanced Flow and Instability Phenomena", *Arch. Mech.* (Warsaw), 32, 3, 1980.

Pippin, H.G., "Analysis of Materials Flown on the Long Duration Exposure Facility: Summary of Results of the Materials Special Investigation Group", NASA CR, 1995.

Poloso, A. and Bradley, M.B., "High-Impact Polystyrene (PS, HIPS)", *Engineering Materials Handbook, Vol. 2, Engineering Plastics*, ASM International, Metals Park, OH, 1988, p 194-199.

Popelar, C. F., Popelar, C. H. and Kenner, V. H., "Viscoelastic material characterization and modeling for polyethylene", Poly. Engr. Sci., 30, 1990, p 577-586.

Popelar, C. F. and Liechti, K. M., "A Distortion-Modified Free Volume Theory for Nonlinear Viscoelastic Behavior", *Mechanics of Time-Dependent Materials,* 7: 89–141, 2003.

Prandtl, *Z. Angew. Math. Mech.*, 9, p. 91 – 100, 1928.

Press, William H. et al. Numerical Recipes in C: The Art of Scientific Computing. Cambridge: Cambridge University Press, 1992.

Read, W.T., "Stress Analysis for compressible viscoelastic materials", *J. Appl. Phys.* 21, 671-674, 1950.

Reiner, M. and Weissenberg, K., "A Thermodynamical Theory of the Strength of Materials", *Rheological Leaflet*, 10, 1939, p.12-20. (See also: "A Thermodynamical Theory of the Strength of Materials", in

Rosen, B., ed., *Fracture Processes in Polymeric Solids* (B. Rosen, ed.), 1964, Wiley-Interscience, NY, 1964)

Reiner, M., *Advanced Rheology*, H.K. Lewis, London, 1971.

Renieri, M.P., *Rate and Time Dependent Behavior of a Structural Adhesives*, Ph.D. Thesis, VPI&SU, 1976.

Renieri MP, Herakovich, CT, Brinson, HF. *Rate and time dependent behavior of structural adhesives,* Virginia Tech Report, VPI-E-76-7, April, 1976.

Richards, C. W., *Engineering Materials Science*, Wadsworth, 1961.

Robotnov, Y. N., *Elements of Hereditary solid Mechanics*, MIR Publishers, Moscow, 1980.

Rochefort, M.A, *Nonlinear Viscoelastic Characterization of Structural Adhesives*, M.S. Thesis, VPI&SU, 1983.

Rochefort, M.A, and Brinson, H.F., *Nonlinear Viscoelastic Characterization of Structural Adhesives*, NASA Contractor Report 172279, Dec. 1983.

Rodriguez, F., *"Principles of Polymer Systems"*, 4th ed., Taylor and Francis, Washington, D.C., 1996.

Rosen, S.L., *Fundamental Principles of Polymeric Materials*, 2nd ed., JW, NY, 1993.

Rybicky, E.F., and Kanninen, M.F., "The Effect of Different Behavior in Tension and Compression on the Mechanical Response of Polymeric Materials". *Deformation and Fracture of High Polymers,* (H. Kausch, et al., Ed.'s.) Plenum Press, NY, 1973, p. 417-427.

Sandhu, R.S., *A survey of failure theories of isotropic and anisotropic materials*, Technical Report, AFFDL –TR-72-71, Jan. 1972.

Sancaktar, E., *Applied Mechanics Reviews*, 1987, p. 1393.

Sakai, M., *Philosophical Magazine*, Vol. 86, 5607–5624, "Elastic and viscoelastic contact mechanics of coating/substrate composites in axisymmetric indentation", (2006).

Sane, S. and Knauss, W.G.: The time-dependent bulk response of Poly (methyl methacrylate), *Mech. Time-Depend. Mat.* **5**, 293-324 (2001)

Segard, E., Benmedakhene, S. Laksimi, A, and Lai, D., "Influence of the fibre–matrix interface on the behaviour of polypropylene reinforced by short glass fibres above glass transition temperature" *Composites Science and Technology*, 62 p. 2029–2036, 2002.

Serpooshan, V. Zokaei, S. and Bagheri., R., "Effect of rubber particle cavitation on the mechanical propertics and deformation behavior of high-impact polystyrene", *J. of Applied Polymer Science,* Vol. 104, Issue 2, Pages 1110-1117, 2007

Shames, I.H. and Cozzarelli, F.A., *Elastic and Inelastic Stress Analysis*, Prentice Hall, 1992.

Schapery, R. A., "Application of Thermodynamics to Thermomechanical Fracture and Birefringent Phenomena in Viscoelastic Media, *J. of Applied Physics*, Vol. 35, No. 5, pp. 1451-1465, 1964.

Schapery, R.A., "A Theory of Non-Linear Thermoviscoelasticity Based on Irreversible Thermodynamics", *Proc. 5th U.S. National Congress of applied Mechanics*, ASME, pp. 511-530, 1966.

Schapery, R.A., "On the Characterization of nonlinear Viscoelastic Materials", *Poly. Engr. and Sc.*, July, 1969, vol. 9, No. 4, p. 295-310.

Schapery, R.A., "Nonlinear Viscoelastic and Viscoplastic Constitutive Equations Based on Thermodynamics" *Mechanics of Time-Dependent Materials* 1: 209–240, 1997.

Schmitz, J.V. and Brown, W.E., *Testing of Polymers*, Vol. 3, JW, NY, 1967.

Schwarz, F. and Staverman, A.J., "Time-Temperature Dependence of Linear Viscoelastic Behavior", *J. of Appl. Physics*, Vol. 23 (1952), p. 838.

Shaw, J.A., Jones, A.S., and Wineman A.S. "Chemorheological response of elastomers at elevated temperatures: Experiments and simulations", *J. Mech. and Phy. of Solids*, 53 (12): 2758-2793 Dec. 2005

Smith, J.O. and Sidebottom, O.M., *Inelastic Behavior of Load-Carrying Members*, JW, NY, 1965.

Smith, T.L., "Deformation and Failure of Plastics and Elastomers", Polymer Engineering and Science, V. 5, No. 5, Oct. 1965, p. 1-10.

Struik, L.C.E., *Physical Aging in Amorphous Poly. and Other Mat'ls*, Elesivier, London, 1969.

Sullivan, R.F., *The Fixation of Photoelastic Stress Patterns by Electron Irradiation And Its Application To Three-Dimensional Photoelasticity*, Ph.D. Dissertation, Stanford University, 1969.

Taylor, R.L., K. Pister, and G. Goudreau, Thermomechanical Analysis of Viscoelastic Solids, *Int. J. Numer. Meth. Engr.*, 1970. 2: p. 45-59.

Thomson, W.T. , Laplace Transformation, Second Edition, Prentice-Hall, Inc., Englewood Cliffs, N.J. (1960).Timoshenko, S.P. and Goodier, J.N., *Theory of Elasticity*, McGraw Hill, NY, 1970.

Tobolsky, A.V., *Properties and Structure of Polymers*, JW, NY, 1962.

Tobolsky, A.V., Mark, H.F., *Poly. Sc. and Mat'ls*, J.W./Interscience, NY, 1971.

Treloar, L. R. G., *The Physics of Rubber Elasticity*, 3rd Ed. Clarendon Press, Oxford, 1975.

Tschoegl, N. W., *The Phenomenological Theory of Linear Viscoelastic Behavior*, Springer Verlag, 1989.

Tschoegl, N.W., Knauss, W. G., and Emri, I., "Poisson's Ratio in Linear Viscoelasticity – A Critical Review", *Mech. of Time-Dependent Materials* 6: 3–51, 2002.

Tuttle, M.E., *Accelerated Viscoelastic Characterization of T300/5208 Graphite-Epoxy Laminates*, Ph.D. Thesis, VPI&SU, 1985.

Tuttle, M.E. and Brinson, H.F., *Accelerated Viscoelastic Characterization of T300/5208 Graphite-Epoxy Laminates*, NASA-CR 3875, March 1985.

Tuttle, M.E., and Brinson, H.F., "Prediction of the Long-Term Creep Compliance of General Composite Laminates", *Experimental Mechanics*, Vol 26 (1), March 1986.

Tuttle, M.E., Pasricha, A., and Emery, A.F., "The Nonlinear Viscoelastic-Viscoplastic Behavior of IM7/5260 Composites Subjected to Cyclic Loading", J. Composite Materials, Vol 29 (15), pp 2025-2046 (1995).

Wang, N., Isothermal Physical Aging Characterization and Test Methodology for Neat and Nanotube-Reinforced Poly(Methyl Methacrylate) (PMMA) Near the Glass Transition Temperature, MS Thesis, University of Louisville, 2007

Ward, I.M. and Hadley, D.W., *An Introduction to the Mechanical Properties of Solid Polymers*, JW, 1993.

Warner, S.B., *Fiber Science*, Prentice-Hall, NJ, 1995.

Weitsman, Y.J. A continuum damage model for viscoelastic materials. *J. of Applied Mechanics* 55. 1988, p773-780.

Westergaard, H.M., *Theory of Elasticity and Plasticity*, Dover, 1964.

Wing G, Pasricha A., Tuttle M., and Kumar, V. "Time-Dependent Response Of Polycarbonate and Microcellular Polycarbonate", *Polymer Engineering And Science* 35 (8): 673-679 Apr 1995.

Wong, E.S.W., Lee, S.M. and Nelson, H.G., *Physical Aging in Graphite Epoxy Composites, NASA TM* 81273, 1981.

Williams, J.W., *Ultracentrifugation of Macromolecules*, Academic Press, NY, 1972.

Williams, M.L. and Arenz, R.J., "The Engineering Analysis of Linear Photoviscoelastic Materials", *Experimental Mechanics*, 4, 9, Sept, 1964, p. 249.

Williams, M.L., Landel, R.F., and Ferry, J.D., "The Temperature Dependence of Relaxation Mechanisms in Amorphous Polymers and Other Glass Forming Liquids", *J. Am. Chem. Soc.*, 77, 3701-3707, 1955.

Yeow, Y.T., "Creep Rupture Behavior of Unidirectional Advanced Composites", *Materials Science and Engineering*, 45, 1980, p. 237-245.

Yu, M. "Advances in strength theories for materials under complex stress state in the 20th Century", *Applied Mechanics Reviews*, Vol 55, no 3, May 2002.

Zhang, M.J. and Brinson, H.F., "Cumulative Creep Damage of Polycarbonate and Polysulfone", *Experimental Mechanics*, June, 1986, p. 155-162.

Zhou, J.-J., Liu, J.-G., Yan, S.-K., Dong, J.-Y, Li, L., Chan, C.M. and Schultz, J. M. "Atomic force microscopy study of the lamellar growth of isotactic polypropylene", *Polymer*, Volume 46, Issue 12, 26 May 2005, Pages 4077-4087

Zhurkov, S.N., Kinetic Concept of the Strength of Solids", *Int. J. of Fracture Mech.*, Vol. 1, 1965.

Author Index

Index